T0173332

CSEC®

PHYSICS

Peter DeFreitas
Reviewers: Lenore Dunnah, Raphael Johnson

Collins

William Collins' dream of knowledge for all began with the publication of his first book in 1819. A self-educated mill worker, he not only enriched millions of lives, but also founded a flourishing publishing house. Today, staying true to this spirit, Collins books are packed with inspiration, innovation and practical expertise. They place you at the centre of a world of possibility and give you exactly what you need to explore it.

Collins. Freedom to teach.

Published by Collins
An imprint of HarperCollins*Publishers*
The News Building
1 London Bridge Street
London
SE1 9GF

HarperCollins*Publishers*
Macken House,
39/40 Mayor Street Upper,
Dublin 1, D01 C9W8, Ireland

Browse the complete Collins Caribbean catalogue at
www.collins.co.uk/caribbeanschools

British Library Cataloguing in Publication Data

A catalogue record for this publication is available from the British Library.

The publishers gratefully acknowledge the permission granted to reproduce the copyright material in this book. Every effort has been made to trace copyright holders and to obtain their permission for the use of copyright material. The publishers will gladly receive any information enabling them to rectify any error or omission at the first opportunity.

Author: Peter DeFreitas
Reviewers: Lenore Dunnah, Raphael Johnson
Publisher: Dr Elaine Higgleton
Development editor: Tom Hardy
Project leaders: Julianna Dunn, Gillian Bowman, Peter Dennis
Illustrator: Ann Paganuzzi
Copy editor: Aidan Gill
Proofreaders: Mitch Fitton, Lauren Reid
Typesetter: Siliconchips Services Ltd
Cover designers: Kevin Robbins and Gordon MacGilp
Cover photos: Jason Winter/Shutterstock,
SomjaiJaithieng/Shutterstock
Printed and Bound in the UK by
Ashford Colour Press Ltd.

MIX
Paper | Supporting
responsible forestry
FSC™ C007454

This book contains FSC™ certified paper and other controlled sources to ensure responsible forest management.

For more information visit: www.harpercollins.co.uk/green

Contents

Acknowledgments

P6 Peter DeFreitas, P106 Checubus/Shutterstock, P106 Romolo Tavani/Shutterstock, P154 Smit/Shutterstock, Menno Schaefer/Shutterstock, serato/Shutterstock,P167 Chiyacat/Shutterstock, P168 INSTANT photography/Shutterstock, P169 Peter DeFreitas, P173 Wikimedia Commons/Meganbeckett27, P175 Peter DeFreitas, P176 Peter DeFreitas, P179 GIPHOTOSTOCK / SCIENCE PHOTO LIBRARY, P181 Peter DeFreitas, P182 Peter DeFreitas, P183 ktsdesign/Shutterstock, P188 GIPHOTOSTOCK / SCIENCE PHOTO LIBRARY, P190 natural.sound.design/Shutterstock, P207 Peter DeFreitas, P220 durk gardenier/Alamy Stock Photo, P223 BERENICE ABBOTT / SCIENCE PHOTO LIBRARY, P223 Eugen Thome/Shutterstock, P233 Cristi Kerekes/Shutterstock, P235 wavebreakmedia/Shutterstock, GagliardiPhotography/Shutterstock, P245 DarkMediaMotion/Shutterstock, P245 Anita van den Broek/Shutterstock, P246 Everything You Need/Shutterstock, P247 Tyler Olson/Shutterstock, Lena Si/Shutterstock, P252 haryigit/Shutterstock, P255 Peter DeFreitas, P282 Peter DeFreitas, P293 Mrs_ya/Shutterstock, P293 BetterTogether/Shutterstock, P29 Quayside/Shutterstock, P296 Peter DeFreitas, P297 Craig Russell/Shutterstock, P299 Peter DeFreitas, P350 Maurice Savage / Alamy Stock Photo

Getting the best from the book

Welcome to *CSEC® Physics*

This textbook has been written as a comprehensive course designed to help you achieve **maximum success** in your CXC® CSEC® Physics examination. Facts are presented in an easily understandable way, using **simple** and **clear language**. A variety of formats are used, including diagrams and tables, to make the facts easy to **understand** and **learn**. Colour **photographs** of real-life situations are also included to enliven your learning, and the relationship between **structure** and **function** is continually highlighted.

Key terms are highlighted in **bold** type, and important **definitions** which you must learn, are written in *italics* and highlighted in **colour**. These definitions are appropriately positioned within the text itself. No key information is given in boxes to the side as it may be easily missed. All the information needed to fully cover the syllabus is given within the text, and only the required information is given.

Practical activities identified by the syllabus are fully covered within the **text** itself or in the **analysing data** questions at the end of the relevant chapters. In the case of the latter, the method used in each activity is outlined, possible results are given, and you are provided with a set of questions to help you analyse and interpret those results.

A variety of methods are used to make the facts easy to **understand** and **learn**.

Useful techniques on tackling certain types of questions are outlined to provide you with the confidence required to always produce the correct answer. **Worked examples** are provided that demonstrate how to answer the **spectrum of questions** on any topic which may confront you in the examination.

The CSEC® Physics syllabus is divided into **five sections**, outlined below. A **preliminary section** is included to provide information fundamental to the processing of scientific data. Each section has been colour coded and Sections A to E have been divided into chapters presenting its constituent topics.

The book has been divided as follows.

- **Ch 1**   **Scientific measurement and manipulation of data**
- **Ch 2–6**   **Section A: Mechanics**
- **Ch 7–14**   **Section B: Thermal physics and kinetic theory**
- **Ch 15–22**   **Section C: Waves and optics**
- **Ch 23–31**   **Section D: Electricity and magnetism**
- **Ch 32–34**   **Section E: The physics of the atom**

Practice exercises are provided at the end of each chapter to help you test your knowledge and comprehension, and to improve your ability to use your knowledge by developing your thinking, investigative and analytical skills. The exercises are presented in three distinct levels of difficulty as outlined below, and are colour-coded for clarity.

- **Recalling facts:** These exercises are designed to help you **assess your knowledge and understanding** of the facts, concepts and principles covered in the unit. You should be able to find the answers to these questions within the chapter and answering them should help you to learn this information.

- **Applying facts:** These exercises are designed to test how you **use** the facts, concepts and principles covered in the chapter to answer questions about unfamiliar or novel situations. They should help you to develop your ability to **apply** knowledge and develop **critical thinking skills.**

- **Analysing data:** These exercises are designed to develop your **investigative and analytical skills**. You are given data or information to analyse, usually in the form of tables or graphs. You may be asked to read information directly from the table or graph, use data from a table to draw a graph, use data from a graph to construct a table, determine the slope of a graph or perform other calculations using information extracted from a table or graph. You may also be asked to identify patterns or trends, to make predictions and to draw conclusions from the data given.

To adequately prepare yourself for a topic, you must start by **carefully studying** the chapter, paying close attention to the definitions, laws, rules and techniques of dealing with specific calculations. You should also make a dedicated effort to thoroughly understand how each of the worked examples in the chapter has been approached. Every question at the end of the chapter should then be tackled.

A list of *important relations and formulae* and a list of *important scientists relevant to the topics of the course* have been provided in this introduction to assist you in answering the questions.

On completion of the questions, you should visit www.collins.co.uk/caribbean to find their **correct answers**. This will enable you to assess your knowledge, to develop your thinking, investigative and analytical skills, and to place you on the path to achieve **maximum success** in your examination.

Key features of the book

Learning objectives inform you of what you are expected to learn ▶

Key terms in bold for emphasis ▶

Learning objectives

- Investigate **magnetic field patterns** around **current-carrying conductors**.
- Apply the **right-hand grip** rule.
- Investigate the forces existing between wires carrying currents.
- Describe the structure of an **electromagnet** and explain the function of some of its applications.
- Investigate the **force on a current-carrying conductor** immersed in a magnetic field and describe the factors affecting the size of the force produced.
- Apply **Fleming's left-hand rule** to currents in wires and to charged particles moving in magnetic fields.
- Explain the function of the **moving coil loudspeaker**, the simple **DC motor** and the **moving coil ammeter**.

Real-life situations displaying applications in physics ▶

Full colour photos to enhance maximum interest ▶

Prenatal scanning (Figure 21.7), **measuring blood flow**, as well as the **examination** of the **eye** and internal organs including the heart, can be carried out using ultrasound. Ultrasonic waves are **non-ionising** and are therefore **relatively safe** in **comparison to X-rays**. X-rays are high energy, **ionising** waves which damage and kill body cells and which can cause cancers to develop.

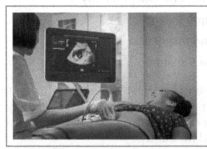
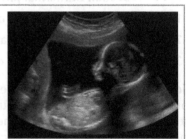

Figure 21.7 *Prenatal scanning*

Annotated diagrams
provide relation between
form and function ►

The solar water heater

The features of the solar water heater are illustrated in Figure 14.18.

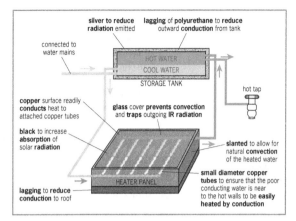

Key terms in bold for
emphasis ►

Common example in
everyday life ►

Experiments important for
school-based assessments
(SBA) fully described ►

Experiment to verify Snell's law using pins

- Place a rectangular glass block on a sheet of paper near its centre and carefully outline its perimeter.
- Remove the block and construct a line NOM to act as a normal as shown in Figure 16.17.
- Construct six lines on the paper from point O to act as rays with incident angles ranging from 10° to about 60°. Mark each of the lines sequentially, 1 through 6.
- Secure the paper to a **pinboard** using thumb tacks and place the block in its outline on the paper.
- Insert two large pins into the pinboard, **vertically** and **far apart** on one of the numbered lines, to act as objects.
- By viewing from a position as shown in Figure 16.17, insert two **search pins**, vertically and far apart, so that they are aligned with the **images** of the two **object pins** as **seen through the block**.
- Mark the positions of the search pins by small crosses, labelled with the same number as the corresponding incident ray.
- Repeat the procedure for the other numbered lines.
- Remove the block.
- Join each incident ray to its corresponding emergent ray through the outline of the block.
- Measure and record the angles of incidence, i, and the angles of refraction, r, at the point where the rays enter the block.
- Record the corresponding values of sin i and sin r.
- Plot a graph of sin i against sin r. The **gradient** (slope) of the graph is the **refractive index** of the material of the block.

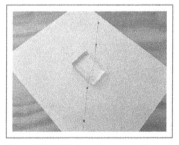

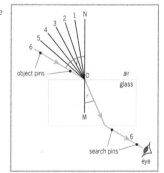

Photos to assist in **difficult**
to grasp concepts ►

Figure 16.18 *Aligning pins as seen through a glass block*

Tackling problems involving coplanar forces in equilibrium

- Sketch a free body diagram showing all the **forces** acting **ON** the **body in equilibrium**. These are
  1. the **weight** (acts through the **centre of gravity**)
  2. all **mechanical forces** acting on the body (act at **points of contact**)
- Use the rules for translational and rotational equilibrium to formulate equations and solve for unknown forces and distances.
  - If there is only **one** unknown force, use the equation for **translational equilibrium**.
  - If there are **two** unknown forces, use the equation of **rotational equilibrium** to find one of them. **Select a point** about which **one of the unknown forces acts**. By taking moments about that point, the unknown force is excluded from the calculation since there is no distance between its line of action and the point, and therefore it creates no moment.

Example 2

Figure 2.6 shows a uniform metre rule of weight 1.8 N resting on a fulcrum. A block of weight 8.0 N hangs from the pulley and a block of weight, x, hangs from the 25 cm mark. Calculate:

a the weight x

b the normal reaction R.

Notes:

- The system is at rest and is therefore in translational equilibrium.
- Since the metre rule is uniform, its centre of gravity is at its midpoint.
- The downward force of the rule on the pivot and the downward force of the block on the pulley are **not considered** since they are **not on the rule**, the body whose equilibrium we are investigating.
- The block hanging from the pulley exerts an upward force **on the rule**.
- A suitable point must be selected from which to take moments in order to eliminate one of the two unknown forces from the equation. By taking moments about the fulcrum, we can eliminate R.

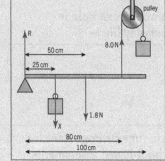

Figure 2.6 *Example 2*

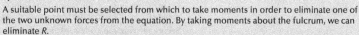

a $\sum$ anticlockwise moments $= \sum$ clockwise moments

$$(8.0 \times 80) = (x \times 25) + (1.8 \times 50)$$
$$640 = 25x + 90$$
$$640 - 90 = 25x$$
$$550 = 25x$$
$$x = 22 \text{ N}$$

b $\sum$ upward forces $= \sum$ downward forces

$$8.0 + R = 22 + 1.8$$
$$R = 22 + 1.8 - 8.0$$
$$R = 15.8 \text{ N} = 16 \text{ N to 2 sig.fig.}$$

Recalling facts

Recalling facts questions help you to assess your knowledge and understanding of facts, concepts and principles ▶

1
a Which scientist is often referred to as the father of scientific methodology?

b List the FOUR steps of the methodology proposed by the person mentioned in **a**.

2 Distinguish between an independent variable, a dependent variable and a control variable as used in the experimental process.

3
a What is a simple pendulum?

b How do each of the following factors affect the period of a simple pendulum?

   i mass of bob

   ii length of string

Applying facts

Applying facts questions test your use of the facts, concepts and principles ▶

4 Andrea is aware that the temperature of a gas may cause its volume to change. She has decided to carry out an experiment to investigate how the pressure of a gas is affected by changing its volume. Name:

a the independent variable

b the dependent variable

c a control variable.

Analysing data

Analysing data questions improve your **SBA** performance. ▶

22 Figure 33.16 shows the apparatus used to determine the half-life of radon-220 (Rn-220), a radioactive gas of short half-life which emits α-particles. A freshly prepared sample of the gas is forced into the vessel to take the place of air which is forced out, and the taps are immediately closed as the Rn-220 immediately begins to ionise the air. A pair of oppositely charged electrodes connected to a power source and a milliammeter register the ionisation current. As the Rn-220 decays, readings of the ionisation current, I, and the corresponding time, t, are taken at intervals of 20 s. Table 33.5 shows the readings taken.

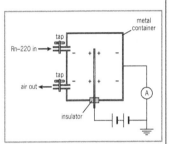

Figure 33.16 *Question 22*

Drawing graphs and tables develops your **Observation, Recording and Reporting (ORR) skills** and investigating and analysing the results develops your **Analysis and Interpretation (AI) skills** ▶

Develops your ability to tackle the **data analysis** question of **Paper 2** which is worth 25% of that examination ▶

Table 33.5 *Readings of ionisation current, I, and corresponding times, t*

Ionisation current I/mA	1.40	1.10	0.82	0.68	0.50	0.44	0.28	0.22	0.22	0.16	0.08
Time t/s	0	20	40	60	80	100	120	140	160	180	200

a Plot a graph of ionisation current versus time, using the graph paper in landscape format.

b Determine the time t_1 for the current to fall from 1.40 mA to half its value.

c Choose two other points on the graph and find the times t_2 and t_3 for their corresponding currents to fall to half their values.

d Determine the half-life of Rn-220 by calculating the mean value of t_1, t_2 and t_3.

e Why are the data points not falling exactly on the curve although the milliammeter and power source are not faulty?

f Read from the graph the current when $t = 50$ s.

g Why do the experimental points deviate more from the curve for larger values of t?

Important relations and formulae

One of the most important things to learn for a Physics examination is the relations and formulae necessary for its calculations. These have been summarised below using symbols defined in the text.

MECHANICS

Density $\rho = \dfrac{m}{V}$

Weight $W = mg$

Force extending spring $F = kx$

Work on extending spring   *work done or potential energy stored in stretching a spring is given by the area between the graph line and x-axis of a F-x graph.*

Moment $T = Fd_{\perp}$

                                 sum of anticlockwise moments = sum of clockwise moments

Displacement–time graphs $v = \dfrac{\Delta s}{\Delta t}$ (i.e. gradient)

Velocity–time graphs $a = \dfrac{\Delta v}{\Delta t}$ (i.e. gradient)

    Distance d = area between graph line and t-axis (all areas positive)

    Displacement s = area between graph line and t-axis (areas above axis are positive)
(areas below axis are negative)

Acceleration $a = \dfrac{v - u}{t}$

Resultant force $F_R = ma$ $F_R = \dfrac{m(v - u)}{t}$

Momentum $p = mv$

                                 sum of moments before = sum of moments after

Work and energy $W = Fd_{\parallel}$ $E = Fd_{\parallel}$

Kinetic energy $E_K = \dfrac{1}{2}mv^2$

Gravitational potential energy $\Delta E_{GP} = mg\Delta h$

Power $P = \dfrac{W}{t}$ $P = \dfrac{E}{t}$

                                 $P = \dfrac{Fd_{\parallel}}{t}$ $P = Fv$ ..... *since* $v = \dfrac{d}{t}$

Efficiency $\eta = \dfrac{\text{useful work or energy output}}{\text{work or energy input}} \times 100\%$

                                 $\eta = \dfrac{\text{load} \times \text{distance moved by load}}{\text{effort} \times \text{distance moved by effort}} \times 100\%$ ... *in relation to machines*

                                 $\eta = \dfrac{\text{useful power output}}{\text{power input}} \times 100\%$

Pressure	$P = \dfrac{F}{A}$ $\qquad$ $P = h\rho g$ ... (for fluid pressure)

Pressure $\qquad$ $P = \dfrac{F}{A}$ $\qquad\qquad$ $P = h\rho g$ ... (for fluid pressure)

Pascal's principle:
(applied to hydraulic lift)

pressure at smaller piston = pressure at larger piston

$$\frac{f}{a} = \frac{F}{A}$$

Archimedes' principle:

upthrust = weight of fluid displaced

$$U = m_f g \qquad U = \rho_f V_f g \qquad \text{... since } m = \rho V$$

Flotation:

weight of object = weight of fluid displaced = upthrust

$$m_o g = m_f g \qquad m_o = m_f \qquad \rho_o V_o g = \rho_f V_f g \qquad \rho_o V_o = \rho_f V_f$$

SECTION B $\qquad$ THERMAL PHYSICS AND KINETIC THEORY

Celsius to kelvin $\qquad$ $T = 273 + \theta$ ... where T is in K and θ is in °C

Heat energy $\qquad$ **Temperature change** $\quad E_H = mc\Delta T$ $\qquad$ **Phase change** $\quad E_H = ml$

Heat power $\qquad$ **Temperature change** $\quad P_H = \dfrac{mc\Delta T}{t}$ $\qquad$ **Phase change** $\quad P_H = \dfrac{ml}{t}$

Ideal gas laws $\qquad$ $\dfrac{P_1 V_1}{T_1} = \dfrac{P_2 V_2}{T_2}$ $\quad$ ... T must be in kelvins

SECTION C $\qquad$ OPTICS AND WAVES

Shadows $\qquad$ $\text{magnification} = \dfrac{\text{size of shadow}}{\text{size of object}} = \dfrac{\text{distance of shadow from light source}}{\text{distance of object from light source}}$

Pinhole camera $\qquad$ $\text{magnification} = \dfrac{\text{size of image}}{\text{size of object}} = \dfrac{\text{distance of image from pinhole}}{\text{distance of object from pinhole}}$

Reflection $\qquad$ angle of incidence = angle of reflection

Wave propagation $\qquad$ **Within a medium:** $\qquad v = \lambda f$ $\qquad\qquad T = \dfrac{1}{f}$

$\qquad$ **Refraction to different medium:** $\qquad \dfrac{\sin \theta_x}{\sin \theta_y} = \dfrac{v_x}{v_y} = \dfrac{\lambda_x}{\lambda_y} = \dfrac{\eta_y}{\eta_x}$

Critical angle $\qquad$ If one medium is air and the calculation involves refractive index: $\sin c = \dfrac{1}{\eta}$

$\qquad$ Otherwise: $\dfrac{\sin c_x}{\sin 90°} = \dfrac{v_x}{v_y} = \dfrac{\lambda_x}{\lambda_y} = \dfrac{\eta_y}{\eta_x}$

$$\sin c_x = \frac{v_x}{v_y} = \frac{\lambda_x}{\lambda_y} = \frac{\eta_y}{\eta_x} \quad \text{... since } \sin 90° = 1$$

Note: All ratios in this equation have the smaller value in the numerator.

$\qquad$ c is always in the medium of greater refractive index.

Lenses $\qquad$ $\dfrac{1}{u} + \dfrac{1}{v} = \dfrac{1}{f}$

$\qquad$ $\text{magnification, } \quad m = \dfrac{v}{u} \qquad\qquad m = \dfrac{I}{O}$

SECTION D   ELECTRICITY AND MAGNETISM

Electrical charge $Q = It$ $Q = Nq$ (... N is the number of particles of charge, q)

Energy $E = QV$ $E = VIt$ $E = I^2Rt$ $E = \dfrac{V^2 t}{R}$

Power $P = VI$ $E = I^2R$ $E = \dfrac{V^2}{R}$

Ohm's law $V = IR$

Adding resistance   **Series:** $R = R_1 + R_2 + R_3 \ldots$ **Parallel (general):** $\dfrac{1}{R} = \dfrac{1}{R_1} + \dfrac{1}{R_2} + \dfrac{1}{R_3} \ldots$

    Parallel (2 resistors): $R = \dfrac{R_1 R_2}{R_1 + R_2}$

For resistors in series $\dfrac{R_A}{R_B} = \dfrac{V_A}{V_B}$ $\dfrac{R_A}{R_{Total}} = \dfrac{V_A}{V_{Total}}$

Emf   emf = sum of pd's in circuit

Transformers $\dfrac{V_p}{V_s} = \dfrac{N_p}{N_s}$ **efficiency:** $\eta = \dfrac{P_s}{P_p} \times 100\%$ $\eta = \dfrac{V_s I_s}{V_p I_p} \times 100\%$

Ideal transformers $\dfrac{V_p}{V_s} = \dfrac{N_p}{N_s} = \dfrac{I_s}{I_p}$ ... 100% efficient, so no need to calculate efficiency

SECTION E   THE PHYSICS OF THE ATOM

Mass number (nucleon number), proton number (atomic number) and neutron number $A = Z + N$

Einstein's mass–energy equivalence $\Delta E = \Delta mc^2$

Converting atomic mass units to kilograms $1\ u = 1.66 \times 10^{-27}\ \text{kg}$

Scientists related to the topics of the CSEC® examination

Scientist	Scientists related to topics of the CSEC® examination
Galileo	Father of scientific methodology.
Hooke	Showed that the force applied to a spring is proportional to its extension provided the proportional limit is not exceeded.
Aristotle	Incorrectly proposed that the force applied to a body is proportional to its velocity.
Newton	• Proved Aristotle wrong by experimentally demonstrating that force is proportional to acceleration. • Proposed the three laws of motion. • Demonstrated that white light is a physical combination of coloured light.
Pascal	Showed that pressure applied to an enclosed fluid is transmitted uniformly throughout the fluid.
Archimedes	Showed that the upthrust on a body is equal to the weight of the fluid it displaces.

Scientist	Scientists related to topics of the CSEC® examination
Benjamin Thompson	Also known as Count Rumford – argued against the caloric theory of heat.
Joule	Established the principle of conservation of energy.
Kelvin	Proposed the Kelvin scale of temperature
Boyle	Showed that the pressure of a fixed mass of an ideal gas is inversely proportional to its volume provided its temperature remains constant.
Charles	Showed that the volume of a fixed mass of an ideal gas is directly proportional to its absolute temperature provided its pressure remains constant.
Huygens	Proposed a wave theory of light.
Young	Demonstrated, through his 'double-slit experiment', that light can undergo interference and therefore propagates as a wave.
Einstein	• Demonstrated by the photoelectric effect that light also has a particle nature and therefore exhibits a wave–particle duality. • Provided the equation of mass–energy equivalence, $\Delta E = mc^2$.
Ohm	Showed that the current through a metallic conductor is proportional to the potential difference between its ends, provided physical conditions remain constant.
Faraday	Showed that whenever there is a relative motion between a conductor and a magnetic field, an emf is produced proportional to the motion.
Lenz	Showed that an induced current is always in a direction opposite to the motion creating it.
Democritus	Proposed that matter was made of fundamental indivisible units (atoms).
Thomson	Proposed a *plum pudding* concept of the atom in which negative *corpuscles* floated around in a mass of positive charge. These corpuscles were later called electrons.
Rutherford	Proposed that the atom is mainly empty space with a small, concentrated positive nucleus at its centre around which electrons orbit.
Geiger and Marsden	Worked with Rutherford in carrying out the experiment where alpha particles were shot through gold foil. This experiment is the basis for Rutherford's conclusion on the structure of the atom.
Bohr	Proposed that electrons orbit the nucleus of an atom in shells with discrete orbit radii and discrete energies.
Chadwick	Proposed that the atom contains neutral particles (neutrons) within its nucleus.
Marie Curie	• Proposed that radioactivity is an atomic phenomenon. • Shared in a Nobel prize for her work in radioactivity and was awarded a second Nobel prize for isolating the elements radium and polonium. • Opened the field of radiotherapy and nuclear medicine.

Examination tips

Format of the CSEC® Physics examination

The examination consists of **two papers** and your performance is evaluated using the **two** profiles:

- **Knowledge and comprehension**
- **Use of knowledge**

Paper 01 (1¼ hours)

Paper 01 is worth 60 marks and represents 30 per cent of the total examination mark. It consists of **60 multiple choice questions**, each worth **1 mark**. Four **choices** of answer are provided for each question.

- Make sure you read each question **thoroughly**; some questions may ask which answer is **incorrect**.
- Some questions may give two or more correct answers and ask which answer is the **best**; you must consider each answer very carefully before making your choice.
- If in doubt, try to **eliminate** incorrect answers. Never leave a question unanswered.

Paper 02 (2½ hours)

Paper 02 is worth 100 marks and represents 50 per cent of the total examination mark. It is divided into Sections A and B, and consists of **six compulsory questions**. The answers are to be written in spaces provided on the paper. These spaces indicate the length of answer required and answers should be restricted to them. Take time to **read the entire paper** before beginning to answer any of the questions.

Section A consists of three questions. Answers are to be written in the spaces provided for each part of the question.

- Question 1 involves **data analysis** and is the most heavily weighted question worth **25 marks**. It usually involves completing a table of readings, plotting a graph and calculating its slope, extracting information from the graph, forming conclusions from the results, and testing basic knowledge on the topic. You should designate **35 minutes** to this question.
- Questions 2 and 3 are each worth **15 marks**. These are **structured questions** whose parts require short answers, usually a word, a sentence or a short paragraph. They may also involve labelling a diagram, filling a table, or performing a calculation. You should designate **20 minutes** to each of these questions.

Section B consists of three **extended response questions**, each worth **15 marks**. The answers are to be written in the space provided at the end of each question.

- Usually, you will be asked to describe an experiment, describe and/or explain physical phenomena, or perform calculations involving physics laws and equations. Your writing skills and your ability to think outside the box are more relevant in this section. You should designate **20 minutes** to each of these questions.

If you use the times designated above to answer the questions, you should have 15 minutes to fully read the paper before you begin, and time to check over your answers when you have finished.

Interpreting examination questions

It is essential that you fully **understand** what is being asked in each question before you begin to answer. Always look at the **number of marks** allocated for each question and make sure you include at least as many **points** in your answer as there are **marks**. The following **key instruction words** tell you the **type of detail** expected in your answers.

Annotate	provide a short note to a label
Apply	solve problems using knowledge and principles
Assess	give reasons for the importance of particular structures, relationships or processes
Cite	quote or refer to
Classify	partition into groups according to characteristics
Comment	state an opinion or point of view validated by reason
Compare	state similarities and differences
Construct	represent data using a specific format to draw a graph, scale drawing or other device
Contrast	state differences
Deduce	logically connect pieces of information; arrive at a conclusion from data
Define	concisely state the meaning of a word or term
Demonstrate	show
Derive	logically determine a relationship, formula or result from data
Describe	give detailed information on the form of a structure or the sequence of a process
Determine	find the value of a physical quantity
Design	plan and present with relevant practical detail
Develop	expand on an idea or argument by using supportive reasoning
Distinguish	briefly identify differences between items that place them in separate categories
Discuss	present the relative merits of a situation with reasons for and against
Draw	produce a line illustration depicting an accurate relation between components
Estimate	arrive at an approximate quantitative result
Evaluate	judge based on given criteria from supporting evidence
Explain	make an account giving reasons based on recall
Formulate	develop a hypothesis
Identify	point out specific constituents or characteristics
Illustrate	clearly present using examples, diagrams or sketches
Investigate	observe, log data, and logically conclude by using basic systematic procedure
Note	record observations
Observe	examine and record changes that occur during a particular situation
Outline	write an account with the main points only
Plan	make preparation for the operation of a particular exercise
Predict	suggest a possible conclusion based on given information
Record	give an accurate account of all observations made during a procedure
Relate	show connections between sets of facts or data
Sketch	a freehand diagram illustrating relevant approximate proportions and characteristics
State	give facts in concise form without explanations
Suggest	explain, based on given information or on recall
Test	discover using a set of procedures

Successful examination technique

- **Read the instructions** at the start of each paper very carefully and do **precisely** what is required.
- **Read through the entire paper** before you begin to answer questions.
- **Read each question at least twice** before beginning your answer to ensure you **understand** what is being asked.
- **Study diagrams**, **graphs** and **tables** in detail and make sure that you **understand** the information given before answering the questions that follow.
- **Underline important words** in each question to help you in precisely answering it.
- **Reread** the question when you are **part way through** your answer to check that you are answering what it asks.
- **Give precise** and **factual answers**. You will not get marks for 'padded out' or irrelevant information.
- **Use correct terminology** throughout your answers.
- Present **numerical answers** with their appropriate **unit** (if any) using correct symbols e.g., cm^3, g, °C.
- If asked to give a **specific number of points**, use **bullets** to make sure each point is clearly separate.
- If a question requires you to give **similarities** and **differences**, you must make it clear which points you are proposing as similarities and which points as differences. The same applies if you are asked to give **advantages** and **disadvantages**.
- **Watch the time** as you work. Know the time available for each question and stick to it.
- **Check over your answers** when you have completed all the questions.
- **Remain in the examination room** until the **end** of the examination and, if there is time, recheck your answers to ensure you have done your very best.

School-Based Assessment (SBA)

School-based assessment (SBA) is an integral part of your CSEC® Examination. It assesses you in the **experimental skills** and the **analysis and interpretation skills** that are involved in laboratory and field work. It is intended to assist you in acquiring certain knowledge, skills and attitudes that are critical in your field of study. The important points that you should note about SBA are summarised below.

- The assessments are carried out in your school by **your teacher** during Terms 1 to 5 of your two-year period of study.
- The assessments are carried out during **normal practical classes** and not under examination conditions, so you have every opportunity to gain a high score in each assessment.
- SBA is worth **20** per cent of your final examination mark.
- You will be assessed in the following **four skills**:
  - Manipulation and Measurement
  - Observation, Recording and Reporting
  - Planning and Designing
  - Analysis and Interpretation
- Each skill will be assessed at **least three times** over the two-year period, and you will be awarded a mark between **0 and 10** for each assessment made.
- You will be **taught** the skills and be given enough opportunity to **develop** them before you are assessed. You will do a minimum of **sixteen** practical experiments over the two-year period.
- All your experimental reports are recorded in a **practical notebook** which is subject to moderation by the CXC® Examination Board to assess the standard of marking in your school.
- As an integral part of your SBA, you will also carry out an **Investigative Project** during the second year of your two-year study period (see p. xxi). This project will assess your Planning and Designing, and Analysis and Interpretation skills. If you are studying two or more of the single science subjects, Biology, Chemistry and Physics, you may choose to carry out ONE investigation only from any one of these subjects.

Skills assessed for SBA

Manipulation and Measurement (MM)

Manipulation and Measurement is assessed whilst you are conducting experiments in the laboratory. In assessing this skill, your teacher will be looking to see if you fulfill the following **general criteria**:

- Use basic laboratory equipment with competence and skill.
- Take precise and accurate readings when using laboratory equipment to make measurements.
- Show mastery of laboratory techniques.
- Carefully follow instructions.
- Carefully handle the apparatus.
- Use materials economically.

The criteria to be met for this skill can be specific to the apparatus used in the experiment. Your teacher will be looking to see if you fulfill the following **specific criteria:**

Use of balance to measure mass

- Place the instrument on a level surface.
- Ensure that there is no zero error.
- Read the scale to avoid parallax error (if the scale is analogue).

Use of vernier caliper and micrometer

- Ensure that there is no zero error.
- Lightly close the caliper/micrometer onto the object.
- Read the scale to avoid parallax error.

Use of measuring cylinder

- Select a cylinder of appropriate size.
- Place the cylinder to stand vertically on a level surface.
- Take readings from the midpoint of the curve of the meniscus. For most liquids, this is the bottom of the meniscus, but for mercury, it is the top.

Use of burette

- Initially rinse the inside of the burette with the liquid to be measured.
- Adjust the burette so that it stands vertically.
- Correctly use the funnel when adding liquid to the burette.
- Eliminate air bubbles and ensure that the lower tip of the burette is filled with liquid.
- Remove excess drops from the lower tip of the burette before each measurement; allow excess drops to fall into the measured volume after each measurement.
- Avoid parallax error by taking eye-level readings.
- Take readings from the bottom of the meniscus of most liquids. For mercury, the readings are taken from the top of the meniscus.

Use of stopwatch

- Ensure that there is no zero error.
- Use a 'count-down' method where appropriate. For example, to start the timing of the oscillations of a pendulum, count from 5 to 0, decrementing by 1 each time the pendulum swings through the vertical position from the same direction. On reaching 0, start the stopwatch.
- Operate the stopwatch correctly.
- Avoid parallax error when reading analogue scales.

Use of thermometer to measure temperature of liquid

- Completely submerge the bulb of the thermometer.
- Ensure that the bulb of the thermometer does not touch the walls of the container.
- Stir the liquid to ensure uniform temperature.
- Allow sufficient time for the thermometer to acquire the temperature of the liquid.

Use of Bunsen burner

- Check that the connections of the hose to the gas supply and to the Bunsen burner are good.
- Wear safety goggles.

- Close the air hole on the burner before igniting it.
- Carefully bring lit splint or match over the top of the burner.
- Turn on gas flow producing luminous flame at top of barrel.
- Open the air hole and adjust it to produce a clear burning flame.
- Adjust the gas supply to regulate the desired rate of combustion.

Plotting light rays using optical pins

- Place the pins to stand firmly and vertically along each ray.
- Place pins to stand far apart along each ray.
- Circle and label the pinholes to identify the position of the pins.
- Accurately draw lines through the pinholes to represent rays.
- Correctly place the baseline and centre of the protractor when measuring angles.
- For experiments showing refraction through a transparent block,
  - accurately outline the block with a pencil
  - carefully position the block within its outline when aligning pins.
- For experiments showing reflection from a mirror, accurately place the silver back surface of the mirror on the line when aligning pins.

Setting up electrical circuits

- Correctly draw the circuit diagram (if required).
- Select ammeter and/or voltmeter of suitable range (if applicable).
- Check ammeters and voltmeters to ensure that there is no zero error.
- Correctly arrange the components of the circuit.
- Connect the appropriate scale of the meters if they have more than one scale.
- Ensure that all connections are tight in the circuit.
- Ensure that ammeters, voltmeters, cells, and diodes are connected with correct polarity (see Chapter 26, Figure 26.2).
- Set up circuit with switch initially off.
- Set variable resistor initially to maximum resistance to avoid possible excessively high current.
- Read ammeter and/or voltmeter with the line of sight perpendicular to the scale (if analogue) to avoid parallax error. Use mirror behind pointer if applicable (see Chapter 1, Figure 1.3).

Observation, Recording and Reporting (ORR)

In assessing this skill, your teacher will look to see that:

- You use the **correct format** for your report with appropriate headings: Aim, Apparatus and Materials, Method/Experimental Procedure, Results/Observations, Discussion, Conclusion.
- Your report is written in the **past tense** and **passive voice** with correct **spelling** and **grammar**.
- Your report of the method is **complete, concise,** and **logically sequenced**.
- You use the **correct terminology** and **expressions**.
- You draw an appropriate **diagram** which
  - is **fully labelled** and **accurately** represents the apparatus as it was set up for use
  - is drawn with a **sharp pencil**, using a **ruler for all straight lines,** and is **not shaded**
  - bears an appropriate **title** below it
  - is **two-dimensional, adequately sized** and **proportioned.**

- You make appropriate use of prose, if applicable, to **accurately** record your observations.
- Your table recording data
  - is **neatly enclosed** and appropriately **titled**
  - has correct **column headings** indicating **quantity**, **symbol**, and **unit** (if applicable)
  - indicates a **repetition** and **averaging** of readings if applicable, e.g. when finding the period of a simple pendulum or the force constant of a spring
  - has values in each column to the **same precision**, i.e. the precision of the measuring instrument
  - has accurate **values in line with expected results**
  - has an **adequate range** of readings
  - has an **adequate number** of readings (usually six when related to straight line graphs, but more when related to curved graphs).
- Your graph
  - is **appropriately titled**
  - has data points marked by **small crosses** or **fine points** surrounded by a **small circle**
  - has **appropriate** and **easy to work with** scales that use **2/3** or more of the graph paper
  - has **axes** which are **correctly labelled** with quantity or symbol, and unit (if applicable)
  - has its data points **accurately plotted**
  - shows a **smooth curve or straight line** passing between the data points such that the mean deviation from them is minimum (a '**line of best fit**').

Planning and Designing (PD)

Your teacher will supply you with a **scenario** or **problem statement**. You will be expected to suggest an appropriate hypothesis for the scenario or problem and then design a suitable experiment to test your hypothesis. A hypothesis is a proposal based on limited evidence intended to explain certain facts or observations. In some cases, you may be asked to carry out the experiment. You will be expected to include the following in your design:

- A clear, concise statement of your **hypothesis** in a form which is testable, and which relates to the problem statement.
- A clear statement of your **aim** which relates to the problem statement and to your hypothesis. Your aim should also specify the technique to be used.
- A list which contains all the essential **apparatus and materials** to be used in your design.
- A workable or feasible **method** which could be used to test your hypothesis. Your method must include all the important steps written in a logical sequence and be in the present or future tense.
- An appropriate **circuit diagram** for electrical experiments.
- The manipulated, responding and controlled **variables** clearly identified together with how they will be measured.
- A clear statement of the **data** you expect to collect that will **support** or **disprove** your hypothesis.
- An explanation of how the data you expect to collect is to be **treated or interpreted** to fulfill the aim and to support or disprove your hypothesis.
- A statement of any **assumptions** to be made, possible **limitations** of your design and any **non-standard precautions** to be taken.

Analysis and Interpretation (AI)

In assessing this skill, your teacher will look to see if you can:

- **Make accurate calculations from data**
  -) If the values of a column in a table are calculated as the product or quotient of one or more columns, then they must be to the same number of significant figures as the **least number of significant figures** of the items in the calculation (see Chapter 1, page 7); the new unit must also be the logical combination of the units of the quantities from which the item was derived.
- **Analyse and interpret graphs.**
  -) Gradients of graphs must be calculated from a **large gradient triangle** marked on the paper.
  -) The coordinates used for the gradient must be taken from the **line** of the graph, not from the experimental points, and should also be marked on the paper.
  -) The **correct formula** should be used for calculating gradients.
  -) **Coordinates** used for the gradient calculation must be **accurately read** from the graph to the correct number of significant figures.
  -) Calculations of gradients must be **accurate** and to the correct number of significant figures.
  -) Most gradients and intercepts have **units**.
  -) The number of significant figures obtained from the axes or gradient of a graph should be obtained from that used in the plot.
  -) Calculations of other quantities produced by manipulation of a gradient or intercept must be presented to the correct number of significant figures and with the correct derived unit.
- Accurately **identify relationships, patterns and trends** displayed by your results.
- **Make logical predictions and inferences** from your results.
- **Compare actual results with expected results** and suggest reasons for any differences.
- Suggest **alternative methods or modifications** to existing methods.
- **Evaluate data** by identifying limitations and sources of error.
- **Draw conclusions** from data or graph.

The Investigative Project

You will be expected to carry out an Investigative Project during your second year of study. To begin your project, your teacher may supply you with a **scenario** or **problem statement**. Alternatively, you may be asked to come up with a scenario or problem statement of your own. Your Investigative Project will be divided into two parts as follows:

- **The proposal**

  Firstly, you will be expected to suggest an appropriate hypothesis for the scenario or problem and then to design a suitable experiment to test your hypothesis. This is your **proposal**, and your teacher will use it to assess your **Planning and Designing** skills. Your proposal is worth a maximum of 10 marks, and the criteria that will be used in its assessment are outlined below:

  - ✦ Your **hypothesis** is clearly stated and testable .... *2 marks*
  - ✦ Your **aim** clearly relates to your hypothesis .... *1 mark*
  - ✦ The appropriate **materials and apparatus** are given .... *1 mark*
  - ✦ Your **method** is suitable and includes at least one manipulated or responding variable ... *2 marks*
  - ✦ At least one **controlled variable** is stated .... *1 mark*
  - ✦ Your **expected results** are given and linked to the method .... *2 marks*
  - ✦ One **assumption, precaution or possible source of error** is stated .... *1 mark*

  10 marks

- ## The Implementation

  You will then be expected to carry out your experiment. This is your **implementation**, and your teacher will use it to assess your **Analysis and Interpretation** skills. Your implementation is worth a maximum of 20 marks, and the criteria that will be used in its assessment are outlined below:

  + Your **method** is linked to the proposal and is written in the correct tense   *1 mark*

  + Your **results** contain correct formulae and equations, and are accurate   *4 marks*

  + Your **discussion** explains and interprets your results and states any trends   *5 marks*

  + At least one **source of error**, one **precaution** and one **limitation** are stated   *3 marks*

  + A written **reflection** on your work is given using the appropriate scientific language, grammar and clear expression, and which includes the relevance between your experiment and real life, the impact of the knowledge you gained from the experiment on yourself and the justification for any adjustments you made during the experiment   *5 marks*

  + A **conclusion** is given which relates to the aim   *2 marks*

       20 marks

Starter unit

In this unit

Quantities, units and measurement

- Here you will learn of the SI system of units, fundamental and derived quantities and units, the use of scientific notation, prefixes and significant figures, the meaning of accuracy and precision, and the types of errors associated with measurements.

Measuring instruments

- A look at various measuring instruments used in the physics laboratory and their suitability for particular measurements based on sensitivity, accuracy and range.

Scalars and vectors

- This chapter distinguishes between scalar and vector quantities and teaches the special techniques that must be employed to sum vector quantities.

Scientific method

- Here you will learn the scientific methodology to apply to experiments, including the use of different types of variables, the format of presenting experimental data and the techniques necessary to analyse that data.

1 Scientific measurement and manipulation of data

This chapter provides the preliminary tools and knowledge required to fully understand, analyse and synthesise the information provided in the five sections of the syllabus. It is divided into four subsections: quantities, units and measurement; measuring instruments; scalars and vectors; and scientific method.

Quantities, units and measurement

Scientists, engineers and many others are constantly working to better understand the world around us. To do so, they must **measure** various **quantities** and must use their measurements to help in the design of devices, utilising the best materials and techniques available. So that a quantity be adequately measured, it is generally expressed by a **number** with a specific **unit** assigned to it.

Learning objectives

- Be familiar with the **SI system of units**.
- Define the term **fundamental quantity** and list the fundamental quantities with their SI base units.
- Define the term **derived quantity** and determine the SI base units of derived quantities from formulae.
- Understand that some derived units have **special names** and symbols.
- Express numbers in **scientific notation**.
- Express and **convert SI units using prefixes**.
- Distinguish between **accuracy** and **precision**.
- Distinguish between **systematic errors** and **random errors**.
- Understand the phenomenon of **parallax error** and describe ways of **preventing it**.
- Describe a **non-linear** scale and explain why it is difficult to estimate readings on such a scale.
- Understand the use of **significant figures** and **precision** in measurements and calculations.

Fundamental and derived quantities and units

Measurements are easily interpreted internationally by using a standard system of units known as the **SI system of units** (International System of units). It has been developed from the **metric system**, a system in which multiples and sub-multiples of its units are expressed as powers of 10. The SI system of units is comprised of the units of **seven fundamental quantities**.

Fundamental quantities are those which do not depend on any other physical quantity for their measurement.

These are shown in Table 1.1, together with their **fundamental SI base units**.

Table 1.1 *Fundamental quantities and their SI base units*

SI fundamental quantity	SI base unit	Symbol of SI base unit	SI fundamental quantity	SI base unit	Symbol of SI base unit
Mass	kilogram	kg	Temperature	kelvin	K
Length	metre	m	Amount of substance	mole	mol
Time	second	s	Luminous intensity	candela	cd
Current	ampere	A			

Derived quantities are those which depend on other physical quantities for their measurement.

These quantities are obtained by applying physical laws and equations to fundamental quantities. They may be expressed in terms of fundamental units and/or other derived units.

An example of a derived quantity is speed. Speed (*v*) depends on the fundamental quantities length (*l*) and time (*t*) in the relation $v = \dfrac{l}{t}$. The SI base unit of speed is therefore **m s⁻¹**, where **m** and **s** are fundamental base units.

Another derived quantity is force. Force (*F*) depends on the fundamental quantity mass (*m*) and the derived quantity acceleration (*a*) in the relation **$F = ma$**. Using Table 1.2, it can be shown that the SI unit of the derived quantity, force, is therefore **kg m s⁻²**, where **kg**, **m** and **s** are the fundamental base units. A force of 5 kg m s⁻², however, is better expressed as 5 N using the **special name** SI unit, the newton (N).

Not all derived units have special names. See Table 1.2 for a better understanding of derived units.

Table 1.2 *Some derived quantities and their SI units*

SI derived quantity	Formula		Special name SI unit	SI unit using the formula column	Constituent fundamental units
Velocity (*v*)	$v = \dfrac{\Delta s}{\Delta t}$	$\text{velocity} = \dfrac{\Delta \text{displacement}}{\Delta \text{time}}$	-	m s⁻¹	m s⁻¹
Acceleration (*a*)	$a = \dfrac{\Delta v}{\Delta t}$	$\text{acceleration} = \dfrac{\Delta \text{velocity}}{\Delta \text{time}}$	-	m s⁻²	m s⁻²
Force (*F*)	$F = ma$	force = mass × acceleration	newton (N)	kg m s⁻²	kg m s⁻²
Work (*W*) or energy (*E*)	$W = Fd_\parallel$ $E = Fd_\parallel$	work = force × distance energy = force × distance	joule (J)	N m	kg m² s⁻²
Power (*P*)	$P = \dfrac{W}{t}$	$\text{power} = \dfrac{\text{work}}{\text{time}}$	watt (W)	J s⁻¹	kg m² s⁻³
Pressure (*P*)	$P = \dfrac{F}{A}$	$\text{pressure} = \dfrac{\text{force}}{\text{area}}$	pascal (Pa)	N m⁻²	kg m⁻¹ s⁻²
Charge (*Q*)	$Q = It$	charge = current × time	coulomb (C)	A s	A s
Voltage (*V*)	$V = \dfrac{E}{Q}$	$\text{voltage} = \dfrac{\text{energy}}{\text{charge}}$	volt (V)	J C⁻¹	kg m² A⁻¹ s⁻³
Resistance (*R*)	$R = \dfrac{V}{I}$	$\text{resistance} = \dfrac{\text{voltage}}{\text{current}}$	ohm (Ω)	V A⁻¹	kg m² A⁻² s⁻³
Frequency (*f*)	$f = \dfrac{1}{T}$	$\text{frequency} = \dfrac{1}{\text{period}}$	hertz (Hz)	s⁻¹	s⁻¹

Scientific notation or standard form

To express a number in a **shorter way** it can be represented as the product of a **mantissa *M*** and a **power of 10**: $M \times 10^P$.

- *M* is a number in **decimal form** with only **one non-zero digit before the decimal point**.
- *P* (the **exponent**) is an integer whose value is the number of decimal places moved in representing the mantissa.
- *P* is negative when the number being expressed is less than 1.

$$45\,000 = 4.5 \times 10^4 \qquad\qquad 0.000\,52 = 5.2 \times 10^{-4}$$

Prefixes

Another way to express very large or very small quantities in a shorter way is to use a **prefix** with the unit. A prefix **joined to a unit** denotes that the unit is multiplied by a certain power of 10.

The symbols for some prefixes are in **lower case** and others are in **uppercase**.

Table 1.3 *Some useful prefixes*

Prefix	pico	nano	micro	milli	centi	deci	no prefix	kilo	mega	giga	tera
Symbol	p	n	μ	m	c	d	no prefix	k	M	G	T
Power of 10	10^{-12}	10^{-9}	10^{-6}	10^{-3}	10^{-2}	10^{-1}	10^{0}	10^{3}	10^{6}	10^{9}	10^{12}

For each jump in the table there is a shift of decimal places equal to the absolute change in the exponents (**absolute change = larger exponent – smaller exponent**).

a   45 000 W = 45 kW

No prefix (10^0) to kilo (10^3):   Absolute change = **(3) – (0) = 3**

Converting to a **larger unit** always produces a **smaller number**, so the decimal point is shifted by 3 places to make the number smaller.

b   45 MW = 45 000 000 000 mW

Mega (10^6) to milli (10^{-3}):   Absolute change = **(6) – (–3) = 9**

Converting to a **smaller unit** always produces a **larger number**, so the decimal point is shifted 9 places to make the number larger.

Accuracy and precision

Accuracy: The degree of correctness of a measurement.

An accurate result is one that is close to the **true value.**

*Precision: The degree of **discrimination** between values of a measured quantity.*

A precise result is one in which the **uncertainty** of the measurement is small.

An accurate measuring instrument is one that produces readings whose **mean value is close to the true value** of the quantity being measured.

A precise measuring instrument is one which always produces readings **close** to some **mean value** for a given measurement. A measure of the **precision** of an instrument can be obtained from the **place value** of the **least significant figure** of its readings: an instrument which produces a reading of **3.6 N** is more precise than one which produces a reading of **4 N** for the same measurement.

Figure 1.1 shows that a reading can be very precise but still inaccurate. This would be the case if the true value of the measurement of 3.6 N mentioned above is **4.1 N.**

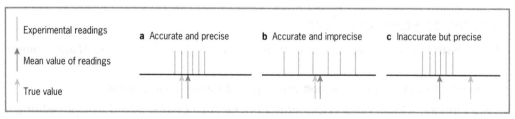

Figure 1.1 *Accuracy and precision*

Systematic errors and random errors

Systematic errors

*A systematic error is one that affects all measurements of a given quantity by the **same amount** or by the **same factor**.*

These errors produce values **either** greater **or** lesser (not both) than the value being measured and may therefore be either positive or negative. This type of error **can be controlled**.

Causes

Usually due to poor **apparatus** but can be due to some physical disability of the experimenter.

- **Zero error** – This occurs when the reading on the instrument is not zero although the quantity being measured is. For example, the pointer of an ammeter may indicate 0.2 A when no current flows through the instrument. It will indicate 0.2 A in excess of each true reading when current is flowing.
- **Poorly calibrated scale**.
- **Faulty apparatus** – for example, a weak spring of a spring balance.
- **Parallax error** (see Figure 1.2) – This is generally a random error but can be a systematic error if the observer always incorrectly reads the instrument from the same angle.

Effect

Reduces **accuracy**.

Reducing systematic errors

- Calibrate the apparatus against a known reliable source.
- Add or subtract a fixed amount from each reading (e.g. for zero error).
- Adjust or service instruments (e.g. for zero error or rusty spring).

Random errors

*A **random error** is one which affects repeated measurements of the **same** value **by random amounts** (uncertainties) which may be greater or lesser than the **mean value**.*

These '*uncertainties*' are generally the value of the nearest ½ division of the markings on the scale above and below the measured value. This type of error **cannot be controlled**.

Causes

- Poor estimations of readings between divisions, particularly of a **non-linear scale** (see Figure 1.4).
- Reaction time in starting and stopping a stopwatch.
- Poor skill in avoiding **parallax** error.
- **Unpredictable** variations of **environmental conditions** (temperature changes/wind draughts).

Effect

Reduces **precision**.

Reducing random errors

- Find the mean of **repeated** readings or of readings taken by **different observers** (see Table 1.8 page 22).
- Construct the **best straight line** or **smoothest curve** through points of a graph (see Figures 1.26 and 1.27 page 23).
- Skillfully take readings, including the application of techniques to reduce **parallax error** (see below).
- Prevent variations in environmental conditions. For example: be careful not to cause vibrations when measuring the time of a pendulum's oscillation; close windows to avoid sudden drafts.

Parallax error and how to reduce it

Parallax error occurs when the marking on a scale is **not coincident in position** with the point to which the measurement should be made and the line of sight of the observer is not perpendicular to the scale (see Figure 1.2).

- Ammeters and voltmeters usually have a **mirror** placed **behind their pointer** (see Figure 1.3). Parallax error is reduced if readings are taken from a position where the pointer is directly in front of its image.
- The scale should be viewed using only **one eye** since the line of sight must be perpendicular to it.
- For instruments with pointers, the scale should be as **close** as possible **to the pointer**. For instruments such as mercury thermometers, the scale should be marked onto the stem so that it is close to the mercury thread.
- If the scale is **vertical**, such as with a measuring cylinder, **eye level** readings should be taken.

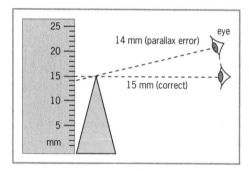

Figure 1.2 *Parallax error*

Figure 1.3 *DC ammeter*

Linear and non-linear scales

A **linear scale** is one which has a uniform separation between its intervals, e.g. the scale on a metre rule.

A **non-linear scale** is one in which the **separation of its intervals** is **non-uniform**, making it **difficult to estimate readings**.

The reading on the scale shown in Figure 1.4 is greater than 25 mA although the pointer is physically midway between the graduations of 20 mA and 30 mA. This occurs since a smaller space is designated on the scale for currents between 20 mA and 25 mA than for currents between 25 mA and 30 mA.

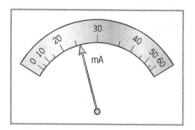

Figure 1.4 *Non-linear scale of AC ammeter*

Significant figures and precision in measurements and calculations

Significant figures

Significant figures are the set of digits, **excluding leading zeros**, used to express a **measured** quantity, or a quantity **calculated** from measurements. Note the following:

- all **non-zero digits** are **significant**
- **leading** zeros are **NOT significant**
- **trailing zeros** in the **decimal section** of the number are **significant**.

Examples:

17.5432   6 significant figures

  27.90   4 significant figures; trailing zeros in the decimal section are significant

  0.045   2 significant figures; leading zeros are not significant

    900   These trailing zeros are not in the decimal section.

            The number could be represented to 1, 2 or 3 significant figures; it is better expressed in scientific notation:

- if expressed as 9×10^2, it is 1 significant figure
- if expressed as 9.0×10^2, it is 2 significant figures
- if expressed as 9.00×10^2, it is 3 significant figures.

Calculations involving products and/or quotients of measured values

The result should be to the **same number of significant figures** as the **measured value** with the **least number of significant figures**.

Note: Only **measured values** are considered. Exact values such as integers, or constants such as π, are treated as though they are expressed with an infinite number of significant figures.

For example:

$$P = \frac{32 \text{ kg} \times 61.25 \text{ m} \times 2 \times \pi}{3.75 \text{ s}^2} = 3284.011521 \text{ kg m s}^{-2} \qquad \dots \text{ as per calculator}$$

32 kg is the measured value expressed to the least number of significant figures – TWO significant figures. The result should therefore be expressed to **TWO significant figures:** $P = 3300 \text{ kg m s}^{-2}$.

Calculations involving addition and/or subtraction of measured values

The result should be to the **same precision** as the **measured value** with the **least precision**.

For example: $Q = 3.76 \text{ m} + 24.1 \text{ m} = 27.86 \text{ m}$ ... as per calculator

The first term is precise to TWO decimal places, but the second term is precise to only **ONE decimal place**. The result should therefore be precise to **ONE decimal place:** $Q = 27.9 \text{ m}$.

Recalling facts

1 **a** Distinguish between fundamental and derived quantities.

b State FIVE fundamental quantities with their corresponding fundamental SI base units.

c State THREE derived quantities with their *special name SI units*.

2 Distinguish between the accuracy and precision of a measurement.

3 Why are readings taken from a *non-linear scale* usually not as accurate as those taken from a linear scale?

Applying facts

4 **a** Distinguish between systematic and random errors.

b Complete the table to show whether the following are random errors or systematic errors.

Reason for error	Type of error
Pointer of ammeter not at zero when no current flows	
Poorly calibrated scale	
Reading a measuring cylinder from a vantage-point which is at times too high and at other times too low	
Unpredictable and unavoidable wind gusts (i.e. uncontrollable)	
Reaction time	
Rusted spring of a spring balance	
Consistently reading the scale at the same incorrect angle	

c List TWO ways of reducing random error and TWO ways of reducing systematic error.

d Which type of error affects precision and which type affects accuracy?

5 Express the following in scientific notation (standard form).

a 297.3 **b** 42 754 **c** 0.0057

6 Express the following in kΩ.

a 7215 Ω **b** 0.045 MΩ **c** 654 321 mΩ

7 The temperature of a liquid is measured by thermometers A and B as 24.67 °C and 23 °C, respectively. The true value of the temperature to 3 significant figures is 23.1 °C. Which thermometer is more precise, and which is more accurate?

8 Express X, Y and Z to the correct number of significant figures.

a i $X = 2.42 \text{ A} + 1.2 \text{ A}$ **ii** $Y = 1.4 \text{ g} + 1.3 \text{ g} + 1.4 \text{ g} + 4 \text{ g}$ **iii** $Z = \dfrac{4.0 \text{ m}^2}{2.304 \text{ m}}$

b Express 4000 in scientific notation to two significant figures.

Measuring instruments

Different measuring instruments are designed to measure different quantities within various ranges and to varying degrees of accuracy and precision.

Learning objectives

- Differentiate between an **analogue** instrument and a **digital** instrument.
- Be capable of **using measuring instruments** to measure length, area, volume, mass and time.
- Be capable of **converting units** between mm², cm² and m², and between mm³, cm³ and m³.
- Discuss the **suitability of measuring instruments** in terms of sensitivity, accuracy and range.

An **analogue** instrument usually produces readings which can be read to the nearest half of an interval on its scale. For a metre rule **graduated in mm**, its **precision is 1 mm**. This means that a reading of 254 mm can be expressed with its possible error as 254.0 mm ± 0.5 mm. The instrument is therefore **sensitive** to the nearest 0.5 mm (**half of the interval size**).

A **digital** instrument produces readings which can differ in amount by a minimum **step-size** in its **display window**. For a voltmeter this may be 0.1 V, 0.2 V etc. If a voltmeter has a step-size of 0.1 V, its precision is 0.1 V. A reading of 4.3 V can be expressed with its possible error as 4.30 V ± 0.05 V. The instrument is therefore sensitive to the nearest 0.05 V (**half of the step-size**).

Measuring length

Table 1.4 *Instruments to measure length*

Instrument	Appropriate application of measurement	Approximate range of common applications	Precision
Measuring tape	length of garage	a few cm ⟶ about 50 m	variable
Ruler (metre rule)	height of desk	1 cm ⟶ 100 cm	1 mm
Vernier calliper	diameter of table tennis ball	1 cm ⟶ 10.0 cm	0.1 mm
Micrometer screw gauge	thickness of writing paper	0.1 mm ⟶ 30.0 mm	0.01 mm

Vernier calliper

The vernier calliper (Figure 1.5) has an extra scale known as a **vernier scale** which is capable of measuring to one extra decimal place over that measured by the **main scale**. This calliper has a main scale which can be read to the nearest **1 mm** and a vernier scale which gives a reading to the nearest **0.1 mm**. Note the following when using the instrument.

- The jaws of the calliper are adjusted to hold the object by sliding the vernier scale over the main scale.
- The **main scale** is read to the marking just **before** the zero marking on the vernier scale.
- The marking on the **vernier scale** that best aligns with a marking on the main scale gives the added significant figure.

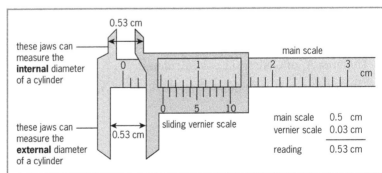

Figure 1.5 *Vernier calliper*

Micrometer screw gauge

Figure 1.6 shows a micrometer screw gauge. This instrument is more precise than a vernier calliper and can be used for measuring lengths as small as the diameter of a thin wire. Like the vernier, it produces a reading with greater precision than that of just the main scale. Each revolution of the thimble moves the jaws by 0.50 mm. The following should be noted when using the instrument.

- The **main scale** is on the sleeve and is read at the marking just before the thimble. The markings below the horizontal line are half-intervals (½ of 1 mm).
- The **thimble scale** is read at the marking that coincides with the horizontal line on the sleeve.
- The two readings are summed.

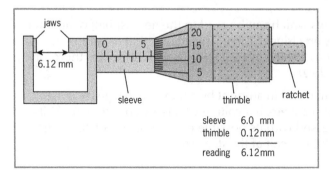

sleeve	6.0 mm
thimble	0.12 mm
reading	6.12 mm

Figure 1.6 *Micrometer screw gauge*

Measuring area

Regular shapes

Area of rectangle: area = length × width $A = l \times w$

Area of triangle: area = $\dfrac{\text{base} \times \text{height}}{2}$ $A = \dfrac{b \times h}{2}$

Area of circle of radius r: $A = \pi r^2$

Surface area of sphere of radius r: $A = 4\pi r^2$

Irregular shapes

- The area being measured is superimposed onto a grid as in Figure 1.7.
- The number of squares, N, covered by the area is estimated. This can be done by finding the sum of the squares that are completely covered and then adding to it the number of complete squares that can be comprised from the remaining partially covered squares.
- The area A_1 of ONE square is calculated.
- The total area A_T is calculated from: $A_T = N \times A_1$

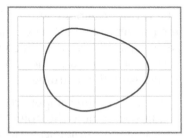

Figure 1.7 *Measuring the area of an irregular shape*

Converting between units of area

From Figure 1.8 we see that:

$$1 \text{ m}^2 = 100 \text{ cm} \times 100 \text{ cm} = 1 \times 10^4 \text{ cm}^2$$

- when converting from m^2 to cm^2, multiply by 10^4
- when converting from cm^2 to m^2, divide by 10^4, (i.e. multiply by 10^{-4}).

Similarly:

$1 \text{ m}^2 = 1000 \text{ mm} \times 1000 \text{ mm}$

$1 \text{ m}^2 = 1 \times 10^6 \text{ mm}^2$

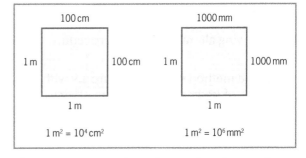

Figure 1.8 *Converting m^2 to cm^2 and to mm^2*

Measuring volume

Regular shaped solids

Rectangular cuboid: volume = length × width × height $V = l \times w \times h$

Sphere of radius, *r*: $V = \dfrac{4}{3}\pi r^3$

Cylinder of radius *r* and length *l*: $V = \pi r^2 l$

Regular or irregular shaped solids

A **displacement can** is filled with water until it overflows from its spout. When the overflow stops, a measuring cylinder is placed below the spout and the object is gently lowered by means of a string into the can until it is submerged (see Figure 1.9a).

The volume of water displaced into the measuring cylinder is the volume of the object.

Alternatively, if the object **can fit into a measuring cylinder**, as in Figure 1.9b, the volume of a given amount of water in the cylinder is measured **before and after** immersing the object into it. The difference in volumes represents the volume of the object.

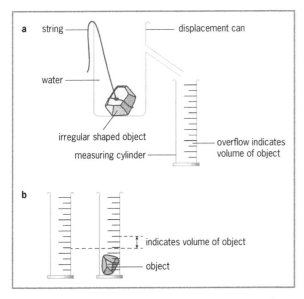

Figure 1.9 *Finding the volume of regular and irregular shaped solids*

For an **irregular shaped solid** that **floats in water** (e.g. a piece of cork or wood) one of the following alterations to the procedure is made.

1. Use the method shown in Figure 1.9 with a liquid of **lesser density** than the object.

2. Find the displacement of water produced as the object is pinned below the surface by a heavier object, and then the displacement produced by the heavier object alone. The **difference** in the two measurements is the volume of the solid which floats (see Figure 1.10).

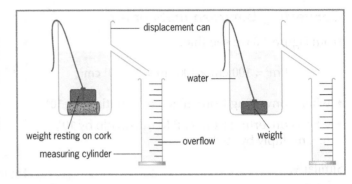

Figure 1.10 *Finding the volume of an irregular solid which floats*

For **irregular solids** that **dissolve in water** (e.g. sugar and salt), use the methods shown in Figure 1.9 with a non-polar solvent (e.g. kerosene) in which the solid is insoluble.

Liquids

Figure 1.11 shows instruments that can be used to measure the volume of a liquid by observation of the position of its meniscus. If the **meniscus is concave**, an **eye-level reading** is taken of its **lowest point** as shown in Figure 1.12. Note that the meniscus of mercury is convex and so is read from its highest point (see Figure 8.4, page 102).

Measuring cylinder – With the cylinder on a **firm level surface**, the volume of its contents is indicated by the meniscus.

Burette – With the burette clamped **vertically**, an initial reading of the liquid's meniscus is taken. The **tap** is then opened, and an amount of liquid is removed. The tap is then closed, and the new reading taken. The volume removed is obtained from the difference in the readings.

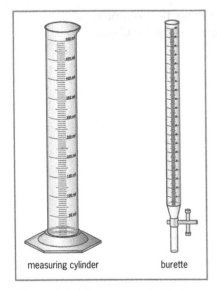

Figure 1.11 *Instruments that measure the volume of a liquid*

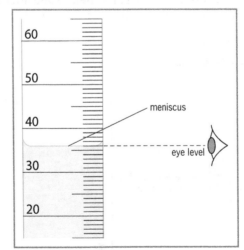

Figure 1.12 *Reading a meniscus*

Converting between units of volume

From Figure 1.13 we see that:

$$1\ m^3 = 100\ cm \times 100\ cm \times 100\ cm = 1 \times 10^6\ \mathbf{cm^3}$$

- when converting from m^3 to cm^3, multiply by 10^6
- when converting from cm^3 to m^3, divide by 10^6, (i.e. multiply by 10^{-6}).

Similarly:

$1\ m^3 = 1000\ mm \times 1000\ mm \times 1000\ mm$

$1\ \mathbf{m^3} = 1 \times 10^9\ \mathbf{mm^3}$

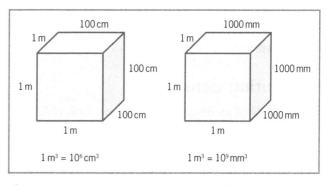

Figure 1.13 *Converting m^3 to cm^3 and to mm^3*

Measuring mass

Mass can be measured using the instruments shown in Figure 1.14.

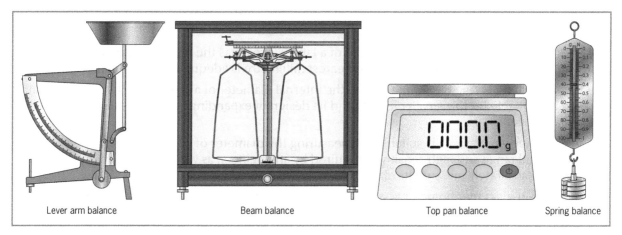

Figure 1.14 *Measuring mass*

The **top pan balance** and **spring balance** really **measure weight**, **not mass**, but can be graduated in units of mass by dividing the weight by a **mean value** of the **acceleration due to gravity on Earth** (see Chapter 2). Of course, since the acceleration due to gravity varies slightly at different points on Earth, and significantly on different planets, the top pan balance and spring balance will give different readings for the same mass at various locations.

A **beam balance** functions by **comparing standard masses** with the mass to be measured. It does require gravity to function, but the effect of gravity on the masses in the two scale pans cancels, and so it can accurately measure mass at various locations.

The **mass of a liquid** may be measured by subtracting the mass of an empty container from the mass of the container with the liquid in it.

Measuring time

Time may be measured by:

- **manually** operated stop clocks or stop watches
- **automatic stop clocks** which start and stop when **triggered** by **sensors** connected to their **start** and **stop terminals**

- by connecting **microphones** (sound sensors) to the start and stop terminals, the time of travel of a sound wave between two points can be measured (see Chapter 21, Figure 21.8, measuring the speed of sound)
- by connecting **motion sensors** to the start and stop terminals, the time of fall of an object between two points can be measured.

Measuring density

The **density** of an object is its mass per unit volume.

$$\rho = \frac{m}{V}$$

The density of a substance can therefore be calculated from its measured mass and volume.

Finding the mass – Any of the instruments shown in Figure 1.14 may be used.

Finding the volume – The volume of regular and irregular solids can be determined by methods shown on pages 11-12.

Suitability of measuring instruments

The **sensitivity (precision)**, **accuracy** and **range** of a measuring instrument determine its suitability for making a given measurement.

- A metre rule is used to measure the height of a table since it has the necessary range. Its scale is in mm and in most cases, accuracy to the nearest millimetre is adequate for such a measurement.
- A vernier calliper is suitable for measuring the internal diameter of a water pipe. Its sensitivity and range are suitable for this smaller length, and its design of expanding jaws make measurement of the diameter accurate.
- A micrometer screw gauge is suitable for measuring the diameter of an electrical wire because it has the small range and high sensitivity required by an engineer, as well as jaws for firmly holding the wire to accurately measure its diameter.
- A mercury-in-glass thermometer is suitable for measuring temperatures in the school laboratory because the experiments generally performed are within its range. It does not have to be as sensitive as a clinical thermometer.
- A clinical thermometer needs to be very sensitive and accurate for the doctor or nurse to make correct decisions regarding the patient's health. It requires only a small range since the body temperature of a living human varies by only a small amount.

Recalling facts

1 List THREE factors that determine the suitability of a measuring instrument for carrying out a given measurement.

2 Explain why a spring balance may produce different readings when measuring the same mass at different locations on Earth, but a beam balance will produce the same reading at those locations.

3 **a** Describe an experiment to determine the density of
  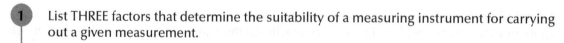
   i the cork of a wine bottle   **ii** sugar
  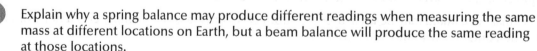
   b A measuring cylinder containing 26 cm³ of water has a mass of 84 g. When a small stone is submerged into the water, the water meniscus rises and indicates a volume of 80 cm³. Calculate the density of the stone if the mass of the cylinder and its contents increases to 192 g.

Applying facts

4 Express: **a** 2 m² as mm²   **b** 4 cm² as m²   **c** 5 m³ as cm³

5 What instrument is most appropriate for measuring each of the following?

  a The diameter of a very small bead of a necklace.

  b The height of a bar stool.

  c The internal diameter of a cylindrical drinking glass.

  d The length of the school hall.

6 Determine the reading given on each of the instruments shown in Figure 1.15.

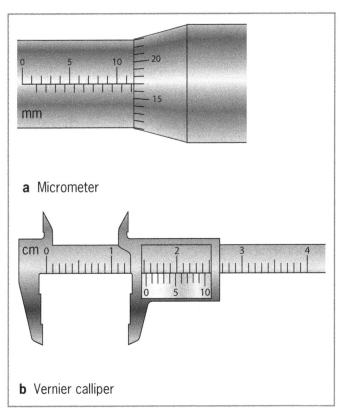

a Micrometer

b Vernier calliper

Figure 1.15 *Question 6*

Scalars and vectors

Some quantities, such as **energy**, are known as **scalar** quantities. They can be added by simply summing their values algebraically. For example, if the energy stored in a battery is 300 kJ, then in two such batteries the total energy is 600 kJ. However, other quantities, such as **force**, can only be summed if their directions are known. Such quantities are known as **vector** quantities. When two forces, each of 300 N and acting in the same direction are combined, they produce a total force of 600 N, but when acting in opposite directions they cancel to zero.

Learning objectives

- Distinguish between **scalar** quantities and **vector** quantities.
- Recall examples of scalars and vectors.
- Use scale diagrams to find the **resultant of two vectors.**
- **Calculate the resultant of vectors** which are parallel, anti-parallel and perpendicular to each other.
- **Resolve a single vector** into two perpendicular directions.
- Analyse situations with **three vectors in equilibrium.**

Table 1.5 *Distinguishing between scalars and vectors*

	Scalar	Vector
Definition	*a quantity that has only magnitude (size)*	*a quantity that has magnitude and direction*
Examples	mass, length, time, temperature, area, volume, speed, pressure, distance, work, energy, power, resistance, current	displacement, velocity, acceleration, force, momentum
How added	algebraic summation	vector summation (see below)

Determining the resultant of two vectors using scale drawings

Figure 1.16 shows four cases of forces A and B pushing or pulling on a particle. In each case, B is horizontal and to the right, and A is acting to the right at an angle θ above the horizontal. The resultant force can be calculated using either of the following methods illustrated in Figure 1.17.

Parallelogram method – The vectors are represented to **scale** in **magnitude** and **direction** by arrows, **starting at the same point** and forming the adjacent sides of a parallelogram. The resultant vector is then represented in magnitude and direction by the arrow forming the **diagonal** of the parallelogram **originating at the same point as the vectors being added.**

Polygon method – The vectors are represented in magnitude and direction to scale by arrows drawn head-to-tail forming a chain. The resultant vector is then represented to scale by the arrow **closing the polygon originating from the beginning of the chain.** This method is also referred to as the **triangular method** if only TWO vectors are being added.

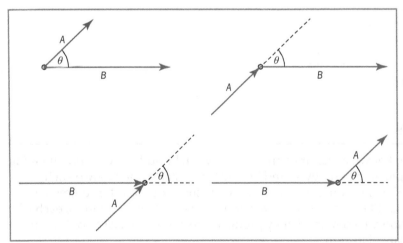

Figure 1.16 *Situations of two vectors acting at a point*

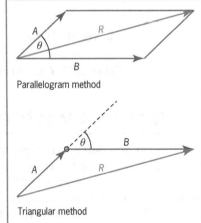

Figure 1.17 *Methods of adding vectors*

Example 1

An object is acted on by forces *A* and *B* of magnitudes 50 N and 40 N, respectively. *B* acts horizontally to the right and *A* acts in a direction 30° above the horizontal. By means of a scale diagram determine the resultant force.

Scale drawings representing the forces can be constructed using a ruler and protractor; a scale of 1 cm = 10 N is used here but a larger scale is more appropriate (see Figure 1.18).

Parallelogram method – Using the scale mentioned above, the **diagonal arrow** representing the **resultant** vector is 8.7 cm in length and therefore the resultant force is of magnitude 87 N. Its direction when measured by a protractor is found to be 17° above the horizontal.

Triangular method – Using the scale mentioned above, the **arrow closing the triangle** and representing the **resultant** vector is 8.7 cm in length and therefore the resultant force is of magnitude 87 N. Its direction is similarly found to be 17° above the horizontal.

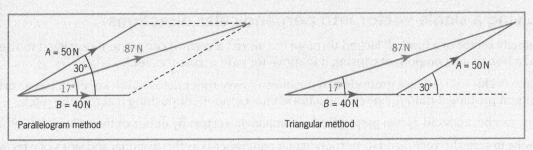

Figure 1.18 *Example 1*

Determining the resultant of vectors by calculation

Parallel and anti-parallel vectors

Since the vectors act along the same line, we assign one direction as positive and the opposite direction as negative. For the case shown in Figure 1.19, forces to the right are assigned positive values and forces to the left are assigned negative values.

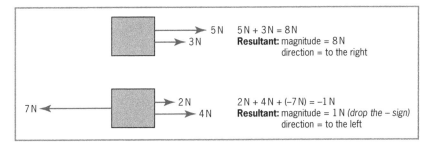

Figure 1.19 *Finding the resultant of parallel and anti-parallel vectors*

Perpendicular vectors

For the case shown in Figure 1.20a, upward forces and forces to the right are assigned positive values, and downward forces and forces to the left are assigned negative values.

- The resultant horizontal and vertical forces are first found as in Figure 1.20a.
- These resultants are then combined by either the parallelogram or triangular methods. A scale drawing or a calculation using Pythagoras' theorem and trigonometry can be applied as shown in Figure 1.20b.

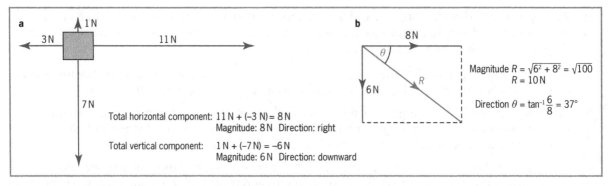

Figure 1.20 *Finding the resultant of perpendicular vectors*

Resolving a single vector into perpendicular directions

The velocity vector of a football kicked through the air has a vertical component causing it to rise and fall and a horizontal component causing it to move forward across the field.

The force vector exerted by a groundsman pushing a heavy roller across a cricket pitch has a vertical component pushing it onto the pitch and a horizontal component pushing it across the pitch.

A vector can be analysed as two **perpendicular component vectors** by either of the following methods.

- **Drawing to scale** the vector and its perpendicular components as the diagonal and sides of a **rectangle**.
- **Sketching** the vector and its perpendicular components as the diagonal and sides of a **rectangle**, and then applying **Pythagoras' theorem** and simple **trigonometry** to calculate the components.

Figure 1.21 shows how the components are calculated for a 5 N force acting at 37° above the horizontal.

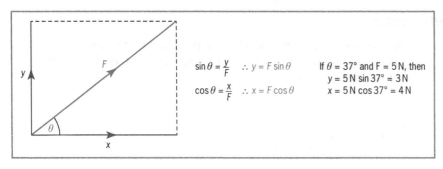

Figure 1.21 *Resolving a vector into perpendicular components*

Three vectors in equilibrium

If **THREE** vectors are in equilibrium (in balance), we can use the **parallelogram method** shown in Figure 1.22a to find the resultant of any **two**, and this would be equal but oppositely directed to the **third**. Alternatively, since the vectors are in **equilibrium**, their **resultant is zero**, and therefore they will form a **CLOSED polygon** (i.e. with no resultant) when joined head-to-tail as shown in Figure 1.22b.

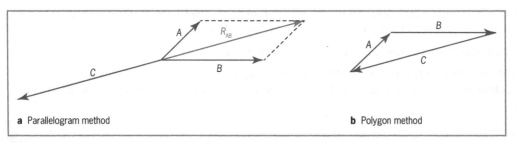

a Parallelogram method

b Polygon method

Figure 1.22 *Three vectors in equilibrium*

An object of weight 16.0 N is suspended by a vertical string. A horizontal force F acts on the object so that the string makes an angle of 30° with the vertical as shown in Figure 1.23. Determine by scale drawing:

a the tension, T, in the string

b the force, F.

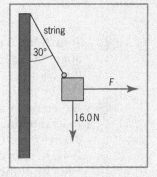

Figure 1.23 *Example 2*

The tension, T, in the string acts along the string away from the block. Since the block is at rest, the forces are in **equilibrium**. Using the polygon method, the three vectors will therefore form a **closed** figure when joined head-to-tail as shown in Figure 1.24.

- Start by drawing the vector arrow for the 16.0 N force since it is the only one for which both the magnitude and direction are given. Use a scale that will produce a **large** drawing.

- From the head of the 16 N force arrow, draw a horizontal line to represent the vector F. Draw it to the edge of the page since its magnitude was not given.

- Since we do not know where the arrow F ends, we do not have a starting point for arrow T. However, we do know that T **ends** at the beginning of the 16 N arrow, since the polygon must be closed. We can therefore place the head of arrow T at the beginning of the 16 N arrow and draw T from there at an angle of 30° from the vertical until it meets the vector F.

- T and F are then found by measurement.
  $F = 9.2 \text{ N}$ $T = 18 \text{ N}$

Note: Since these vectors form a right-angled triangle, the same result can be obtained from a sketch and using simple trigonometry.

$$\tan 30° = \frac{F}{16} \qquad F = 16 \tan 30° = 9.2 \text{ N}$$

$$\cos 30° = \frac{16}{T} \qquad T = \frac{16}{\cos 30°} = 18 \text{ N}$$

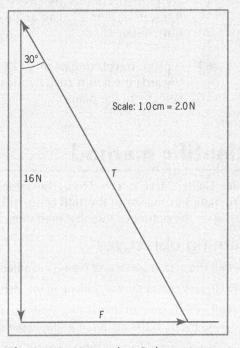

Scale: 1.0 cm = 2.0 N

Figure 1.24 *Example 2 Polygon method*

Recalling facts

1   **a** Distinguish between a scalar and a vector quantity.

   b List:

   i THREE scalar quantities

   ii THREE vector quantities.

Applying facts

2   Tugs A and B pull a ship with forces of 5.0 MN acting due east, and 4.0 MN acting at an angle of N 30° E. Determine by scale drawing the resultant force in magnitude and direction.

3   A man walks across the deck of a ship perpendicular to its motion. The ship is travelling due west at 8.0 m s⁻¹ and the man's speed relative to the ship is 4.0 m s⁻¹ in a direction due north. Calculate the velocity of the man relative to the water (i.e. the resultant velocity of the man).

4   Determine the weight, W, of the object shown in Figure 1.25, which is suspended by tensions of 3.0 N and 4.0 N.

5   Calculate the magnitude of the vertical and horizontal components of the velocity of a football when it travels at 20 m s⁻¹ at an angle of 37° above the horizontal.

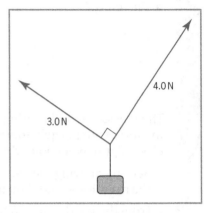

Figure 1.25   *Question 4*

6   A plane travels due north at 72 ms⁻¹ in still air. It meets a wind of 30 m s⁻¹ blowing towards the south-east. By means of a scale drawing, determine the resultant velocity of the plane.

Scientific method

Galileo Galilei, often referred to as the '**father of modern scientific methodology**', was an Italian mathematics professor of the 16th century. He was the first to implement a scientific approach to investigate the nature of the observed world.

Learning objectives

- Recall the significance and basis of **Galileo's scientific methodology**.
- Recall, and correctly use, **independent, dependent and controlled variables** in experiments.
- Recall and be able to investigate how the mass of the bob, the length of the string and the angular displacement of the string of a **simple pendulum** can affect its period.
- Use the required format for presenting experimental observations in **tables and graphs**.
- Be competent in plotting experimental points and drawing a **line of best fit** for straight-line graphs and for curved graphs.
- Determine the **gradient**, the **intercepts**, the **area** between the graph line and the time axis, and other information from the coordinates of a graph.

Galileo's scientific methodology

Today's scientists have the advantage of added tools for investigation, such as **calculus mathematics** developed for scientific research by **Sir Isaac Newton** and **Gottfried Wilhelm Leibniz**, the extremely high processing **speed of the modern computer** and the web of pooled information available on the **internet**. Despite these advantages however, the basis of Galileo's methodology – listed below – is still adopted by modern-day scientists.

Also,

$$m = \frac{4\pi^2}{g}$$

$$\therefore g = \frac{4\pi^2}{m}$$

$$= \frac{4\pi^2}{4.0 \text{ s}^2 \text{ m}^{-1}}$$

$$= 9.9 \text{ m s}^{-2}$$

Determining the intercepts of a graph

If the axes of the graph **begin at the origin** (coordinates (0,0)), the y-intercept is the value where the line cuts the y-axis and the x-intercept is the value where the line cuts the x-axis. The y-intercept in Figure 1.28 is 5.0 V.

If one or more of the axes do not start at the origin, as in Figure 1.29, the following procedure can be taken.

- The **y-intercept**, *c*, can be found by substituting the gradient, *m*, and the x and y values of a coordinate from any point on the line into the equation $y = mx + c$.

- The **x-intercept** is the x coordinate when the **y coordinate is zero**. It is calculated by substituting the gradient, *m*, the y-intercept, *c*, and the value $y = 0$ into the equation $y = mx + c$.

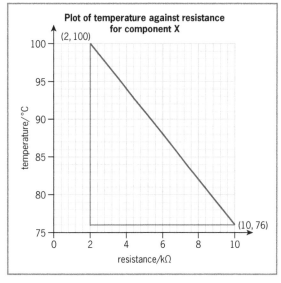

Figure 1.29 *Determining the intercepts of a straight-line graph*

Analysing the graph of Figure 1.29:

Gradient: $m = \dfrac{y_2 - y_1}{x_2 - x_1}$

$$m = \frac{76 - 100}{10.0 - 2.0} \frac{°C}{k\Omega} = -3.0 \frac{°C}{k\Omega}$$

y-intercept: Using coordinate (6.0, 88)

$y = mx + c$

$88 = -3.0 \times 6.0 + c$

$88 = -18 + c$

$\therefore 106 = c$ y-intercept = 106 °C

x-intercept: Using coordinate (x, 0)

$y = mx + c$

$0 = -3.0x + 106$

$3.0x = 106$

$x = 35.3$ x-intercept = 35 kΩ (2 sig. fig.)

Note: Intercepts usually have units.

Recalling facts

1   **a**   Which scientist is often referred to as the father of scientific methodology?

       b   List the FOUR steps of the methodology proposed by the person mentioned in **a**.

2   Distinguish between an independent variable, a dependent variable and a control variable as used in the experimental process.

3   **a**   What is a simple pendulum?

       b   How do each of the following factors affect the period of a simple pendulum?

          i   mass of bob

          ii   length of string

Applying facts

4   Andrea is aware that the temperature of a gas may cause its volume to change. She has decided to carry out an experiment to investigate how the pressure of a gas is affected by changing its volume. Name:

       a   the independent variable

       b   the dependent variable

       c   a control variable.

Analysing data

5   With reference to Figure 1.26, page 23, calculate:

       a T when $l = 50$ cm

       b l when $T = 2.0$ s

6   Examine the graph shown in Figure 1.28.

       a   Calculate the x-intercept, given that the gradient is 5.0 V A^{-1}.

       b   Assuming the relation between the voltage and current remains unchanged, determine the current when the voltage is:

          i   2.0 V

          ii   27.0 V

7 The graph of Figure 1.30 indicates how the area, A, of a triangle of FIXED HEIGHT, h, varies with the length of its base, b.

a Suggest a suitable title for the graph.

b What is the area when the base is of length 15 cm?

c Calculate the gradient of the graph.

d The equation of the graph is $A = \frac{1}{2} \times h \times b$. Calculate the height of the triangle.

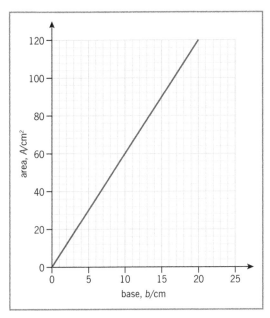

Figure 1.30 *Question 7 Graph of how the area of a triangle of fixed height varies with the length of its base*

Section A
Mechanics

In this section

Statics

- The study of bodies at rest under the action of several forces.

Kinematics

- The study of motion in relation to distance, displacement, speed, velocity and acceleration.

Dynamics

- The effects of forces on the motion of bodies including situations of collisions and explosions.

Work, energy and power

- This teaches of the interrelation between work, energy, and power. Energy is the ability to do work and is transformed from one type to another at a rate known as the power.

Pressure and Archimedes' principle

- Here you will learn of pressure – the force acting per unit area on a surface – and about Archimedes' principle, which equates the upthrust on a body to the weight of fluid it displaces.

2 Statics

Statics is the study of how forces keep systems in equilibrium: a study of situations in which the resultant force and torque on a system is zero and therefore the system does not accelerate. In this course it is concerned with **systems at rest**.

Learning objectives

- Define **force** and identify various **types of force**.
- Define **mass** and **weight** and use equations relating them.
- Define **centre of gravity** and describe experiments to locate the centre of gravity of a body.
- Use **free-body force diagrams** for analysing the behaviour of bodies acted on by forces.
- Define the **moment** of a force about a point and state the **conditions** required for a system of **coplanar forces** to be **in equilibrium**.
- State and verify the **principle of moments** and use it to solve problems.
- Explain the action of **levers** as it relates to common tools and devices.
- Define and explain **stable, unstable, and neutral equilibrium.**
- Relate the **stability** of an object to the **height of its centre of gravity** and to the **width of its base**.
- State **Hooke's law** and describe an experiment to verify the law.
- Use equations and graphs to solve problems involving Hooke's law.

Force is an action which changes, or tends to change, the size, shape, and/or motion of a body.

It is simply a push or a pull that one body exerts on another.

The Newton (N) is the force that will cause a mass of 1 kg to accelerate at 1 m s^{-2}.

A force can be exerted on body A due to **contact** with body B, or it can be exerted on body A **from a distance** if it is in a **field** of body B. Force is a **vector** quantity as explained in Chapter 1.

Contact forces (mechanical forces)

- **Thrust** is the force exerted by a person or system (e.g. a motor) in pushing an object.
- **Tension** and **compression** are forces produced within a spring or elastic material being stretched or compressed. These forces act in directions which oppose the deformation of the material by pulling or pushing respectively on objects connected at their ends.
- **Normal reaction** is the contact force acting **perpendicularly from a surface** as a reaction to a force acting **onto the surface** from another body.
- **Buoyancy** is an upward force from a fluid onto a body submerged in the fluid – see Chapter 6.
- **Friction** is the **opposing** force which acts along the surfaces of bodies moving or tending to move relative to each other.
- **Drag** is the **opposing** force which acts along the surface of a body as it moves through a **fluid**.

Non-contact forces (field forces)

- **Gravitational force** is the **attractive** force between bodies due to their masses and to the distance of separation between their centres.
- **Nuclear force** is the strong attractive force which binds subatomic particles in an atomic nucleus.
- **Electric force** is the **attractive** or **repulsive** force existing between bodies of **opposite charge**, or of **similar charge**, respectively.
- **Magnetic force** is the **attractive** or **repulsive** force existing between bodies due to the interaction of opposite magnetic poles, or of similar magnetic poles, respectively. **Electric currents** also have magnetic fields around them. Magnetic material or conductors carrying electric currents may be attracted or repelled if placed in these fields.

Mass and weight

Mass is the quantity of matter making up a body.

Weight is the gravitational force of a planet or other large object on a body.

Gravitational force

The weight of a body depends on the gravitational field strength that acts on it:

$$\text{weight} = \text{mass} \times \text{gravitational field strength}$$

$$W = mg$$

On Earth, the generally accepted value for the **gravitational field strength**, g, is **10 N kg^{-1}**.

A body falling in this field has an **acceleration due to gravity**, g, of **10 m s^{-2}**.

Gravitational field strength and acceleration due to gravity are given the same symbol, g, and their units are equivalent.

An object of mass **1 kg** has a weight of **10 N** since:

$$W = mg = 1 \text{ kg} \times 10 \text{ N kg}^{-1} = 10 \text{ N}$$

The **gravitational force** between two masses increases if

- the **mass** of either of the bodies **increases**
- the **distance** between the centres of gravity (see below) of the bodies **decreases**

A breadfruit of mass 2.0 kg has a weight of 20 N on Earth but on the Moon its weight would be much lower. This is because the **mass** of the Moon is much smaller than the mass of the Earth.

If the breadfruit is transported away from the Earth's surface, its weight will decrease as a result of the increasing **distance** between it and the planet. The breadfruit would be completely **weightless** in outer space where there is no large mass nearby to attract it.

The **mass** of the breadfruit **is the same** at each location since it is comprised of the **same quantity of matter**.

Centre of gravity

*The **centre of gravity** of a body is the point through which the **resultant** gravitational force on the body may be **considered** to act.*

Each atom or molecule has its own weight, and we can consider the sum of those weights (the resultant force) as acting from the centre of gravity of the body (see Figure 2.1).

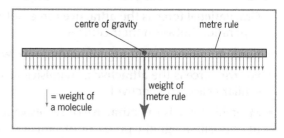

Figure 2.1 *Centre of gravity*

Centre of gravity of regular shapes and three-dimensional bodies

The centre of gravity (C of G) of a **regular shape**, such as a circle or square, and a **regular solid body**, such as a sphere or cube, is at its **geometric centre**. Its location on various shapes and in various bodies is as described below.

Shapes

- **Rectangle** – intersection of the diagonals.
- **Triangle** – intersection of its medians (lines drawn from the vertices to **bisect** the opposite sides).
- **Circle** – mid-point of its diameter.

Bodies

- **Uniform thin rod** – mid-point of its length.
- **Uniform cube or cuboid** – intersection of its diagonals from opposite vertices.
- **Uniform sphere** – mid-point of its diameter.
- **Non-uniform thin rod** – point where a fine edge placed below the rod keeps it in balance.

Experiment to find the centre of gravity of an irregular shaped lamina

- The lamina is hung so that it swings freely from a pin placed through a small hole near its edge as shown in Figure 2.2.
- A plumb line (a small weight attached at one end of a string) is hung from the pin and its position marked by small crosses on the lamina.
- The procedure is repeated by suspending the lamina from another hole near its edge.
- The point of intersection of the drawn lines is the centre of gravity.
- The procedure can be repeated from a third point near its edge as a check that the third line drawn across the lamina intersects at the same point as the first two.

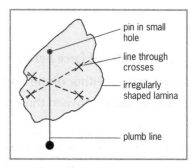

Figure 2.2 *Finding the centre of gravity of an irregular lamina*

Free-body force diagrams

A free-body force diagram is important in analysing how a body behaves under the action of several forces. It is a sketch showing the **forces** acting **ON** the **body being investigated**. These are

1. **field forces** – usually the weight acting through the **centre of gravity**

2. **mechanical forces** of the environment acting at **points of contact**

A free-body diagram cannot include both forces of a *Newton's 3rd law pair* (Chapter 4, page 54). For example, when investigating the behaviour of the rod shown in Figure 2.3, we are not concerned with the downward force of the rod ON THE PIVOT or the upward force of the rod ON THE BLOCK.

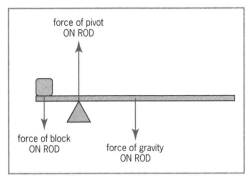

Figure 2.3 *Free-body diagram of a rod in equilibrium*

Forces producing moments

When we sit on a seesaw, use a spanner or push open a door, we are using forces that have a **turning effect** or **moment** about some pivot point.

*The **moment of a force** about a point is the product of the force and the perpendicular distance of its line of action from the point.*

$$\text{moment about a point} = \text{force} \times \text{perpendicular distance from the point}$$
$$T = F \times d_{\perp}$$

The **SI unit** of moment is **N m**.

Work (see Chapter 5) is the product of a *force* and the *distance moved in the direction* of the force. Its unit is therefore also the product, **N m**. The unit **joule (J)** is assigned to *work* to indicate the difference between a *moment* of a force and *work*.

The moment about a point tends to produce a clockwise or anticlockwise rotation about that point.

Example 1

Figure 2.4 shows two circumstances of similar trapdoors being pulled on by strings. The tension is 30 N in each string, but the strings pull in different directions. For each case, calculate the moment about the hinge caused by the tension.

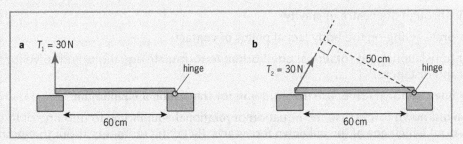

Figure 2.4 *Example 1*

Note: To calculate the moment caused by a tension, T, about the hinge we must use the **perpendicular distance** from the **line of action** of T to the hinge.

a moment of T_1 about hinge = 30 N × 60 cm = 1800 N cm (clockwise)

b moment of T_2 about hinge = 30 N × 50 cm = 1500 N cm (clockwise)

Conditions necessary for a system of coplanar forces in equilibrium

1. **Translational equilibrium** – the sum of the **forces** in any direction is equal to the sum of the **forces** in the opposite direction.

2. **Rotational equilibrium** – the sum of the clockwise **moments** about any point is equal to the sum of the anticlockwise **moments** about that same point. This is known as *the principle of moments*.

Experiment to verify the principle of moments

- Balance the metre rule by placing the pivot (fulcrum) directly below its centre of gravity. The weight of the metre rule then has no moment about the pivot since there is no distance between its line of action and the pivot (it passes through the pivot).

- Without moving the position of the pivot, rest slotted weights, W_1, W_2, W_3 and W_4, at distances x_1, x_2, x_3 and x_4 as shown in Figure 2.5 to obtain equilibrium. Slotted weights are used so that the position on the scale can be viewed through the slots.

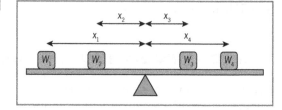

- Tabulate weights, W_1, W_2, W_3 and W_4, and distances x_1, x_2, x_3 and x_4, and calculate the sum of the anticlockwise moments and the sum of the clockwise moments.

Figure 2.5 *Verifying the principle of moments*

sum of anticlockwise moments = $W_1x_1 + W_2x_2$
sum of clockwise moments = $W_3x_3 + W_4x_4$

- Repeat the procedure by varying the positions of the weights.

It will be found that for each set of readings, the sum of the clockwise moments about the pivot is equal to the sum of the anticlockwise moments about the pivot.

Tackling problems involving coplanar forces in equilibrium

- Sketch a free body diagram showing all the **forces** acting **ON** the **body in equilibrium**. These are
  1. the **weight** (acts through the **centre of gravity**)
  2. all **mechanical forces** acting on the body (act at **points of contact**)

- Use the rules for translational and rotational equilibrium to formulate equations and solve for unknown forces and distances.
  - If there is only **one** unknown force, use the equation for **translational equilibrium**.
  - If there are **two** unknown forces, use the equation of **rotational equilibrium** to find one of them. **Select a point** about which **one of the unknown forces acts**. By taking moments about that point, the unknown force is excluded from the calculation since there is no distance between its line of action and the point, and therefore it creates no moment.

Example 2

Figure 2.6 shows a uniform metre rule of weight 1.8 N resting on a fulcrum. A block of weight 8.0 N hangs from the pulley and a block of weight, x, hangs from the 25 cm mark. Calculate:

a the weight x

b the normal reaction R.

Notes:

- The system is at rest and is therefore in translational equilibrium.

- Since the metre rule is uniform, its centre of gravity is at its midpoint.

- The downward force of the rule on the pivot and the downward force of the block on the pulley are **not considered** since they are **not on the rule**, the body whose equilibrium we are investigating.

- The block hanging from the pulley exerts an **upward** force **on the rule**.

- A suitable point must be selected from which to take moments in order to eliminate one of the two unknown forces from the equation. By taking moments about the fulcrum, we can eliminate R.

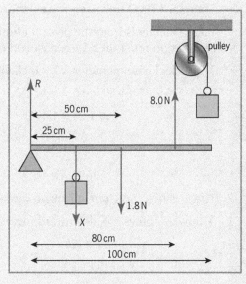

Figure 2.6 *Example 2*

a $\sum$ anticlockwise moments = $\sum$ clockwise moments

$$(8.0 \times 80) = (x \times 25) + (1.8 \times 50)$$
$$640 = 25x + 90$$
$$640 - 90 = 25x$$
$$550 = 25x$$
$$x = 22 \text{ N}$$

b $\sum$ upward forces = $\sum$ downward forces

$$8.0 + R = 22 + 1.8$$
$$R = 22 + 1.8 - 8.0$$
$$R = 15.8 \text{ N} = 16 \text{ N to 2 sig.fig.}$$

Example 3

Determine the normal reaction forces from fulcrums A and B shown in Figure 2.7, given that the weight of the uniform plank, PQ, is 120 N and the weight of the block X is 40 N.

A useful way to redraw the diagram is to label the key positions on PQ as if it were a measuring stick of length 3.0 m, starting with 0 m at end P (see Figure 2.8).

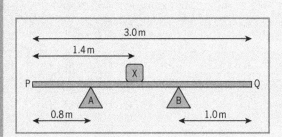

Figure 2.7 *Example 3*

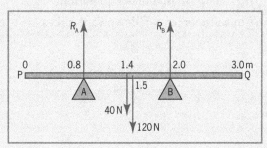

Figure 2.8 *Example 3 solution*

There are **TWO unknown forces**, therefore start with the equation of **rotational equilibrium**.

Taking moments from the pivot B eliminates R_B from the calculation. All distances used in the calculation must then be measured from B.

$\sum$ **anticlockwise moments** = $\sum$ **clockwise moments**

$$120 \times 0.50 + 40 \times 0.6 = R_A \times 1.2$$

$$60 + 24 = 1.2\,R_A$$

$$\frac{84}{1.2} = R_A$$

$$70\,\text{N} = R_A$$

There is now only **one unknown force**, so use the equation for translational equilibrium.

$\sum$ **upward forces** = $\sum$ **downward forces**

$$70 + R_B = 40 + 120$$

$$R_B = 160 - 70$$

$$R_B = 90\,\text{N}$$

Figure 2.9 shows a tree trunk AB lying horizontally on the road. Its centre of gravity is 1.2 m from B and a force of 4000 N is required to raise it as shown. Calculate:

a the weight, W, of the tree trunk

b the force, R, on the tree trunk at point A.

Hint: Whenever there are **THREE parallel** forces in equilibrium, the one between the other two is always opposite in direction to them (see Figure 2.10).

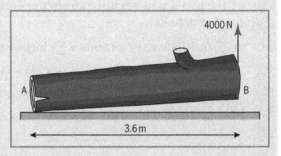

Figure 2.9 *Example 4*

Taking moments **about the end A** eliminates R from the calculation. All distances are then measured from A.

$\sum$ **anticlockwise moments** = $\sum$ **clockwise moments**

$$4000 \times 3.6 = W \times 2.4$$

$$\frac{4000 \times 3.6}{2.4} = W$$

$$6000\,\text{N} = W$$

There is now only **one unknown force**, so apply the rule of **translational equilibrium**.

$\sum$ **upward forces** = $\sum$ **downward forces**

$$4000 + R = 6000$$

$$R = 6000 - 4000$$

$$R = 2000\,\text{N}$$

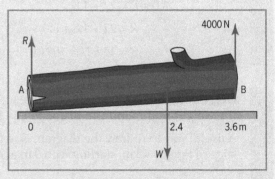

Figure 2.10 *Example 4 solution*

Example 5

a A uniform metre rule of weight 4.0 N is suspended by strings as shown in Figure 2.11. Determine the tension in B if the tension in A is 2.4 N.

b A small mass of 800 g is now hung from the rule to the right of B in such a position that A just becomes slack.

  i What is the new tension in B?

  ii How far from B is the 800 g mass?

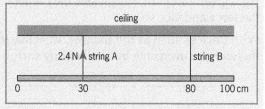

Figure 2.11 *Example 5*

a There is only **ONE unknown force**, so apply the rule of **translational equilibrium** (see Figure 2.12).

$$\sum \text{upward forces} = \sum \text{downward forces}$$

$$2.4 + T_B = 4.0$$

$$T_B = 4.0 - 2.4$$

$$T_B = 1.6 \text{ N}$$

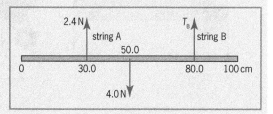

Figure 2.12 *Example 5 solution a*

b i Since another force has been added, a new diagram is required to include it (see Figure 2.13). Since string A is **slack**, it provides **no force**. Let the distance from B to the point from which the added mass is hung be x.

  weight of added mass:
  $$W = mg = 0.800 \times 10 = 8.0 \text{ N}$$

There is only **ONE unknown force**, so apply the rule of **translational equilibrium**.

$$\sum \text{upward forces} = \sum \text{downward forces}$$

$$T_{B2} = 4.0 + 8.0$$

$$T_{B2} = 12.0 \text{ N}$$

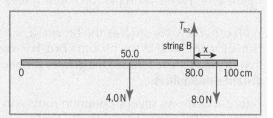

Figure 2.13 *Example 5 solution b*

  ii Taking moments about the 80 cm mark eliminates T_{B2} from the calculation.

$$\sum \text{anticlockwise moments} = \sum \text{clockwise moments}$$

$$4.0 \times 30 = 8.0x$$

$$\frac{4.0 \times 30}{8.0} = W$$

$$15 \text{ cm} = x$$

The 800 g mass is hung 15 cm to the right of B.

Common tools and devices utilising the principle of moments

Figure 2.14 shows three devices illustrating the lever, a simple machine which utilises the principle of moments. For each example, an effort is used to create a moment about a pivot in order to overcome the moment created by a load about the same pivot. In each case:

effort × perpendicular distance from effort to pivot = load × perpendicular distance from load to pivot

$$E \times d_E = L \times d_L$$

For a given moment, the **greater the perpendicular distance** from the **pivot P** to the force, the **smaller the force** and vice versa.

With devices such as the **spanner, wheelbarrow and crowbar** $d_E > d_L$, and so $L > E$; **large loads can therefore be overcome by small efforts** and so these devices are known as **force multipliers.**

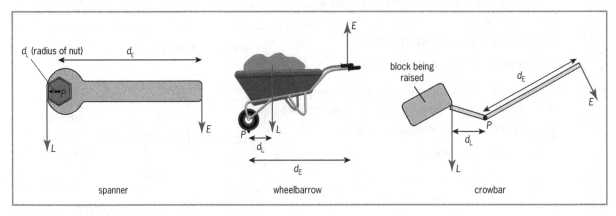

Figure 2.14 *Levers acting as force multipliers*

With other devices such as the **broom**, $d_E < d_L$, and so $L < E$; the **effort is then greater than the load**! However, the load at the broom's brush is small, making this a small sacrifice to pay for the advantage that the **load** moves through a **greater distance** than the effort. These devices are therefore known as **distance multipliers**.

Figure 2.15 shows several common tools and devices that act as levers.

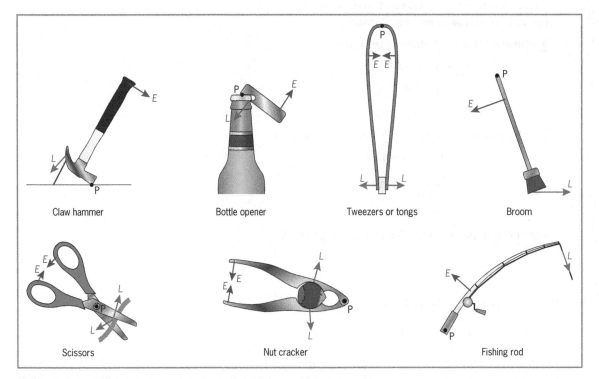

Figure 2.15 *Other common tools and devices acting as levers*

Stable, unstable and neutral equilibrium

*A body is in **stable equilibrium** if when slightly displaced, its centre of gravity rises, providing a restoring moment that returns the body to its original equilibrium position.*

*A body is in **unstable equilibrium** if when slightly displaced, its centre of gravity falls, providing a toppling moment that moves the body away from its original equilibrium position.*

*A body is in **neutral equilibrium** if when slightly displaced, its centre of gravity remains at the same level, providing no moment and therefore leaving the body in the displaced position.*

Figure 2.16 illustrates these types of equilibrium.

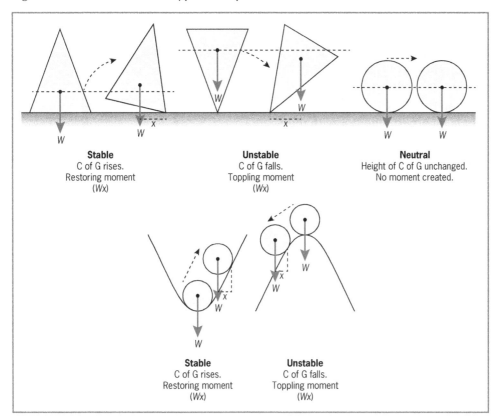

Figure 2.16 *Stable, unstable, and neutral equilibrium*

Factors affecting stability of an object

The stability of an object **increases** with

- decreased height of its centre of gravity
- increased width of its base.

See Figure 2.17. Examples:

- Low centres of gravity and wide wheelbases make go-carts more stable. When tilted, restoring moments are created which return them to their original position.
- Large cargo buses have their luggage compartments below the floor to lower their C of G and therefore increase their stability.

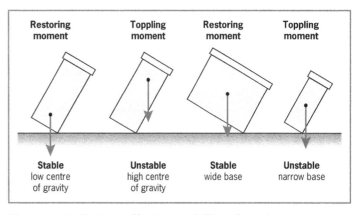

Figure 2.17 *Factors affecting stability of an object*

Deformation and Hooke's law

When we **deform** a spring by stretching it, we increase the spacing of its molecules, causing bonds to be under tension as attractive forces are set up to resist the separation. Similarly, on compressing a spring, we force the molecules closer together, creating repelling forces to resist the compression.

Hooke's law: Up to some maximum load, the force applied to a spring is proportional to its extension.

$$F \propto x \quad \text{i.e.} \quad F = kx$$

The constant of proportionality, k, is a property of the spring known as its **force constant**.

Experiment to verify Hooke's law

- The apparatus is set up as shown in Figure 2.18 and the position of the pointer on the metre rule is recorded.
- A mass of 20 g is placed on the mass holder and the new position of the pointer is recorded together with the corresponding mass.
- The process is repeated by adding masses, one at a time, until six pairs of readings are taken.
- The masses are then removed, one at a time, each time recording the position of the pointer together with the corresponding mass.
- The load (weight) corresponding to each mass is recorded.
- The average scale reading and the extension for each load is then calculated and tabulated.
- A graph of load against extension is plotted.

Provided the proportional limit is not exceeded (see Figure 2.19), a **straight-line graph through the origin** should be obtained, thus **verifying Hooke's law**. The **slope** of the graph is the **force constant** of the spring.

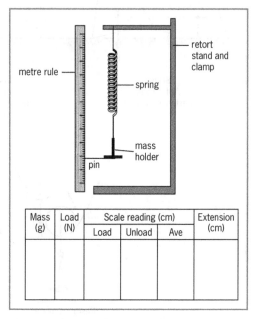

Mass (g)	Load (N)	Scale reading (cm)			Extension (cm)
		Load	Unload	Ave	

Figure 2.18 *Verifying Hooke's law*

Precautions to minimise errors

- The pointer is securely fixed to the mass holder.
- Eye level readings are taken to avoid parallax error.
- Readings are taken only when the mass is at rest.
- To reduce random error, the spring is loaded and unloaded and the average reading for each load is calculated.
- It is ensured that the spring is securely supported so that it does not shift during the experiment.

Analysing force–extension graphs

Springs

Figure **2.19a** shows the relation between force and extension for a spring which has been stretched to the extent that it is damaged; on removing the load, the spring remains with a permanent stretch.

The **potential energy** stored in a spring (**strain energy**) is equal to the **work done** in stretching it. It can be calculated by finding the **'area'** between the **graph-line** and the **extension axis** of the force–extension graph. If the proportional limit has not been exceeded, this is simply the 'area' of a triangle as shown in Figure **2.19b**.

*The **proportional limit** (P) is the point beyond which any further increase in the load applied to a spring, produces an extension that is no longer proportional to the force.*

*The **elastic limit** (E) is the point beyond which any further increase in the load applied to a spring, produces a permanent stretch.*

Note: For many materials, the proportional limit occurs at loads that are only slightly less than the elastic limit.

Elastic deformation occurs for loads within the elastic limit; if the load is removed, the spring returns to its original length and there is **no permanent stretch.**

Plastic deformation occurs if the elastic limit is exceeded; the material is then **permanently stretched.**

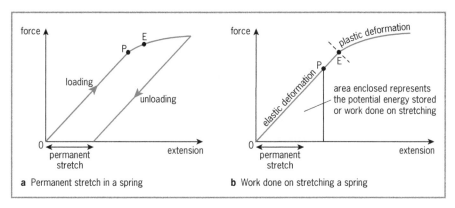

a Permanent stretch in a spring   **b** Work done on stretching a spring

Figure 2.19 *Stretching a spring*

Elastic bands

Unlike springs, elastic rubber bands **never obey Hooke's law**; Figure 2.20 shows that the graph is not a straight line at any point. As the elastic band is unloaded, the extension for any given load is greater than when it was being loaded. Heat energy is dissipated in the process.

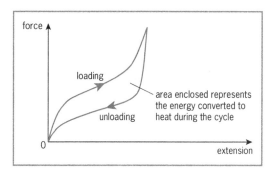

Figure 2.20 *Loading and unloading an elastic band*

A force of 10 N stretches a spring by 20 cm. The length of the relaxed spring is 25 cm.

a Assuming that the proportional limit is not exceeded, calculate:

 i the force constant of the spring

 ii the load which will stretch it to 65 cm.

b In relation to the spring being stretched to 65 cm, sketch graphs of:

 i force against extension of the spring

 ii force against length of the spring.

c Calculate:

   i the work done in stretching the spring to 65 cm

   ii the strain energy (elastic potential energy) stored in stretching the spring to 65 cm.

a i $F = ke$

   $10 = k(0.20)$

   $k = \dfrac{10}{0.20}$

   $k = 50 \text{ N m}^{-1}$

ii $F = ke$

   $F = 50(0.65 - 0.25)$

   $F = 20 \text{ N}$

b

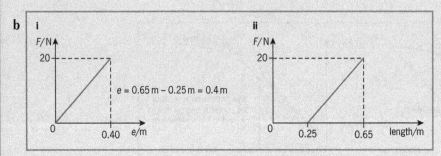

Figure 2.21 *Example 6 b*

c i work done = area under force–extension graph: $W = \dfrac{20 \times 0.40}{2} = 4.0 \text{ J}$

  ii work done = strain energy stored = 4.0 J

Recalling facts

1 Define:

  a force

  b weight

  c mass

  d centre of gravity

2 List three examples of each of the following types of forces.

  a Contact forces.

  b Forces that act at a distance due to fields.

3 **a** Define the *moment of a force about a point*.

  b State the conditions required for a system of coplanar forces to be in equilibrium.

  c What is the *principle of moments*?

4 **a** What is meant by the term *stable equilibrium*?

  b State TWO ways of increasing the stability of an object.

5 **a** State Hooke's law.

   b What is meant by the term *elastic limit*?

   c Distinguish between *elastic deformation* and *plastic deformation*.

Applying facts

6 Figure 2.22 shows a wheelbarrow ('pan cart') and its load which are of total weight 400 N. To raise the load, a person exerts a total force X at the ends of the handles. A third force, Y, acts on the wheelbarrow to maintain equilibrium.

   a Make a simple sketch of the diagram, showing by means of labelled arrows, the forces acting on the wheelbarrow.

   b Calculate:

   i X

   ii Y

   c Explain the effect on the required force, X, if

   i the majority of the load is placed closer to the front of the pan that holds it

   ii the handles are longer.

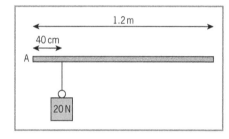

Figure 2.22 *Question 6*

7 A 20 N weight is hung from the uniform bar of weight 30 N and length 1.2 m as shown in Figure 2.23.

   a Redraw the diagram indicating all forces acting on the bar when a fulcrum is placed under it to maintain equilibrium.

   b Determine the reaction force on the bar from the fulcrum.

   c Determine the distance from A to the fulcrum.

Figure 2.23 *Question 7*

8 Figure 2.24 shows a model of a skyscraper which Omar made for the school exhibition. It is mounted on a firm plastic base and has a total weight, W, of 4.0 N. Omar applies a force F as shown to test the stability of his model.

Calculate the minimum value of F required to cause the model to rotate about point P.

9 A load of 6.0 N stretches a spring by 12 cm without exceeding its proportional limit. Calculate:

   a the force constant of the spring

   b the load which could produce an extension of 7.2 cm.

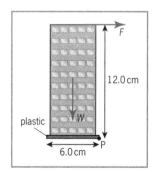

Figure 2.24 *Question 8*

10 The graph of Figure 2.25 shows how the length of a spring varies with the applied load.

a What is the length of the spring before the load is applied?

b What is the extension when the load is 4 N?

c Determine:

   i the extension when stretched to the proportional limit

   ii the force constant of the spring

   iii the potential energy stored in the spring when it is stretched by 2.0 cm.

d Estimate the work done in stretching the spring to a length of 10 cm.

e What type of deformation occurs after the elastic limit?

f What does the gradient of the graph represent for loads up to 8.0 N?

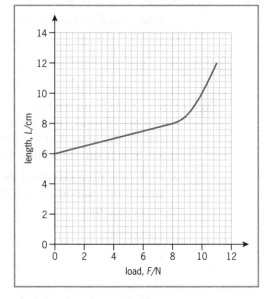

Figure 2.25 *Question 10*

Displacement–time and velocity–time graphs

Displacement–time (s–t) graph

- **Velocity** $= \dfrac{\text{change in displacement}}{\text{change in time}}$

  $= \left(\dfrac{\Delta s}{\Delta t}\right)$

Velocity = slope of the graph.

Examples of obtaining velocity from displacement–time graphs are shown in Figure 3.5.

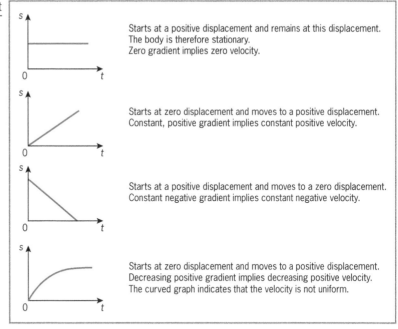

Starts at a positive displacement and remains at this displacement. The body is therefore stationary. Zero gradient implies zero velocity.

Starts at zero displacement and moves to a positive displacement. Constant, positive gradient implies constant positive velocity.

Starts at a positive displacement and moves to a zero displacement. Constant negative gradient implies constant negative velocity.

Starts at zero displacement and moves to a positive displacement. Decreasing positive gradient implies decreasing positive velocity. The curved graph indicates that the velocity is not uniform.

Figure 3.5 *Displacement–time graphs*

Velocity–time (v–t) graph

- **Acceleration** $= \dfrac{\text{change in velocity}}{\text{change in time}}$

  $= \left(\dfrac{\Delta v}{\Delta t}\right)$

Acceleration = slope of the graph.

Examples of obtaining acceleration from velocity–time graphs are shown in Figure 3.6.

- **Distance** = area between the graph line and the time axis (**all areas** treated as **positive**) – see Example 5.

- **Displacement** = area between the graph line and the time axis (areas **above the time axis** treated **as positive**; those **below the time axis** treated as **negative**) – see Example 5.

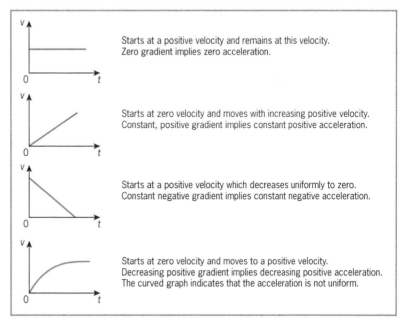

Starts at a positive velocity and remains at this velocity. Zero gradient implies zero acceleration.

Starts at zero velocity and moves with increasing positive velocity. Constant, positive gradient implies constant positive acceleration.

Starts at a positive velocity which decreases uniformly to zero. Constant negative gradient implies constant negative acceleration.

Starts at zero velocity and moves to a positive velocity. Decreasing positive gradient implies decreasing positive acceleration. The curved graph indicates that the acceleration is not uniform.

Figure 3.6 *Velocity–time graphs*

Determine the velocity during each stage, A, B and C, shown in the graph of Figure 3.7.

$$v = \frac{\Delta s}{\Delta t} \; .... \; gradient$$

$$v_A = \frac{4.0 - 0}{5.0 - 0} = \frac{4.0}{5.0} = 0.80 \text{ m s}^{-1}$$

$$v_B = \frac{4.0 - 4.0}{10.0 - 5.0} = \frac{0}{5.0} = 0 \text{ m s}^{-1}$$

$$v_C = \frac{10.0 - 4.0}{15.0 - 10.0} = \frac{6.0}{5.0} = 1.2 \text{ m s}^{-1}$$

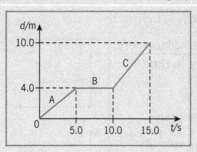

Figure 3.7 *Example 3*

Distance–time and speed–time graphs

- The gradient of a **distance–time** graph is the **speed**.
- The gradient of a **speed–time** graph is the **acceleration** *if the motion is in one direction along a straight line.*

Average speed and average velocity

From a speed–time graph, we can determine the distance travelled from the **area between the graph line and time axis.** From a velocity–time graph we can calculate the distance travelled and the displacement using the areas between the graph line and the time axis. The average speed or velocity can then be calculated by dividing these values by the corresponding times.

$$\text{average speed} = \frac{\text{total distance}}{\text{time}} \qquad \text{average velocity} = \frac{\text{total displacement}}{\text{time}}$$

If the motion is represented by a **single line** of **uniform gradient**, then the average speed or average velocity can also be obtained from the **mean of the highest and lowest speeds or velocities**, respectively.

If an object travels in **one direction** along a **straight line**:

- The **magnitude of its displacement** is the same as the **distance** it travels.
- The **magnitude of its velocity** is the same as its **speed**.

Figure 3.8 shows the motion of an object along a straight path. Determine:

a   the acceleration during stage A
b   the acceleration during stage B
c   the acceleration during stage C
d   the velocity 7.0 s after starting
e   the displacement during stage A
f   the displacement during stage B
g   the displacement during stage C
h   the total displacement
i   the average velocity during stage A
j   the average velocity during stage B
k   the average velocity during stage C
l   the average velocity for total trip

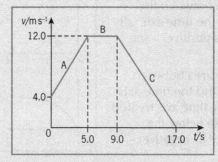

Figure 3.8 *Example 4*

m the average speed for total trip

n the total distance travelled.

a $a_A = \dfrac{12.0 - 4.0}{5.0 - 0} = \dfrac{8.0}{5.0} = 1.6 \text{ m s}^{-2}$

b $a_B = \dfrac{12.0 - 12.0}{9.0 - 5.0} = \dfrac{0}{4.0} = 0 \text{ m s}^{-2}$

c $a_C = \dfrac{0 - 12.0}{17.0 - 9.0} = \dfrac{-12.0}{8.0} = -1.5 \text{ m s}^{-2}$ ... deceleration = 1.5 m s^{-2}

The negative sign indicates that the acceleration is oppositely directed to the initial acceleration. Since the speed (**magnitude of the velocity**) is now **decreasing**, we say that the object is **decelerating**.

d Inspection of the graph indicates that the velocity 7.0 s after starting is 12.0 m s^{-1}.

e Displacement during stage A is given by the area between the graph line and t-axis for the first 5.0 s.

$\text{displacement}_A = (4.0 \times 5.0) + \left(\dfrac{8.0 \times 5.0}{2}\right) = 20 + 20 = 40 \text{ m}$

f $\text{displacement}_B = (12.0 \times 4.0) = 48 \text{ m}$

g $\text{displacement}_C = \left(\dfrac{12.0 \times 8.0}{2}\right) = 48 \text{ m}$

h Total displacement = 40 + 48 + 48 = 136 m

i Average velocity during stage A $= \left(\dfrac{\text{disp}}{\text{time}}\right)_A = \dfrac{40}{5.0} = 8.0 \text{ m s}^{-1}$

j Average velocity during stage B $= \left(\dfrac{\text{disp}}{\text{time}}\right)_B = \dfrac{48}{4.0} = 12 \text{ m s}^{-1}$

k Average velocity during stage C $= \left(\dfrac{\text{disp}}{\text{time}}\right)_C = \dfrac{48}{8.0} = 6.0 \text{ m s}^{-1}$

l Average velocity for total trip $= \left(\dfrac{\text{disp}}{\text{time}}\right)_{\text{total}} = \dfrac{136}{17} = 8.0 \text{ m s}^{-1}$

m Since the **velocity is always positive**, there is **no change in direction**. The average speed is therefore equal to the magnitude of the average velocity, i.e. 8.0 m s^{-1}.

n No change in direction also implies that the total distance travelled is equal to the magnitude of the total displacement, i.e. 136 m.

Example 5

Figure 3.9 shows the velocity–time graph of an object moving to the right and then to the left. Determine:

a the initial acceleration

b the final acceleration

c the distance travelled during the final 6.0 s

d the displacement during the final 6.0 s

e the average speed during the final 6.0 s

f the average velocity during the final 6.0 s

g the time at which it changes direction.

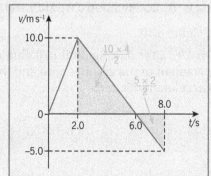

Figure 3.9 *Example 5*

Inspection of the graph indicates that velocity directed to the right has been assigned a positive value, and velocity to the left, a negative value.

a Initial acceleration $a = \dfrac{10.0 - 0}{2.0 - 0} = 5.0 \text{ m s}^{-2}$

b Final acceleration $a = \dfrac{-5.0 - 10.0}{8.0 - 2.0} = \dfrac{-15.0}{6.0} = -2.5 \text{ m s}^{-2}$

Since the speed is decreasing between $t = 2.0$ s and $t = 6.0$ s, the deceleration is 2.5 m s^{-2} for that period. Note that for the period $t = 6.0$ s to $t = 8.0$ s, although the **acceleration is still negative** (negative gradient of -2.5 m s^{-2}), the object is **not decelerating** since it is **not slowing down** (its **speed** is **increasing**).

c The distance travelled is found from the area between the graph line and time axis. It is helpful to work your areas directly on the graph as shown here in green. Since distance is a scalar quantity, the areas of both triangles are taken as positive.

$$d = \frac{10.0 \times 4.0}{2} + \frac{5.0 \times 2.0}{2} = 20 + 5.0 = 25 \text{ m}$$

d The displacement during the final 6.0 s is also found from the area of the triangles, but since displacement is a vector quantity, areas below the t-axis are taken as negative.

$$s = \frac{10.0 \times 4.0}{2} - \frac{5.0 \times 2.0}{2} = 20 - 5.0 = 15 \text{ m}$$

The positive answer infers that the total displacement during the final 6.0 s is directed to the right.

e Average speed $= \dfrac{\text{distance}}{\text{time}} = \dfrac{25}{6.0} = 4.2 \text{ m s}^{-1}$

f Average velocity $= \dfrac{\text{displacement}}{\text{time}} = \dfrac{15}{6.0} = 2.5 \text{ m s}^{-1}$

The positive answer infers that the average velocity during the final 6.0 s is directed to the right.

g The object changes direction at $t = 6.0$ s, when the velocity changes from a positive value to a negative value.

Example 6

An object starts from rest and accelerates uniformly for a period of 4.0 s along a straight road to a velocity of 20 m s^{-1}. It remains at this velocity for a further 6.0 s and then decelerates uniformly to rest in the next 8.0 s.

a Sketch a velocity–time graph of the motion.

b What is the acceleration 6.0 s after starting?

a See Figure 3.10.

b From 4.0 s after starting until 10.0 s after starting, the gradient of the graph is zero and therefore the acceleration is zero.

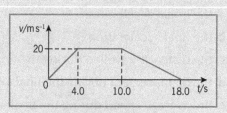

Figure 3.10 *Example 6*

An object starts at 4.0 m s^{-1} and accelerates uniformly at 2.0 m s^{-2} for a period of 3.5 s. Sketch a v–t graph clearly showing the final velocity attained.

First calculate the final velocity and then sketch the graph.

$$a = \frac{\Delta v}{\Delta t} \qquad 2.0 = \frac{v - 4.0}{3.5} \qquad 7.0 = v - 4.0 \qquad v = 7.0 + 4.0 = 11.0$$

Final velocity, $v = 11.0$ m s^{-1}

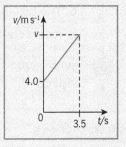

Figure 3.11 *Example 7*

Experiment to determine the acceleration due to gravity by a free fall method

The apparatus shown in Figure 3.12 can be used to determine the acceleration due to gravity.

- The switch at A completes the circuit to the electromagnet and therefore holds the iron ball in position.

- Placing the switch at B, breaks the circuit of the electromagnet and releases the ball. At the same time, the circuit to the millisecond timer is completed and timing starts.

- As the ball strikes the platform, the connection at C separates, the timer circuit breaks and the timing stops, leaving the time of fall, t, to be read from its digital screen.

Using initial velocity = u and final velocity = v

- average velocity of ball = $\dfrac{\text{displacement}}{\text{time}} = \dfrac{h}{t}$

- but average velocity is also = $\dfrac{u + v}{2} = \dfrac{0 + v}{2} = \dfrac{v}{2}$

$\therefore \dfrac{v}{2} = \dfrac{h}{t}$ and so final velocity, $v = \dfrac{2h}{t}$

and therefore acceleration, $a = \dfrac{v - u}{t} = \dfrac{\frac{2h}{t} - 0}{t} = \dfrac{2h}{t^2}$

The acceleration can therefore be calculated from just h and t.

Note: When an object **falls freely from rest**, the **final velocity** is always **twice** the **average velocity**.

Figure 3.12 *Determining the acceleration due to gravity by a free fall method*

Recalling facts

1

   a Distinguish between:

      i distance and displacement

      ii speed and velocity.

   b Define acceleration.

2 How can we determine each of the following concerning the motion of an object?

  a Its velocity from a *displacement–time* graph of its motion.

  b The distance travelled from a *velocity–time* graph of its motion.

  c Its acceleration from a *velocity–time* graph of its motion.

3 When a coconut falls from a tree, what is the relation between its velocity just before striking the ground and its average velocity during free fall?

Applying facts

4 An object starts from rest and accelerates uniformly for a period of 2.0 s to acquire a velocity of 8 m s⁻¹. It then changes its acceleration and after a further 4.0 s reaches a velocity of 12 m s⁻¹. Sketch a *velocity–time* graph of its motion and determine:

  a the initial acceleration

  b the final acceleration

  c the distance travelled within the last 2.0 s

  d the average velocity for the total time.

5 For each of the *displacement–time* graphs shown in Figure 3.13, sketch and explain the velocity–time graph that can represent its motion.

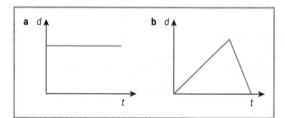

Figure 3.13 *Question 5*

6 For each of the *velocity–time* graphs shown in Figure 3.14, sketch and explain the displacement–time graph that can represent its motion.

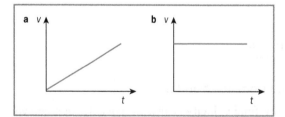

Figure 3.14 *Question 6*

7 Describe each stage of the motion of the object whose velocity varies with time as shown in Figure 3.15.

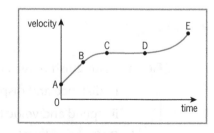

Figure 3.15 *Question 7*

Analysing data

8 The tapes shown in Figure 3.16 were produced by a ticker-tape timer connected to trolley X and then to trolley Y. The distance between each successive pair of dots in X is the same.

a State the type of motion displayed by each of the trolleys.

b Given that the timer produces 50 dots per second, calculate the average velocity of Y.

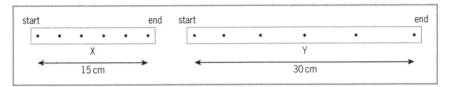

Figure 3.16 *Question 8*

9 The motion of an object moving along a straight line is described by the graph shown in Figure 3.17.

a Calculate:

i the acceleration during the first 12 seconds

ii the deceleration during the last 8 seconds

iii the total distance travelled

iv the average velocity.

b State with reason, whether or not the object changes direction.

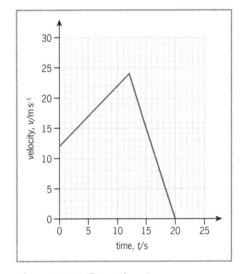

Figure 3.17 *Question 9*

10 The motion of an object moving along a straight line is described by the graph shown in Figure 3.18.

a Calculate:

i the total distance travelled

ii the total displacement

iii the average speed

iv the average velocity.

b How long after starting does the object change direction?

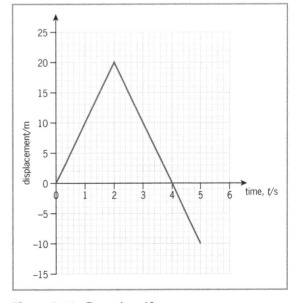

Figure 3.18 *Question 10*

4 Dynamics

Dynamics is the study of the effects of forces on the motion of bodies.

Aristotle, a renowned Greek philosopher, proposed a '**law of motion**' suggesting that the force applied to a body is proportional to its velocity ($F \propto v$). His arguments were based on **observations**, rather than on **experiment.**

Newton discredited the theory by proving **experimentally** that the force on a body is instead proportional to its acceleration ($F \propto a$).

Learning objectives

- Discuss **Aristotle's** arguments in support of his '**law of motion**' and explain why the 'law' was discredited.
- State **Newton's laws of motion** and use them to explain and solve problems on dynamic systems.
- Define **linear momentum** and state the law of **conservation of linear momentum.**
- Apply the law of conservation of linear momentum.

Aristotle's arguments in support of his theory of motion

- To increase the speed of a chariot required more horses and so he concluded that a greater force was necessary to produce a greater velocity.
- If the horses were to disengage from the chariot it would soon come to rest, and so he concluded that forces were necessary to keep a body in motion.

Aristotle failed to consider the frictional forces acting. If friction is negligible, a trolley will **accelerate** when pushed along a level surface by a **constant force**. Aristotle's hypothesis was therefore incorrect since the velocity will continuously increase and there is no force which results in a unique velocity. Also, in the absence of friction, a moving trolley will continue to move with uniform velocity when no resultant external force acts on it.

Newton's laws of motion

1. *A body continues in its state of rest or uniform motion in a straight line unless acted on by a resultant force.*

2. *The rate of change of momentum of a body is proportional to the applied force and takes place in the direction of the force.*

$$F_R = \frac{mv - mu}{t}.$$

3. *If body A exerts a force on body B, then body B exerts an **equal** but **oppositely directed** force on body A. (Every action has an equal but oppositely directed reaction.)*

Note law 1: If the resultant force on a body is zero, then one of the following is true:

- the body is at rest
- the body is moving at constant velocity (it is not accelerating).

Note law 2: F_R = resultant force, m = mass, u = initial velocity, v = final velocity, t = time of action of force, a = acceleration

$$F_R = \frac{mv - mu}{t} \qquad F_R = \frac{m(v - u)}{t} \qquad F_R = ma$$

Momentum (the product of the mass and velocity of an object) is discussed later in the chapter.

Example 4

A body of mass 400 g moves to the right at 8.0 m s⁻¹ and collides head-on with a lighter body of mass 100 g that moves to the left at 40.0 m s⁻¹. After the collision, the heavier rebounds at 11.2 m s⁻¹. Determine the velocity of the lighter body immediately after the collision.

Total momentum before collision = total momentum after collision

$$(400 \times 8.0) + (100 \times -40.0) = (400 \times -11.2) + 100v$$

$$3200 - 4000 = -4480 + 100v$$

$$-800 = -4480 + 100v$$

$$-800 + 4480 = 100v$$

$$3680 = 100v$$

$$\frac{3680}{100} = v$$

$$37 = v$$

Figure 4.5 *Example 4*

Velocity of lighter object after collision:
magnitude = 37 m s⁻¹   direction = towards the right

Example 5

A stationary gun of mass 800 g is loaded with a bullet of mass 20 g. If the recoil velocity of the gun is 10.0 m s⁻¹, determine the forward velocity of the bullet.

Total momentum before firing = total momentum after firing

$$(800 \times 0) + (20 \times 0) = (800 \times -10.0) + 20v$$

$$0 = -8000 + 20v$$

$$8000 = 20v$$

$$\frac{8000}{20} = v$$

$$400 = v$$

Figure 4.6 *Example 5*

Velocity of bullet after firing:
magnitude = 400 m s⁻¹   direction = opposite to recoil velocity of gun

Recalling facts

1

  a What did Aristotle propose in his '*Law of motion*'? Support your answer with an equation.

  b State TWO observations he used in support of his law.

  c What factor did he overlook in coming to his conclusion?

  d Which scientist discredited Aristotle's law, and how did he modify Aristotle's equation of force?

2

  a State Newton's THREE laws of motion.

  b For EACH of the laws, give an example that demonstrates it.

3 **a** Define momentum.

b State an SI unit that can be used to express the momentum of a body.

c State the law of conservation of linear momentum.

d Is momentum a scalar or a vector quantity?

e Can two bodies be moving if their total momentum is zero? Explain your answer.

Applying facts

4 Select from the items below the ones in which the resultant force on the body is zero.

a A raindrop falling at constant velocity (terminal velocity).

b A cricket ball at its highest point when thrown vertically into the air.

c A ball fastened to a string and whirled in a circle at constant speed.

d A car parked in a garage.

e A car moving at a steady speed of 80 km h^{-1} on a straight road.

f A stone falling in a vacuum chamber in a research laboratory on Earth.

g A rocket travelling at a constant velocity of 4×10^3 m s^{-1} through deep space.

5 Nathan's racing car is of mass 2.0×10^3 kg and is powered by two identical engines. It can accelerate from rest to a velocity of 90 km h^{-1} in 5.0 s.

a Calculate:

i the acceleration in m s^{-2}

ii the minimum force that each engine must provide to produce the acceleration.

b Explain why, in practice, the force provided by the engines must be greater.

6 Javan of weight 500 N balances on a stationary log of mass 25 kg floating near the edge of the bank of a river. He steps from the log onto the bank with a velocity of 4.0 m s^{-1}. Calculate the speed at which the log initially moves away.

7 Samara of mass 40 kg runs at 5.0 m s^{-1} and jumps onto a stationary trolley of mass 10 kg. Calculate their common velocity as soon as she lands on the trolley, if they move on together.

8 Jenna of mass 50 kg is skating in a westerly direction and collides head-on with Hagen of mass 75 kg who is skating in the opposite direction at 5.0 m s^{-1}. They are brought instantly to rest on colliding. Determine:

a the magnitude of Hagen's momentum before the collision

b the total momentum before the collision and explain your answer

c Jenna's speed before the collision.

9 A raindrop of mass 0.030 g falls with increasing velocity until the frictional drag on it increases to such a value that it acquires a maximum velocity of 9 m s^{-1}. It then continues to fall at this terminal velocity.

a Sketch a velocity–time graph of the motion.

b What is the frictional force when moving at this terminal velocity?

c Calculate the magnitude of the momentum at terminal velocity.

d Sketch a diagram showing the forces on the raindrop.

10 The graph shown in Figure 4.7 illustrates how a constant force changes the momentum of an object with time.

  a Determine:

    i the change in momentum

    ii the impulse

    iii the force producing the motion

    iv the momentum after 4.8 s.

  b Prepare a table using the experimental points plotted.

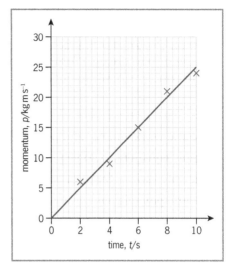

Figure 4.7 *Question 10*

5 Work, energy and power

Energy is needed to do **work**, and whenever work is done, energy is transformed from one type to another. The greater the rate of doing work, the greater is the **power**. Physics explains how things work and so to obtain a firm grasp of this field of science, it is essential to have a vivid understanding of the concepts of work, energy and power.

Learning objectives

- Define work, energy and power, and state their SI units.
- Apply **equations** to solve problems involving work, energy and power.
- Identify various **forms of energy**.
- Define **potential energy** and **kinetic energy**, and apply equations to solve related problems.
- State the principle of **conservation of energy** and cite cases demonstrating energy transformations.
- Explain the term efficiency and **calculate efficiency** in various situations.
- Distinguish among **renewable**, **non-renewable** and **alternative** sources of energy.
- Discuss various examples of renewable and alternative sources of energy used in the Caribbean.

Table 5.1 *Work, energy and power*

Quantity	Equation	
Work is the product of a force and the distance moved by its point of application in the direction of the force.	work = force × distance$_\parallel$	$W = F \times d_\parallel$
Energy is the ability to do work.	energy = force × distance$_\parallel$	$E = F \times d_\parallel$
Power is the rate of doing work (or rate of using energy).	power = $\dfrac{\text{work}}{\text{time}} = \dfrac{\text{energy}}{\text{time}}$	$P = \dfrac{W}{t}$ or $P = \dfrac{E}{t}$
	power = force × velocity	$P = \dfrac{F \times d_\parallel}{t}$ $P = F \times v_\parallel \ldots$ since $v = \dfrac{d_\parallel}{t}$

Note the following when calculating work:

- If the **force** is **perpendicular** to the **distance** moved, **no work** is done.
- Work is positive when the directions of the force and distance are parallel and is negative when the force and distance are antiparallel.

Figure 5.1 shows a block pulled across a surface through a distance, *d*, by a force, *F*. Other forces acting on the block are friction, *f*, the normal reaction, *R*, and the weight of the block, *mg*.

- *R* and *mg* do **no work** since they act perpendicular to the distance, *d*, moved by the block.
- The work done **by** *F* on the block is **positive** since *F* acts in the **same** direction as the distance, *d*, through which its point of application moves.
- The work done **by** *f* on the block is **negative** since *f* acts in the **opposite** direction to the distance, *d*, through which its point of application moves.
- The work done **against** the force, *f*, refers to the work done **by** a force equal in magnitude to *f* but acting in the same direction as *d*. It is therefore **positive**.

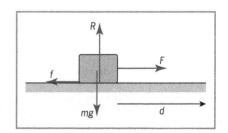

Figure 5.1 *The work done by a force can be positive, negative or zero*

SI units of work, energy and power

1 joule (J) is the work done (or energy used) when the point of application of a force of 1 N moves through a distance of 1 m in the direction of the force.

(1 J = 1 N m)

1 Watt (W) is the power supplied when the work done (or energy used) is at a rate of 1 J per 1 s.

(1 W = 1 J s^{-1})

Amelia pushes a block of weight 150 N through 15 m across a yard by applying to it a horizontal force of 40 N in a time of 5.0 s. Determine:

a the work done by Amelia

b the energy she uses

c the power she supplies.

a $W = F \times d_{\parallel}$ **b** $E = 600$ J   **c** $P = \dfrac{W}{t}$

$W = 40 \times 15$ The energy used is $P = \dfrac{600}{5.0}$

$W = 600$ J   equivalent to the work done.

$P = 120$ W

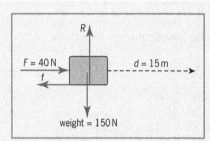

Figure 5.2 *Example 1*

A frictional force, *f*, and a normal reaction force, *R*, also act on the block. The **weight** of the block is **not used in the calculation** since it acts **perpendicularly** to the distance through which it is pushed. Recall that work = force × distance moved in the **direction of the force.**

Forms of energy

Table 5.2 *Some forms of energy*

Energy	Description
Gravitational potential energy	Energy **stored** by a body due to its position in a **gravitational force field.** It increases as the body rises above the Earth's surface.
Elastic potential energy	Energy **stored** by a body due to its **stretched or compressed state.**
Nuclear potential energy	Energy **stored** in an atomic nucleus due to its **physical state.**
Chemical potential energy	Energy **stored** by a body due to its **chemical state.**
Electrical potential energy	Energy **stored** due to the **position of electrical charges.** This energy can be transferred by charged particles, e.g. electrons in metal wires or ions in an electrolyte.
Magnetic potential energy	Energy **stored** by a body due to the **alignment of magnetic atomic dipoles** within it (see Chapter 29).
Kinetic energy	Energy possessed by a body due to its **motion.**
Internal energy	The **sum** of the **potential** and **kinetic** energies of the **particles** (atoms, molecules, ions) of a body.

(Continued)

Table 5.2 *Continued*

Energy	Description
Thermal energy	Energy possessed by a body due to the **motion of its particles** (atoms, molecules, ions). Increased temperature produces increased motion, i.e. **increased kinetic energy**.
Heat	Thermal energy **transferring** from **hotter to cooler** bodies.
Sound energy	Energy transported as particle **vibrations** through a material in the form of a **wave**.
Electromagnetic energy	Energy transported as **electric** and **magnetic field vibrations** in the form of a **wave** (radio, TV, microwaves, infrared, visible light, ultraviolet, X-rays, gamma waves).

A closer look at potential and kinetic energy

Kinetic energy is the energy possessed by a body due to its motion.

The **kinetic energy** of a body of mass, *m*, and speed, *v*, is given by:

$$E_K = \frac{1}{2}mv^2$$

Potential energy is the energy stored by a body due to:

- *its position in a force field*

- *its condition or state (e.g. stretched/compressed/chemical).*

The change in **gravitational potential energy**, ΔE_{GP}, of a body of mass, *m*, as it moves through a height, Δh, in a gravitational field of strength, *g*, is given by:

$$\Delta E_{GP} = mg\Delta h$$

Examples:

- A fast-moving cricket ball has **kinetic energy** mainly due to its high velocity.
- A coconut in a tree has **gravitational potential** energy due to its **position** (height) in the gravitational force field.
- A stretched elastic band has **elastic potential** energy due to its stretched **state**.
- A compressed spring has **elastic potential** energy due to its compressed **state**.
- Gasoline, food and batteries have **chemical potential** energy due to their chemical **state**.

The law of conservation of energy

The law of conservation of energy states that energy cannot be created or destroyed but can be transformed from one type to another.

For example, **chemical energy** stored in batteries and **gravitational potential energy** stored in water behind a river dam can be transformed into **electrical energy** as needed.

Whenever **energy is transformed**, an equal amount of **work is done**. This work can put objects into motion if the transformation is to **kinetic energy**, or it can raise objects if the transformation is to **gravitational potential energy**. If work is done against **friction**, the transformation is to **heat and sound energy**. Several examples of energy transformations are outlined below.

Main energy transformations occurring during various situations

- **Gasoline fueled car moving at constant velocity on a level road**

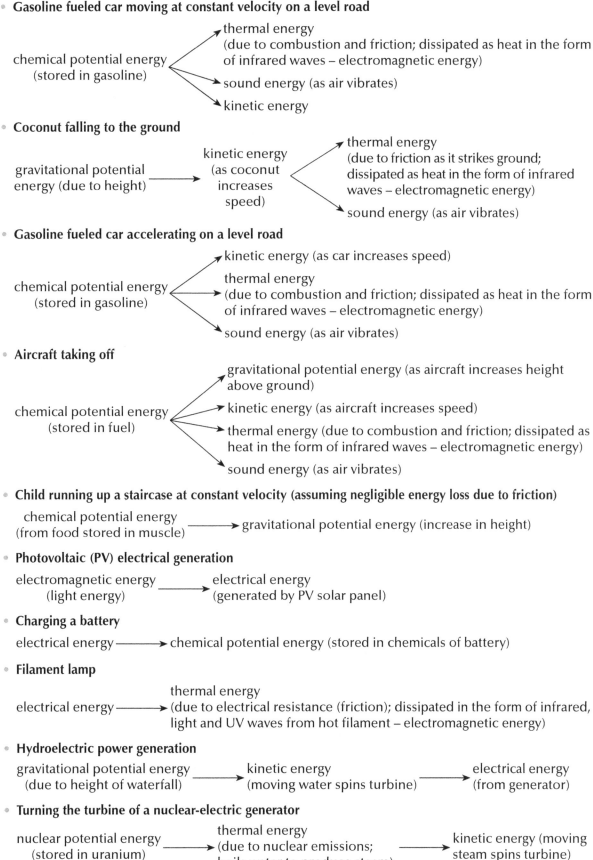

chemical potential energy (stored in gasoline) →
- thermal energy (due to combustion and friction; dissipated as heat in the form of infrared waves – electromagnetic energy)
- sound energy (as air vibrates)
- kinetic energy

- **Coconut falling to the ground**

gravitational potential energy (due to height) → kinetic energy (as coconut increases speed) →
- thermal energy (due to friction as it strikes ground; dissipated as heat in the form of infrared waves – electromagnetic energy)
- sound energy (as air vibrates)

- **Gasoline fueled car accelerating on a level road**

chemical potential energy (stored in gasoline) →
- kinetic energy (as car increases speed)
- thermal energy (due to combustion and friction; dissipated as heat in the form of infrared waves – electromagnetic energy)
- sound energy (as air vibrates)

- **Aircraft taking off**

chemical potential energy (stored in fuel) →
- gravitational potential energy (as aircraft increases height above ground)
- kinetic energy (as aircraft increases speed)
- thermal energy (due to combustion and friction; dissipated as heat in the form of infrared waves – electromagnetic energy)
- sound energy (as air vibrates)

- **Child running up a staircase at constant velocity (assuming negligible energy loss due to friction)**

chemical potential energy (from food stored in muscle) → gravitational potential energy (increase in height)

- **Photovoltaic (PV) electrical generation**

electromagnetic energy (light energy) → electrical energy (generated by PV solar panel)

- **Charging a battery**

electrical energy → chemical potential energy (stored in chemicals of battery)

- **Filament lamp**

electrical energy → thermal energy (due to electrical resistance (friction); dissipated in the form of infrared, light and UV waves from hot filament – electromagnetic energy)

- **Hydroelectric power generation**

gravitational potential energy (due to height of waterfall) → kinetic energy (moving water spins turbine) → electrical energy (from generator)

- **Turning the turbine of a nuclear-electric generator**

nuclear potential energy (stored in uranium) → thermal energy (due to nuclear emissions; boils water to produce steam) → kinetic energy (moving steam spins turbine)

- **Photosynthesis**

    electromagnetic energy ———→ chemical energy (stored in carbohydrate molecules of green
    (light waves from Sun)   plants)

- **Swinging pendulum (assuming negligible friction)**

    gravitational potential ———→ kinetic ———→ gravitational potential
    energy at A   energy at B   energy at C

    At A and C: $E_K = 0$ since the ball is momentarily at rest.

    E_{GP} is maximum since it is now at its highest position.

    At B: E_K is maximum since the ball moves fastest

    E_{GP} is minimum since it is at its lowest position.

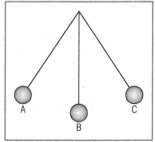

Figure 5.3 *Energy transformation of a swinging pendulum*

- **Microphone**

    sound energy   kinetic energy   electrical energy
    (kinetic energy of vibrating ——→ (air vibrates speaker ——→ (a.c. induced in coil with
    air of sound wave)   coil and cone)   same frequency as sound
                                                                        wave – see Chapter 31)

- **Loudspeaker**

                                          kinetic energy
                                          (oscillation of speaker   sound energy
    electrical energy   coil and cone with same ——→ (air vibrates as it is pushed
    (a.c. fed to speaker coil) ——→ frequency as a.c. fed   and pulled by speaker cone)
                                          to it – see Chapter 30)

- **Vehicle braking and coming to rest**

                                          thermal energy
    kinetic energy   (due to friction; dissipated as heat from the brakes in the form of
    (due to vehicle's motion) ——→ infrared waves – electromagnetic energy)

- **Microwave warming food**

                                          thermal energy
    electromagnetic energy   (due to molecules in food vibrating more vigorously as they
    (microwaves) ——————→ resonate with the frequency of the microwaves; dissipated as heat
                                          in the form of infrared waves – electromagnetic energy)

- **Turning the turbine of a diesel-electric generator**

                                          thermal energy
    chemical potential energy (due to combustion; boils ——→ kinetic energy
    (stored in diesel) ——————→ water to produce steam)   (moving steam spins turbine)

Example 3

Figure 5.4 shows a heavy steel ball on a construction site being used to demolish a building. Determine the maximum velocity acquired by the ball as it swings from a crane through a height of 7.2 m before striking its target. (Gravitational field strength = 10 N kg⁻¹.)

Assuming friction is negligible, E_p at the highest point transforms completely to E_k at the lowest point.

$$\frac{1}{2}mv^2 = mg\Delta h$$

$$\frac{1}{2}v^2 = g\Delta h \ldots. (m \text{ cancels})$$

$$\frac{1}{2}v^2 = 10 \times 7.2$$

$$v^2 = 2 \times 10 \times 7.2$$

$$v^2 = 144$$

$$v = 12 \text{ m s}^{-1}$$

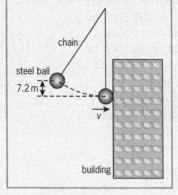

Figure 5.4 *Example 3*

Example 4

Calculate the force exerted by the engine of a boat if 1.2 kW of power drives it across the water at a speed of 3.0 m s⁻¹.

$$P = \frac{W}{t} = \frac{F \times d_{\parallel}}{t} = F \times v_{\parallel}$$

$$\therefore \quad P = F \times v_{\parallel}$$

$$1200 = F \times 3.0$$

$$\frac{1200}{3.0} = F$$

$$F = 400 \text{ N}$$

Example 5

A boy of mass 75 kg runs up a flight of 15 steps, each 20 cm high, at constant speed in a time of 3.0 s (see Figure 5.5).

a By means of an arrow diagram, show the single MAIN energy transformation occurring.

b Calculate the energy he uses, stating any assumption.

c Calculate his personal power.

a Chemical potential energy from food stored in muscles ⟶ gravitational potential energy as he runs upward.

b Since the boy is running at constant speed, kinetic energy is constant and so is not taking part in the transformation. Assuming the heat and sound energy lost due to friction are negligible, the chemical energy used is equal to the gravitational potential energy acquired.

$$E = mg\Delta h = 75 \times 10 \times 3.0 = 2250 \text{ J}$$

c $P = \dfrac{E}{t} = \dfrac{2250}{3.0} = 750 \text{ W}$

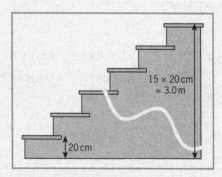

Figure 5.5 *Example 5*

Example 6

Determine the gravitational potential energy and the kinetic energy of a ball of mass 200 g, at each of the marked points shown in Figure 5.6 as it rolls from rest at A down the incline to E.

First find the potential energy E_{GP} at each point by using the formula $E_{GP} = mg\Delta h$.

Since the object is a **ball**, it **rolls** (in contrast to a **block** which would **slide**) and therefore friction is negligible. Any loss or gain in potential energy produces an equal gain or loss in kinetic energy, which can be found by subtraction as shown in the table.

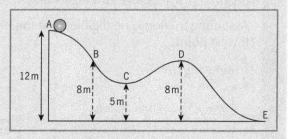

Figure 5.6 *Example 6*

	Potential energy/J	Kinetic energy/J
A	$0.200 \times 10 \times 12 = 24$	0 … (at rest)
B	$0.200 \times 10 \times 8 = 16$	$24 - 16 = 8$
C	$0.200 \times 10 \times 5 = 10$	$24 - 10 = 14$
D	$0.200 \times 10 \times 8 = 16$	$24 - 16 = 8$
E	$0.200 \times 10 \times 0 = 0$	$24 - 0 = 24$

Example 7

A bullet of mass 20 g and velocity 250 m s^{-1} bores a hole through a piece of wood 5.0 cm thick and emerges from the other side with a velocity of 100 m s^{-1}.

Calculate:

a the kinetic energy transformed as the bullet penetrates the wood

b the energy lost to heat and sound

c the work done in boring the hole

d the average frictional force of the bullet on the wood.

a kinetic energy $\longrightarrow$ heat and sound energy

Only part of the initial kinetic energy is transformed.

$$\Delta E_k = \frac{1}{2}mv_1^2 - \frac{1}{2}mv_2^2$$

$$\Delta E_k = \frac{1}{2}m(v_1^2 - v_2^2)$$

$$\Delta E_k = \frac{1}{2} \times 0.020(250^2 - 100^2)$$

$$\Delta E_k = 525 \text{ J}$$

b The 525 J of kinetic energy transforms to 525 J of heat and sound energy.

c Whenever energy is transformed, an equal amount of work is done. The work done is therefore 525 J.

d $W = F \times d_{\parallel}$

$525 = F \times 0.050$

$\dfrac{525}{0.050} = F$

$F = 10\,500 \text{ N}$

Example 8

A block of mass 200 g slides from rest at A and reaches point B with a velocity of 4.0 m s^{-1} (see Figure 5.7).

a Describe the energy transformation occurring by means of an arrow diagram.

b Determine the heat and sound energy produced and the work done against friction.

c Determine the frictional force.

a gravitational potential energy
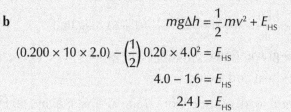
→ kinetic energy
→ heat and sound energy

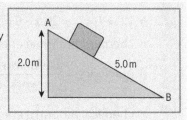

Figure 5.7 Example 8

b
$$mg\Delta h = \frac{1}{2}mv^2 + E_{HS}$$

$$(0.200 \times 10 \times 2.0) - \left(\frac{1}{2}\right) 0.20 \times 4.0^2 = E_{HS}$$

$$4.0 - 1.6 = E_{HS}$$

$$2.4\,J = E_{HS}$$

The 2.4 J of heat and sound energy was the result of 2.4 J of work done against friction.

c Since the frictional force acting along the 5.0 m slope produces the heat and sound energy of 2.4 J:

$$E = F \times d_{\parallel}$$

$$2.4 = F \times 5.0$$

$$\frac{2.4}{5.0} = F$$

$$F = 0.48\,N$$

Efficiency

Machines are devices designed to make work easier for a user. This usually means decreasing the force and increasing the distance moved in the direction of the force. The force to be overcome (usually the **weight**) is referred to as the **load** and the associated work is the **useful work output**. The force exerted by the user of the machine is the **effort** and the associated work is the **work input**.

In practice, the useful energy output is always less than the energy input since thermal energy produced due to friction is wasted. The efficiency of a system is commonly represented as a percentage and can be calculated as follows:

$$\text{efficiency} = \frac{\text{useful work or energy output}}{\text{work or energy input}} \times 100\% \quad \text{OR} \quad \text{efficiency} = \frac{\text{useful power output}}{\text{power input}} \times 100\%$$

Example 9

Figure 5.8 shows the forces acting as Jide pushes a block of weight of 20 N through 8.0 m up an inclined plane at constant speed by using an effort of 12 N. The normal reaction R and the frictional force f are also shown. Calculate:

a the work input (work done by the effort)

b the useful work output (work done against the force of gravity i.e. against the weight; this is also the gravitational potential energy acquired)

c the chemical potential energy used by Jide

d the efficiency and show an arrow diagram of the energy transformation

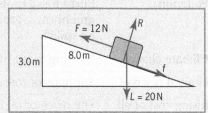

Figure 5.8 Example 9

e the energy lost as heat and sound due to friction

f the work done against the frictional force

g the frictional force

h the work done by the normal reaction, R.

---

a $W_i = F \times d_F = 12 \times 8.0 = 96$ J

b $W_o = L \times d_L = 20 \times 3.0 = 60$ J

c chemical potential energy of 96 J was used by Jide's muscles to provide the work input

d efficiency $= \dfrac{\text{useful work output}}{\text{work input}} \times 100\% = \dfrac{60}{96} \times 100\% = 62.5\%$ (63% to 2 sig.fig.)

chemical potential energy **96 J** $\longrightarrow$ gravitational potential energy **60 J**
$\searrow$ heat and sound energy

work input
(work done by effort) **96 J** $\longrightarrow$ useful work output (work done against gravity) **60 J**
$\searrow$ wasted work (work done against friction)

The second arrow diagram has been included so that you can understand the process in terms of work.

e From the arrow diagram, the energy lost as heat and sound = 96 J – 60 J = 36 J

f From the arrow diagram, the work against friction = 96 J – 60 J = 36 J

g To calculate the frictional force, we use the work done against friction and the distance through which the friction occurs:

$W_f = f \times d_f$

$36 = f \times 8.0$

$f = \dfrac{36}{8.0} = 4.5$ N

h Since R is **perpendicular to the distance** through which the block moves, it does NO WORK (see page 62).

Fossil fuels and alternative sources of energy

*Fossil fuels are buried combustible deposits of decayed plant and animal matter that have been converted to **crude oil**, **natural gas** and **coal** by subjection to heat and pressure in the Earth's crust for hundreds of millions of years.*

Problems associated with the use of fossil fuels are shown in Table 5.3.

Table 5.3 *Problems associated with the use of fossil fuels*

Problem	Details
Pollution	Burnt fossil fuels contaminate the environment with pollutants, including greenhouse gases. Unburnt fossil fuels also pollute the environment (example: oil spills).
Climate change	Burning fossil fuels enhances climate change.
Limited reserves	Supplies of fossil fuels are diminishing.
Rising cost of oil	The price of oil increases as it becomes less abundant.
Health care costs	Government funding of health care facilities to deal with illnesses associated with pollutants from fossil fuels is a burden to economies.

Energy sources may be either renewable or non-renewable.

Renewable sources of energy are those that are readily replenished by natural processes.

Examples are solar, hydroelectric, geothermal, tidal and wind energy.

Non-renewable sources of energy are those that are not readily replenished.

Examples are fossil fuels and nuclear fuel.

Due to the problems associated with fossil fuels, it is necessary to use alternative sources of energy.

An alternative source of energy is one which is not a fossil fuel.

Various alternative sources of energy are shown in Table 5.4. Most alternative sources:

- are renewable
- produce zero or minimum net environmental pollution
- require minimum operational costs, although initial plant costs may be high.

Table 5.4 *Various alternative sources of energy*

Energy source	Uses
Solar energy	**Water heaters** heat water directly and are relatively cheap to install (see Chapter 14 page 161). **Photovoltaic (PV) panels** convert solar radiation into electrical energy.

Advantages	Disadvantages
• Low maintenance and operating costs.	• High start-up cost.
• Clean source of energy.	• Poor performance on cloudy days.
• Energy can be stored in batteries.	• Large production requires much space.
• Infinite supply of free sunshine.	• Low efficiency of conversion to electricity.

	Solar cookers use mirrors to reflect solar radiation to a furnace or pot. **Solar driers** absorb solar radiation through a glass cover. Air below the cover is heated and rises through the drying chamber by convection. The warm saturated air exits through a vent at the top of the drier and unsaturated air is drawn in at the bottom to be heated.
Biofuels or biomass	These provide energy from **plant** or **animal matter**. Although biofuels produce carbon dioxide when burnt, the plants from which they are formed removed carbon dioxide when they were alive, producing a canceling effect on environmental pollution. **Wood** can be burnt directly to be used as a fuel for cooking. **Biogas** is obtained from the decay of **plant** and **animal wastes** in the **absence of oxygen**. It is used for cooking or to drive generators on farms. Disadvantages of using biogas are that it is mainly methane, a greenhouse gas, and forested areas as well as agricultural lands and crops are used in producing it. **Biodiesel** is produced from **vegetable oils, animal oils/fats or waste cooking oils**. It produces less pollution than petroleum diesel. **Gasohol** is made from **gasoline mixed with ethanol** produced from the fermentation of crops such as sugar cane. It is cheaper than regular gasoline, emits less harmful gases when burnt, gives better engine performance and its production provides jobs.

(Continued)

Table 5.4 *Continued*

Energy source	Uses		
Wind energy	**Kinetic energy** from the wind can be used to turn the turbines of electrical generators.		
	Advantages		**Disadvantages**
	• Caribbean islands generally experience strong winds. • Clean energy. • Can be stored in batteries.		• High costs in constructing the plant. • Causes noise pollution and unpleasant scenery. • Wind is variable between seasons. • Vulnerable to stormy weather.
Hydro-electric energy	Water collected behind a dam can be released to turn the turbines of electrical generators.		
	Advantages		**Disadvantages**
	• The ability to release the water as needed is a clear advantage of these systems over the less controllable solar and wind systems. • Clean source of energy.		• High cost in constructing the plant. • Disturbs the ecology. • Danger of possible flooding.
Tidal energy	Water from the ocean can be collected at high tide and then released at low tide to produce powerful pressures that turn the turbines of electrical generators. High costs in constructing the plant, together with the negative impact on tourism due to the disturbance of the natural beauty of the coast, are the main disadvantages.		
Wave energy	Energy of wave motion and ocean currents can be harnessed to turn the turbines of electrical generators **near shorelines** or floating **far offshore**. Advantages of low operational cost and consistent wave power are offset by disadvantages of the high costs in constructing the plant and the negative effect on the marine ecology.		
Geo-thermal energy	It is common in volcanic islands for **hot water** and **steam** from within the Earth to be released in hot springs. The heat can be used directly for industrial processes or to warm buildings. **Geothermal energy plants** drill and install pipes so that hot water and steam can rise through them and be used to turn the turbines of electrical generators. The water is then cooled and returned to the geothermal reservoir in the Earth where it is reheated.		
Nuclear energy	This is an **alternative source** of energy that is **non-renewable**. Energy released by the fission of **uranium** and **plutonium** (see Chapter 34) is used to produce steam to turn the turbines of electrical generators. A small amount of nuclear fuel produces an enormous amount of energy and greenhouse gases are not produced in the process. However, nuclear fuel, before and after use, is extremely dangerous, and spent nuclear fuel is difficult to dispose of.		

Recalling facts

1 **a** Define:

  i work

  ii energy

  iii power

  iv the joule

  v the watt.

 b Define:

  i gravitational potential energy

  ii elastic potential energy

  iii chemical potential energy

  iv electromagnetic energy.

 c Distinguish between *thermal energy* and *heat*.

2 **a** What type of energy is possessed by each of the following?

  i a stretched spring

  ii a book on a shelf

  iii the active material in a battery.

 b State the law of conservation of energy.

 c Describe the MAIN energy transformations occurring in each of the following situations:

  i a truck applying brakes

  ii a microwave oven warming food

  iii a filament lamp in use

  iv an active hydroelectric power station.

3 **a** Define:

  i fossil fuels

  ii renewable energy sources

  iii alternative energy sources.

 b List ONE alternative energy source which is non-renewable.

4 **a** List THREE biofuels.

 b Briefly describe how we can utilise:

  i tidal energy

  ii geothermal energy.

5 **a** List THREE devices, other than photovoltaic (PV) panels, which utilise solar energy.

 b List TWO advantages and TWO disadvantages of using PV panels.

Applying facts

Acceleration due to gravity = 10 m s^{-2}

6 Determine the kinetic energy of an object of weight 12.0 N which moves at 20 m s^{-1}.

7 Zachary throws a ball of mass 500 g vertically upward to a height of 20 m.

   a By means of an arrow diagram, show the energy transformation occurring as the ball rises, assuming friction is negligible.

   b Calculate the maximum kinetic energy of the ball.

8 Calculate the maximum speed of a pendulum bob which swings through a height of 20 cm.

9 Calculate the power of Emmett's toy car if a force of 16 N drives it at a constant velocity of 0.50 m s^{-1}.

10 Jenna drops a heavy stone from the edge of a cliff. Calculate the height through which it falls if it acquires a maximum speed of 30 m s^{-1}. State any assumption made.

11 Zoe drives 50 km at a steady speed in a small vehicle which she has designed. The total force resisting the motion of the vehicle is 900 N and she uses a full tank of fuel. Determine:

   a the useful work output of the vehicle

   b the chemical energy of the fuel initially in the tank if it is transferred to the motion of the vehicle with an efficiency of 30%.

Analysing data

12 The table shows how the energy input to a system increases with time as a block is pulled by a horizontal force through 8.0 m at constant velocity across a horizontal floor.

Energy E/J	21	38	59	80	102
Time t/s	0.8	1.6	2.4	3.2	4.0

   a Use the table to plot a graph of energy against time.

   b Calculate:

      i the slope (gradient) of the graph

      ii the power input to the system

      iii the average force on the block

      iv the velocity of the block.

6 Pressure and Archimedes' principle

Why is it so difficult to press one's thumb into a notice board, but so easy to drive a thumb tack into it? Why does a heavy cement truck need several broad tyres so that it does not cut tracks into the ground?

Why do our ears hurt when we dive deep below the surface of the ocean? These occurrences are all examples of the relationship between forces and the area on which these forces act.

Learning objectives

- Define **pressure** and its **SI unit**.
- State and apply the characteristics of pressure due to fluids.
- Calculate **pressure** due to **solids** and due to **fluids**.
- Demonstrate the strength of the atmospheric pressure.
- Describe and explain the use of the mercury **barometer** and the **manometer**.
- State and experimentally verify **Archimedes' principle**.
- Relate Archimedes' principle to the **principle of flotation**.
- Perform calculations involving Archimedes' principle.

Pressure

*Pressure is the **force** per unit **area** acting perpendicularly to a surface.*

$$\text{pressure} = \frac{\text{force}}{\text{area}} \qquad P = \frac{F}{A}$$

The equation indicates that the **smaller the area** on which the force acts, the **greater will be the pressure**. The point of a thumb tack has a small area of contact with the notice board to which it is being forced, and so the pressure it creates at that point is high. Several broad tyres on a cement truck implies there is a large area of contact with the road, and this reduces the pressure on the road. As we dive deeper into the ocean, there is a greater weight (force) per unit area of water above us and therefore a greater pressure.

The SI unit of pressure is the pascal (Pa).

1 pascal (Pa) is the pressure exerted by a force of 1 newton acting perpendicularly on an area of 1 m^2.

$$(1 \text{ Pa} = 1 \text{ N m}^{-2})$$

A rectangular box of mass 45 kg is of dimensions 20 cm × 30 cm × 50 cm. Determine the minimum pressure that can be exerted by the box when resting on one of its faces given that the gravitational field strength is 10 N kg^{-1}.

The **minimum** pressure occurs when the block rests on its **largest face**.

$$P = \frac{F}{A}$$

$$P = \frac{mg}{l \times w} \quad \text{... } F \text{ is the weight } (mg) \text{ and acts on the base } (l \times w)$$

$$P = \frac{45 \times 10}{0.50 \times 0.30}$$

$$P = 3000 \text{ Pa}$$

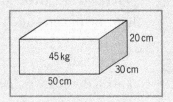

Figure 6.1 *Example 1*

Characteristics of fluid pressure

The pressure at a point in a fluid:

- increases with the **depth** of the point below the surface
- increases with the **density** of the fluid
- increases with the **acceleration due to gravity**
- acts **equally** in **all directions**
- acts **perpendicularly** to a surface within the fluid
- is **transmitted uniformly** through a **confined** fluid when an **external** force is applied (principle of hydraulics).

The **change** in **pressure** between two points in a fluid is calculated from:

$$\Delta \text{pressure} = \Delta \text{depth} \times \text{density} \times \text{acceleration due to gravity}$$
$$\Delta P = \Delta h \rho g$$

Hydrostatic pressure is the pressure exerted by a liquid at rest.

Example 2

A diver is submerged to a depth of 20 m below the surface of the sea. Given that the acceleration due to gravity is 10 m s^{-2}, the density of water is 1000 kg m^{-3}, and the atmospheric pressure on the day is 1.0×10^5 Pa, calculate:

a the hydrostatic pressure on the diver caused by the water

b the total pressure on the diver.

a $P = h\rho g = 20 \times 1000 \times 10 = 2.0 \times 10^5$ Pa

b Total pressure = atmospheric pressure + pressure of water

$P_T = 1.0 \times 10^5 + 2.0 \times 10^5 = 3.0 \times 10^5$ Pa

Note: The pressure increases by approximately **1 atmosphere** for every **10 m** below the surface of the water.

$P = h\rho g = 10 \times 1000 \times 10 = 1.0 \times 10^5$ Pa ... (1 atmosphere)

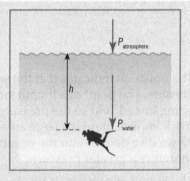

Figure 6.2 *Example 2*

Example 3

Calculate the force created by the hydrostatic pressure of the water on the base of the conical flask shown in Figure 6.3. The diameter of the base is 16.0 cm and the surface of the water is 18 cm above the base. The gravitational field strength is 10 N kg^{-1} and the density of the water is 1000 kg m^{-3}.

First find the pressure of the water on the base:

$P = h\rho g = 0.18 \times 1000 \times 10 = 1800$ Pa

Then, use the pressure to find the force:

$P = \dfrac{F}{A}$

$\therefore F = P \times A = P \times \pi r^2 = 1800 \times \pi (0.080)^2 = 36$ N

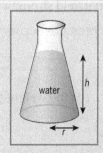

Figure 6.3 *Example 3*

Example 4

Water of density 1000 kg m^{-3} is retained by a dam built across a river of width 100 m and depth 20 m. Determine the total force exerted by the **hydrostatic pressure** of the water on the *wall of the dam*.

Since the pressure due to the water on the wall varies with the depth of the water, we first will find the **average pressure** of the water on the wall. This occurs at **half the depth** of the river, i.e. 10 m below the surface.

$P = h\rho g = 10 \times 1000 \times 10 = 1.0 \times 10^5$ Pa

Using this average pressure, we can then calculate the total force of the water on the dam:

$P = \dfrac{F}{A}$ $\therefore F = PA = 1.0 \times 10^5 \times (100 \times 20) = 2.0 \times 10^8$ N

Note: The base of a dam is constructed much stronger than its top to withstand the **increased pressure** at the **greater depth**. The total force on the dam is greater than this value since it is the result of the sum of the **hydrostatic pressure** and the **atmospheric pressure**.

Pressure in liquids increases with depth and acts equally in all directions from a given point

Figure 6.4 shows a tall jar of water with holes in its sides. Water spouts with **increased strength** from the holes at **greater depths** and with the **same strength** from holes at the **same depth**.

If we dive deep below the surface of the ocean, we experience increased pain in our ears due to the increased pressure. However, at any given depth, turning our head from side to side has no effect on the pain, since pressure at a point in a fluid acts equally in all directions.

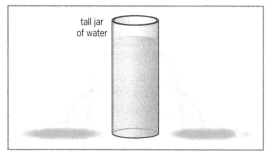

Figure 6.4 *Pressure increases with depth*

The upper surface of a continuous body of fluid always seeks the same level

Figure 6.5 shows two water reservoirs R_1 and R_2 connected by a pipe with an attached tap which is initially closed. When the tap is opened, the pressure on the right end of the horizontal connecting pipe is greater than that on its left end since there is a **taller water column** above the right end of the pipe than above its left end. Water will therefore flow from R_2 to R_1 until the water levels are equal, and hence the pressures at each side of the tap are equal.

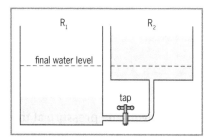

Figure 6.5 *Liquids seek a common level*

Sucking liquid into a drinking straw

Figure 6.6 shows liquid being sucked into a drinking straw by reducing the air pressure above the liquid in the straw to a value, P_X. Atmospheric pressure, P_A, then forces liquid up the straw towards the region of reduced pressure.

Since pressure at any point on a given level in a fluid is the same, the pressure at A is equal to the pressure at B. The reduced air pressure P_X, plus the hydrostatic pressure produced by the liquid column P_L, is therefore equal to the atmospheric pressure P_A.

$$P_X + P_L = P_A$$
$$P_X + h\rho g = P_A$$

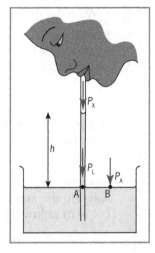

Notes:

- The liquid column above the surface of the liquid in the dish is referred to as a **head** of liquid.

- When investigating the pressure at a point in a liquid, look **above the point** to see what is causing the pressure, or look **above some other point** on the **same level** in the **same body of liquid**.

Figure 6.6 *Sucking liquid into a drinking straw*

Determining the density of a liquid using a concept of fluid pressure

Figure 6.7 shows an arrangement used to determine the density ρ_L of a liquid. An inverted U-tube has its ends immersed into liquid, L, and water, W. With the tap open, some of the air is sucked from the tubes so that the pressure there **decreases** to some value, P_X. The tap is then closed, and the heights h_L and h_W of the liquid heads are measured using a metre rule. The pressure at the surface of the liquid in the dish is **atmospheric pressure**, P_A, and therefore the pressure inside the tubes at this same level is also atmospheric pressure.

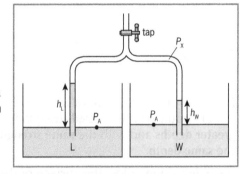

The pressure at the base of each column is caused by the sum of the pressures P_X and the liquid head above it. Knowing the density of water, the density of liquid L can be calculated:

Figure 6.7 *Determining the density of a liquid*

$$P_A = P_X + h_L\rho_L g \quad \text{... pressure at base of left column} \qquad \text{..... (i)}$$
$$P_A = P_X + h_W\rho_W g \text{ ... pressure at base of right column} \qquad \text{..... (ii)}$$

$$\therefore P_X + h_L\rho_L g = P_X + h_W\rho_W g \text{ ..... equating right hand sides of (i) and (ii)}$$

$$\therefore h_L\rho_L g = h_W\rho_W g \text{ ... subtracting } P_X \text{ from both sides}$$
$$\therefore h_L\rho_L = h_W\rho_W \text{ ... dividing both sides by } g$$
$$\therefore \rho_L = \frac{h_W\rho_W}{h_L}$$

Alternatively, the pressure at the base of each column of liquid is atmospheric pressure and the pressure at the upper surface of each is P_X. Both columns therefore have the same **change in pressure** between their ends and therefore:

$$h_L\rho_L g = h_W\rho_W g$$
$$h_L\rho_L = h_W\rho_W$$

Fluid pressure and hydraulics

*Pascal's principle of hydraulics states that whenever an **external force** is applied to the surface of an incompressible fluid in an enclosed vessel, the **pressure** so created is transmitted **uniformly** throughout the fluid.*

It is the **pressure – not the force** – that is transmitted uniformly throughout the fluid. The force will vary depending on the area on which it acts. Hydraulic machines allow us to obtain very large forces from much smaller ones. Examples are hydraulic jacks, hydraulic brakes, excavators, and bulldozers.

Example 5

Figure 6.8 shows a simple **hydraulic lift** (**jack**). A force of just 10 N applied to the small piston of area, 0.10 m² raises the heavy object at the large piston of area, 25 m². Calculate:

a the pressure P_S created at the small piston due to the applied force

b the pressure P_L created at the large piston due to the applied force

c the weight of the object being raised.

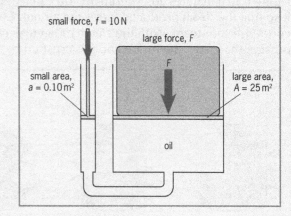

Figure 6.8 *Example 5*

a Pressure at small piston $P_S = \dfrac{f}{a} = \dfrac{10}{0.10} = 100$ Pa

b Pressure at large piston must be the same $P_L = 100$ Pa

c $P_L = \dfrac{F}{A}$ $\qquad \therefore 100 = \dfrac{F}{25}$ $\qquad F = 100 \times 25 = 2500$ N

Notes:

- The area of the larger piston is 250 times the area of the smaller piston, and therefore the force at the larger piston is 250 times the force at the smaller piston.
- The object can be raised at **constant velocity** (acceleration = zero) if the upward force of the oil is equal to its weight since the resultant force is then zero.
- If the force of the oil is greater, the resultant force will cause it to **accelerate** upward.

Hydraulic brakes

Have you ever considered how the enormous braking force necessary to stop a heavily laden truck is produced? When a driver presses on the brake pedal, a **small force** is exerted on the brake fluid inside a **small piston** of the master cylinder. This pressure is then uniformly transmitted through the brake lines to **larger pistons** that press on the wheels of the vehicle, creating correspondingly **larger forces** which reduce its speed.

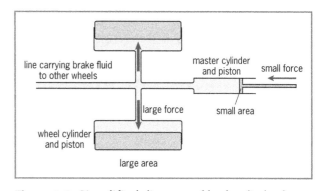

Figure 6.9 *Simplified diagram of hydraulic brakes*

Atmospheric pressure

Atmospheric pressure at sea level varies between **96 kPa** and **104 kPa**. The pressure in the eye of a **category 5 hurricane** is about **92 kPa**, and yet this relatively small fall in pressure produces winds of extremely dangerous force. Imagine the destruction that a hurricane would produce if the pressure in its eye could fall to zero!

Jet aircraft fly at altitudes where the atmospheric pressure is only about 25 kPa. The cabins must therefore be sealed and pressurised to prevent passengers from bleeding due to their excess blood pressure!

The following experiments demonstrate the strength of the atmospheric pressure.

Crushing can experiment

Figure 6.10 illustrates an experiment you can try at home. Place a little water into a metal can and heat it so that the **steam** produced **forces the air out**. Cover the can quickly and pour **cold water** over it to condense the steam. With the steam molecules condensed, more molecules strike the can from the outside than from the inside and it is crushed by the **excess external pressure**.

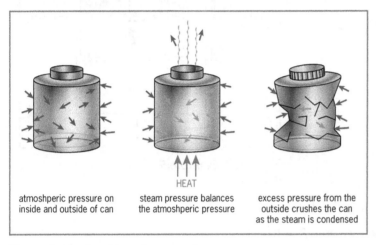

Figure 6.10 *Crushing can experiment*

Magdeburg hemispheres

Otto von Guericke, a German physicist and mayor of the city of Magdeburg, performed an interesting experiment to demonstrate the strength of atmospheric pressure. He joined two metal hemispheres together with a **greased ring** to act as an **air seal** as shown in Figure 6.11. With the tap open, air was pumped from the cavity between the hemispheres; the tap was then closed. Two teams, each of fifteen horses, harnessed to the handles and driven in opposite directions, could not separate the hemispheres.

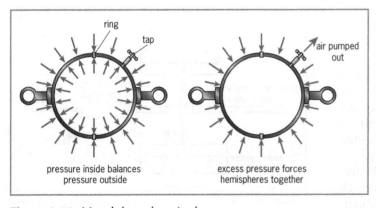

Figure 6.11 *Magdeburg hemispheres*

Rubber sucker

Most of us are familiar with a toy dart having a rubber sucker at its end as shown in Figure 6.12. When the sucker is squeezed onto a surface, **air is expelled** from the cavity between the rubber and the surface. The rubber sucker then returns to its original shape, causing the number of air molecules per unit area striking its inner wall to decrease as the cavity expands. The excess pressure exerted by the air molecules striking the outer wall of the sucker then holds it to the surface. Since the surface is not perfectly smooth, air leaks into the sucker cavity through small crevices, equalises the pressure and causes the dart to detach.

Greasing the edge of the rubber sucker **provides a seal**, preventing air from reentering and therefore allowing the dart to be attached for a longer period.

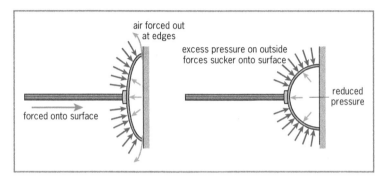

Figure 6.12 *Rubber sucker*

Measuring atmospheric pressure using a mercury barometer

Atmospheric pressure can be measured by the simple mercury barometer shown in Figure 6.13. The region above the mercury in the tube is effectively a **vacuum**, containing only a little mercury vapour, which exerts a **negligible pressure**. The pressure of the atmosphere pushes on the mercury in the dish, forcing it up the tube until the **pressure at Y** created by the column of mercury is **equal to the atmospheric pressure** P_A. **Standard atmospheric pressure** forces the mercury **76 cm** up the tube.

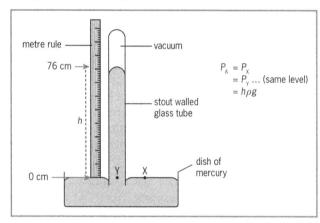

Figure 6.13 *Simple mercury barometer*

Low pressure regions above the ocean result in the formation of updrafts of moisture, producing rainy conditions which can lead to **bad weather**. An approaching **depression** or **tropical storm** in the Caribbean is therefore signaled by a **decrease** in atmospheric pressure.

> **Example 6**
>
> Given that the density of mercury is 13 600 kg m^{-3} and that the gravitational field strength is 10 N kg^{-1}, calculate the pressure of the atmosphere indicated by a barometer having a mercury column 73.6 cm tall.
>
> $$P_A = h\rho g = 0.736 \times 13\,600 \times 10 = 1.0 \times 10^5 \text{ Pa}$$

Increasing the cross-sectional area of the tube of a mercury barometer

Figure 6.14 shows that increasing **the cross-sectional area** of the tube has **no effect** on the height to which the mercury will rise within it. The pressure at any point in a **given liquid** is dependent only on its **depth**, assuming that the acceleration due to gravity is constant. Since the tube is of uniform cross-sectional area, any change in this area will result in a proportional change in weight of the liquid contained. If, for example, the cross-sectional area of B is greater than that of A by a factor of 4, the weight of the liquid column of B will also be greater than that of A by a factor of 4, and this results in the same pressure at the base of each mercury column.

$$P = \frac{F}{A} = \frac{4F}{4A}$$

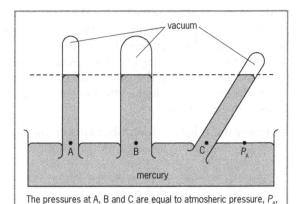

The pressures at A, B and C are equal to atmosheric pressure, P_A, since these points are at the same level as the surface of the liquid outside the tubes where the atmoshperic pressure acts.

Figure 6.14 *Increasing the cross-sectional area or tilting the tube of a mercury barometer*

Tilting the tube of the mercury barometer

Figure 6.14 shows that **tilting the tube** has **no effect** on the height to which the mercury will rise within it. The volume, and hence the weight of mercury in the tilted tube, is greater than that in the vertical tube of same diameter, but a fraction of its weight is now supported by the glass wall of the tube such that the pressure at its base is unchanged.

Measuring the pressure of a gas supply using a manometer

A manometer is used to measure the pressure of a gas supply. It consists of a measuring scale such as a metre rule and a transparent U-tube containing a liquid as shown in Figure 6.15. One arm of the U-tube is exposed to the atmosphere and the other arm is connected to the gas supply being measured.

The difference in the levels of the liquid in the arms indicates the difference in pressure between the gas supply and the atmosphere. This **excess pressure**, represented by the excess liquid column (head), can be found from the equation $P = h\rho g$. For measuring gas pressures close to atmospheric pressure, water may be used as the liquid, but for **higher pressures**, a **denser liquid** such as mercury is necessary.

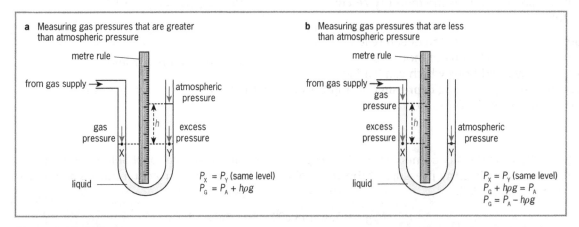

Figure 6.15 *Using a manometer*

Example 7

Given that the density of mercury is 13 600 kg m⁻³ and that the gravitational field strength is 10 N kg⁻¹, calculate the pressure of the gas supply connected to the manometer of Figure 6.16 if the atmospheric pressure on the day is 102 000 Pa.

pressure of gas = pressure of atmosphere + pressure of excess mercury head

$$P_G = P_A + P_M$$
$$P_G = 102\ 000 + (0.20 \times 13\ 600 \times 10)$$
$$P_G = 129\ 200 \text{ Pa}$$

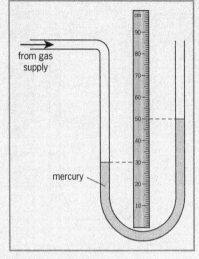

from gas supply

mercury

Figure 6.16 *Example 7*

Expressing pressure as the height of a liquid column

Since the pressure at the base of a barometer tube is proportional to the height of the liquid column it contains, it is common to express **pressure** as the **height of a liquid column.**

Different liquids exert different pressures at the same depth and so when expressing pressure as a height of liquid, **the liquid must be identified.** Atmospheric pressure is often stated in millimetres of mercury (mmHg). The chemical symbol for mercury is Hg.

The equation $P = h\rho g$ can be used to convert this value to pascals, using standard SI base units. A gas pressure of 900 mmHg implies that the pressure in pascals is given by:

$$P = 0.900 \times 13\ 600 \times 10 = 122\ 400 \text{ Pa}$$

Example 8

If, in example 7, the atmospheric pressure on the day was given as 75 cmHg, what would have been the pressure of the gas expressed in:

a pascals?   **b** cmHg?

a Since we are asked to give our answer in pascals, we represent each term in pascals.

$$P_G = P_A + P_M$$
$$P_G = (0.75 \times 13\ 600 \times 10) + (0.20 \times 13\ 600 \times 10)$$
$$P_G = 129\ 200 \text{ Pa}$$

b Since we are asked to give our answer in cmHg, we represent each term as a length in cmHg.

$$P_G = P_A + P_M$$
$$P_G = 75 \text{ cmHg} + 20 \text{ cmHg}$$
$$P_G = 95 \text{ cmHg}$$

Example 9

Determine the pressure in cmHg at each of the following points on the manometer and barometer shown in Figure 6.17.

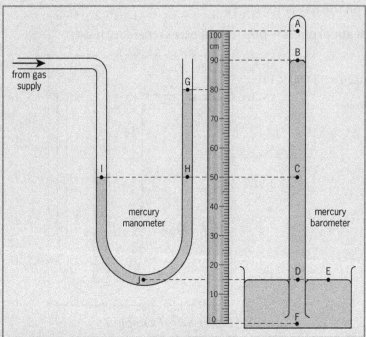

Figure 6.17 *Example 9*

Note: When investigating the pressure at a point in a liquid, **look above the point** to see what is causing the pressure, or **look above some other point** in the **same body of liquid** at the **same level**.

Table 1 *Example 9*

	Pressure in cmHg	Details
A	0	The space above the mercury in a barometer is a vacuum
B	0	As above
C	90 – 50 = 40	40 cm of mercury is above C
D	90 – 15 = 75	75 cm is above D
E	75	Same pressure as D since at same level and in same body of liquid as D (atmospheric pressure since only the atmosphere is above this point)
F	90 – 0 = 90	90 cm of mercury above F
G	75	Atmospheric pressure as indicated by the barometer (pressure at D or E)
H	(80 – 50) + 75 = 105	30 cm of mercury + atmospheric pressure (75 cm of mercury) is above H
I	105	Same pressure as at H since at same level as H and in same body of liquid as H
J	(80 – 15) + 75 = 140	65 cm of mercury + atmospheric pressure (75 cm of mercury) is above J in the right limb

Archimedes' principle

Have you ever wondered why a heavy ship made of steel will float in water, but a small pin made of the same material will sink; or why on taking a deep breath we easily float in the sea, but on exhaling we descend below the surface? The answers to these questions of buoyancy were first explained by the Greek philosopher, Archimedes.

Archimedes' principle states that when a body is wholly or partially immersed in a fluid, it experiences an upthrust equal to the weight of the fluid displaced.

Experiment to verify Archimedes' principle

- A small object is attached by a string to a spring balance and its weight W_1 is measured and recorded.
- It is then completely immersed in the water of a displacement can which has been filled to its overflow spout, and the displaced water is collected by a beaker placed below the spout.
- The new reading, W_2, on the spring balance (the apparent weight) is recorded.
- The weight of the beaker and displaced water, W_3, as well as the weight of the beaker when empty and dried, W_4, are also measured using the spring balance.
- It is observed that:

$$W_1 - W_2 = W_3 - W_4$$

upthrust = weight of fluid displaced

The verification is illustrated in Figure 6.18 by a possible sample of numerical data.

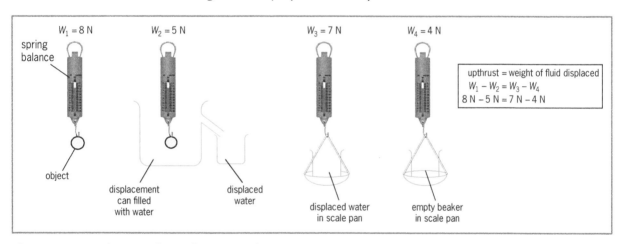

Figure 6.18 *Verifying Archimedes' principle*

Example 10

A and B are cubes of side 20 cm. The densities of A and B are 800 kg m^{-3} and 1500 kg m^{-3}, respectively. The cubes are completely submerged in water of density 1000 kg m^{-3} and released. Calculate:

a the upthrust on (weight of water displaced by) a single cube

b the resultant force in magnitude and direction on each cube

c the magnitude of the resultant acceleration of each cube.

---

a The upthrust on an object has nothing to do with the mass of the object. It depends on the weight of the fluid displaced, and therefore on the **mass, m, of the fluid displaced.**

upthrust = weight of water displaced

$$U = m_w \times g$$

$$U = (\rho_w \times V_w) \times g \ldots \text{(since mass = density} \times \text{volume)}$$

$$U = (1000 \times 0.20^3) \times 10 = 80 \text{ N}$$

b Taking downward as positive:

A: $F_A = \text{weight} - \text{upthrust}$

$$= m_A g - U$$

$$= (\rho_A V_A) g - U$$

$$= (800 \times 0.20^3)10 - 80$$

$$= -16$$

Resultant force on A = 16 N **upward**

B: $F_B = \text{weight} - \text{upthrust}$

$$= m_B g - U$$

$$= (\rho_B V_B) g - U$$

$$= (1500 \times 0.20^3)10 - 80$$

$$= 40$$

Resultant force on B = 40 N **downward**

c

A: $a = \dfrac{F_A}{m_A}$

$$= \dfrac{F_A}{(\rho_A V_A)}$$

$$= \dfrac{-16}{800 \times 0.20^3}$$

$$= -2.5 \text{ m s}^{-2}$$

Acceleration of A = 2.5 m s^{-2} **upward**

B: $a = \dfrac{F_B}{m_B}$

$$= \dfrac{F_B}{(\rho_B V_B)}$$

$$= \dfrac{40}{1500 \times 0.20^3}$$

$$= 3.3 \text{ m s}^{-2}$$

Acceleration of B = 3.3 m s^{-2} **downward**

Flotation

The principle of flotation states that the weight of a floating body is equal to the upthrust on the body.

∴ for a floating body: weight of body = upthrust = weight of fluid displaced

Consider the upthrust U on an object immersed in a fluid of density ρ_f in which a displaced volume V_f of the fluid is of mass m_f:

$$\text{upthrust} = \text{weight of fluid displaced}$$
$$U = m_f g$$
$$U = (\rho_f V_f)g$$

The equation indicates that **upthrust** is **greater in fluids of greater density**.

Table 1 shows that whether an object sinks, floats at the surface, remains suspended at rest totally submerged or rises in a fluid, depends on the relation between the densities of the fluid and of the object.

Table 2 *Effect of density on floating*

weight of object > upthrust	density of object > density of fluid	object sinks through fluid
weight of object = upthrust	density of object < = density of fluid	object can float at surface, partially or totally submerged
	density of object = density of fluid	object can remain at rest or with uniform velocity, totally submerged
weight of object < upthrust	density of object < density of fluid	object rises through fluid

- **Sinking** – An object will sink in a fluid if its density is greater than that of the fluid. The weight of the object is then greater than the weight of fluid displaced (upthrust) and it accelerates downward.
- **Floating at the surface** – A heavy ship made of steel will float, partially submerged. This is because its rooms contain air and therefore the overall density of the ship and its contents is much less than that of steel. The ship displaces just enough water to provide an upthrust equal to its weight.

   A ship travelling from the sea into the mouth of a river moves from salt water to less dense fresh water. The ship displaces more of the fresh water to maintain the upthrust at a value equal to its weight.

   A similar occurrence results when a ship travels from cooler, higher latitude waters to warmer, tropical waters. To remain floating, more of the less dense, warmer water must be displaced to provide the necessary upthrust equal to its weight.
- **Remaining at rest at any depth in a fluid** – By adjusting the amount of water in its ballast tanks, a completely submerged submarine may alter its weight so that it becomes equal to the upthrust on it. If the submarine is at rest, it will remain at rest since the resultant force on it is zero.
- **Rising** – Air is less dense than water, and so an air bubble released under water, will accelerate towards the surface. Its weight is less than the weight of water displaced (upthrust) by it.

Upthrust and submarines

Figure 6.19 shows a toy submarine that has a ballast tank which can hold water. If water is taken into the tank until the submarine's weight exceeds the weight of water displaced (upthrust), the submarine will accelerate downwards. The resultant force on it is downwards and it will descend without the use of its engines.

To accelerate upwards, air under pressure expels water from the ballast tank, decreasing the weight of the submarine. When decreased to a value less than the weight of water displaced, the submarine will accelerate upwards.

Floating occurs if the weight of the submarine is equal to the weight of water displaced.

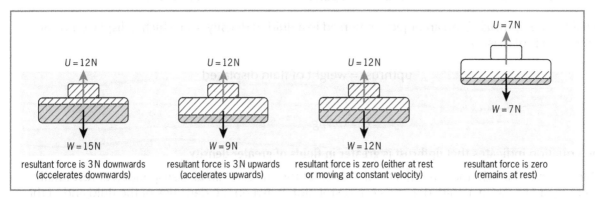

Figure 6.19 *Upthrust and submarines*

Upthrust and balloons

Figure 6.20 (a) shows a balloon containing air. If the air in the balloon is heated, it will become less dense and will expand. The weight of the balloon is unchanged, but the upthrust on it will increase since it now displaces more of the cooler surrounding air. It will accelerate upwards if its weight plus the weight of its contents is less than the weight of the air it displaces (upthrust).

If the balloon contains a gas of low density such as helium, as in Figure 20 (b), its weight will be less than the weight of air it displaces (upthrust), and therefore it will accelerate upwards.

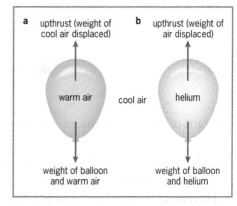

Figure 6.20 *Upthrust and balloons*

Example 11

A sphere of volume 0.20 m³ and of density 3000 kg m⁻³ is suspended below the surface of water of density 1000 kg m⁻³ by means of a string. For the sphere, calculate:

a its weight

b the upthrust of the water on it

c its apparent weight

d the tension in the string.

a Weight of sphere = mass of sphere × acceleration due to gravity

$\quad\quad W$ = (density of sphere × volume of sphere) × acceleration due to gravity

$\quad\quad W = \rho_s V_s g$

$\quad\quad W = 3000 \times 0.20 \times 10$

$\quad\quad W = 6000$ N

---

b Upthrust = weight of water displaced

U = mass of water displaced × acceleration due to gravity

U = (density of water × volume of water displaced) × acceleration due to gravity

$U = \rho_w V_s g$

$U = 1000 \times 0.20 \times 10$

$U = 2000$ N

c Apparent weight = actual weight – upthrust

$W_{app} = 6000 - 2000$

$W_{app} = 4000$ N

d Since the forces are in balance:

$U + T = W$

$2000 + T = 6000$

$\therefore T = 6000 - 2000 = 4000$ N

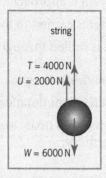

string

$T = 4000$ N

$U = 2000$ N

$W = 6000$ N

Figure 6.21 *Example 11*

Note: The tension is the apparent weight. The upper end of the string will pull on an attached spring balance with a force of 4000 N.

Example 12

The rectangular block shown in Figure 6.22 is of base dimensions 25 cm × 25 cm and floats submerged in a liquid of density 1200 kg m^{-3} to a depth of 20 cm. Calculate the weight of the block given that the gravitational field strength is 10 N kg^{-1}.

Since the block floats:

weight of block = weight of liquid displaced

$W_b = m_w g$

$W_b = (\rho_w V_w)g$ ... since density × volume = mass

$W_b = (1200 \times 0.25 \times 0.25 \times 0.20)10$

$W_b = 150$ N

block

20 cm

volume of liquid displaced

liquid

base dimensions 25 × 25 cm

Figure 6.22 *Example 12*

Recalling facts

 1 Define:

a pressure

b the pascal.

 2 State FOUR characteristics of the pressure acting at a point in a fluid.

 3 Describe, with the aid of a diagram, an experiment to show that the pressure at a point in a liquid increases with depth.

 4 **a** Draw a labelled diagram of a mercury barometer.

b State the height of the mercury column equivalent to standard atmospheric pressure.

5 A mercury barometer supports a column of mercury of length 76 cm when its tube is vertical. Describe the effect on the height of the mercury column if

   a the tube is made of twice the cross-sectional area

   b the tube is tilted so that it is no longer vertical

   c a tropical storm is approaching

   d the barometer is carried to the peak of a mountain

   e a small hole is drilled through the top of the glass tube.

6 **a** State Archimedes' principle.

   b State the principle of flotation.

   c A heavily laden ship floats in water. Describe how this is possible, expressing your answer in terms of:

      i density

      ii upthrust.

7 Describe how a submarine adjusts the quantity of water in its ballast tanks to achieve each of the following:

   a dive, if initially partially submerged

   b rise from the depths of the ocean.

Applying facts

density of fresh water = 1000 kg m^{-3}

density of mercury = 13 600 kg m^{-3}

acceleration due to gravity = 10 m s^{-2}

8 A fishing boat travels from the sea into the mouth of a river and then heads upstream through the less dense fresh water. Describe and explain the change, if any, of each of the following as it makes the journey:

   a the weight of the boat

   b the upthrust on the boat

   c the weight of water displaced by the boat

   d the height of the water line along the side of the boat.

9 Calculate the pressure exerted by a force of 4000 N acting at right angles to an area of 0.20 m².

10 The pressure of the air in a racing bicycle tyre is more than three times the pressure of the air in a car tyre. Explain this phenomenon.

11 **a** Determine the hydrostatic pressure created **by sea water** of density 1200 kg m^{-3} on a fish 5.0 m below the surface.

   b What is the total pressure on the fish given that atmospheric pressure on the day is 1.0×10^5 Pa?

12
  a Anita is of mass 48 kg. Calculate the force she exerts on the ground.

  b Calculate the pressure she exerts as she stands on both feet, if each foot is in contact with 80 cm² of the ground.

  c She raises one foot from the ground. Describe and explain the effect on each of the following:

    i the force she exerts on the ground

    ii the pressure she exerts on the ground.

13 Express:

  a 200 mmHg in pascals

  b 1.0×10^5 Pa in cmHg

14 Calculate the atmospheric pressure in pascals when the mercury column of a barometer is 72 cm tall.

15 A rectangular block of mass 4.0 kg and dimensions 20 cm × 40 cm × 50 cm rests on one of its faces. What are the dimensions of the face on which it rests, if the pressure due to the block is 400 Pa?

16 A ten-wheeler truck of mass of 25×10^3 kg exerts a pressure of 5.0×10^5 Pa on the road surface by means of its tyres. Assuming the weight of the truck is uniformly distributed, calculate the average area of contact between each tyre and the road.

17 Water is sucked 2.0 m into a vertical tube of length 3.0 m. Given that atmospheric pressure is 1.0×10^5 Pa, determine the pressure of the air above the liquid in the tube.

18 If the vapour pressure of water is negligible, to what height will the water column of a *water barometer* rise on a day when the atmospheric pressure is 1.0×10^5 Pa?

19 Determine the force caused by the liquid on the base of the tank of water shown in Figure 6.23.

20 Figure 6.24 shows a hydraulic lift being used to lift an object of weight W by application of a small force of 100 N. The cross-sectional areas of the small and large pistons are respectively 100 cm² and 2.0 m². Determine the weight, W.

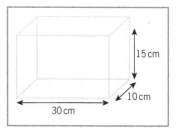

Figure 6.23 *Question 19*

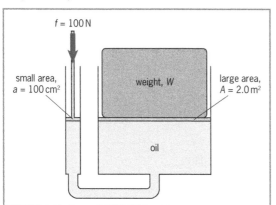

Figure 6.24 *Question 20*

21 A submarine can withstand a pressure of 8.0×10^5 Pa. How deep can it be submerged in fresh water on a day when the atmospheric pressure is 1.0×10^5 Pa?

22 Figure 6.25 shows air trapped by a column of mercury in several glass tubes on a day when atmospheric pressure is 75 cmHg. Determine the pressure at X in cmHg for each case.

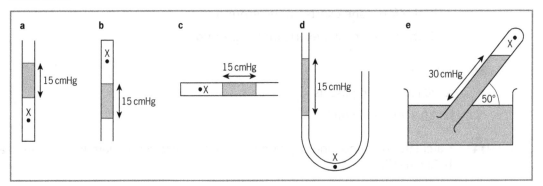

Figure 6.25 *Question 22*

23 **a** Determine the pressure of each gas supply connected to the manometers shown in Figure 6.26, given that atmospheric pressure on the day is 1.0×10^5 Pa.

 b Explain what would occur if manometer (i) contained water instead of mercury.

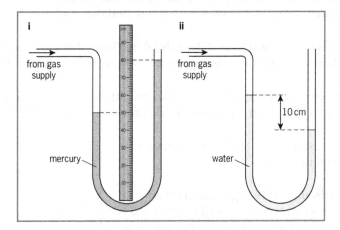

Figure 6.26 *Question 23*

24 With the tap open, liquids are sucked up the tubes shown in Figure 6.27. The tap is then closed. Atmospheric pressure on the day is 1.0×10^5 Pa. Determine:

 a the pressure of the air trapped above the liquid columns

 b the density of liquid L.

 (Tip: The difference in pressure between the top and bottom of each column is the same.)

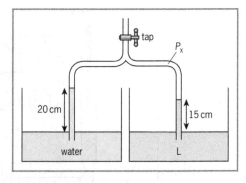

Figure 6.27 *Question 24*

25 A *water* manometer is being used to measure the pressure of a gas supply when the atmospheric pressure is equivalent to 75 cmHg. The surface of the water in the limb connected to the gas supply is 40 cm lower than the surface of the water in the other limb of the manometer. Draw a diagram of the arrangement and determine the pressure of the gas supply in

a Pa

b cmHg

26 An object of weight 50 N accelerates downward through the ocean towards the seabed. The viscous drag force from the water on the object is 12 N and the upthrust caused by the displaced water is 18 N. Draw a diagram showing these forces and determine the acceleration of the object.

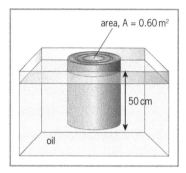

27 Figure 6.28 shows a cylindrical drum of circular cross-sectional area, 0.60 m², floating in oil of density 900 kg m⁻³ with 50 cm of its height below the surface. Calculate the weight of the cylinder.

Figure 6.28 *Question 27*

Analysing data

28 A student carries out an experiment using the apparatus shown in Figure 6.29 together with a metre rule to determine the density of brine. The liquid levels in the tubes can be altered by sucking air from the opening at X and then closing the tap. The table shows the heights of the liquid columns for several pairs of readings of h_W and h_b.

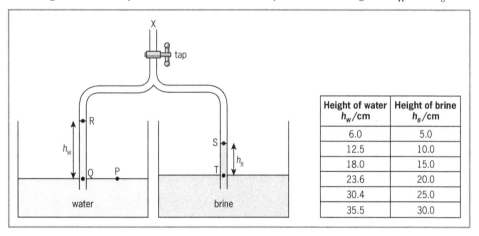

Height of water h_W/cm	Height of brine h_B/cm
6.0	5.0
12.5	10.0
18.0	15.0
23.6	20.0
30.4	25.0
35.5	30.0

Figure 6.29 *Question 28*

a Plot the graph of h_W against h_b starting both axes at zero.

b Calculate the slope, *S*, of the graph.

c Use the graph to determine the height of the water column when the brine rises 35.0 cm.

d Determine the pressure at each of the following points when $h_w = 20$ cm, given that atmospheric pressure is 1.0×10^5 Pa:

   i P

   ii Q

   iii R

   iv S

   v T

e Use the slope of the graph to determine the density of the brine, ρ_b, given that $\rho_b h_b = \rho_w h_w$.

Section B
Thermal physics and kinetic theory

In this section

Nature of heat
- The switch from caloric theory to kinetic theory and the principle of conservation of energy.

Phases of matter and the kinetic theory
- The physical properties of solids, liquids and gases on a particle level.

Thermal expansion
- The effects of temperature change on the volume of a body.

Temperature and thermometers
- This chapter deals with temperature scales and thermometric properties related to the design of various thermometers.

The gas laws
- The relation between the pressure, volume and absolute temperature of an ideal gas.

Heat and temperature change
- The effect on the temperature of a body as it is heated and remains in the same phase.

Heat and phase change
- The effect on the phase of a body as it is heated at its melting point or boiling point.

7 Nature of heat

Heat is forever present in our daily lives. It can **increase the temperature** of our coffee, **change the phase** of water evaporating from the ocean, produce **expansion** of mercury in the bore of a thermometer or cause **chemical decomposition** of limestone. This chapter discusses the changing views which scientists have developed in their understanding of the nature of heat.

Different theories of heat

Learning objectives

- Briefly describe the **caloric theory** of heat and the **properties of 'caloric'**.
- Discuss **arguments for and against** the caloric theory of heat.
- Describe **Count Rumford's experiment** and explain how it **challenged the caloric theory**.
- Describe the **kinetic theory** of heat.
- Describe **Joule's experiment** which established the principle of **conservation of energy**.
- Explain how Joule came to his **conclusions** from the **observations** of the experiment.

Caloric theory of heat

During the 18th century it was believed that heat was an **invisible fluid**, known as 'caloric', which could **combine with matter** and **raise its temperature**.

It was thought that particles of **'caloric'**

- **repelled** each other
- **flowed** from bodies of higher temperature to bodies of lower temperature.

Arguments for the caloric theory

1. Objects **expand** when given 'caloric' because the **increased** numbers of 'caloric' particles **occupy more space**.
2. Heat **flows** from hotter to cooler bodies because **'caloric' particles repel** each other. In hotter bodies, the particles would be more concentrated and therefore have a greater tendency to repel and leave the body.

Arguments against the caloric theory

1. When bodies are **given 'caloric'** so that they change phase (solid to liquid, or liquid to gas), an increase in **'caloric' cannot be detected**.
2. When different materials are given the **same** amount of 'caloric', their temperatures increase by different amounts, suggesting that they receive **different amounts** of 'caloric'.
3. The mass of a body **should increase** as it is heated, since it should then contain more 'caloric' particles. However, the mass **remains the same**.

Rumford's cannon-boring experiments

Benjamin Thompson was a British physicist and inventor. He worked at reorganising the Bavarian army where he acquired the title, **Count Rumford**. In 1798, Rumford experimentally tested the caloric theory. He **immersed a cannon in water** and proceeded to bore its barrel **by friction** with a blunt boring tool.

Observations:

- The water was **heated** and soon came to a boil.
- The fragments produced by boring the barrel had the **same physical properties** as the barrel.
- The release of heat seemed **inexhaustible** so long as the **frictional forces** were continued.

Conclusions:

- It was only the work done due to friction between the relative **motion** of the cannon and the boring tool that produced the heat, since there was **no physical change** in the fragments and therefore **no chemical reaction** had occurred.
- Since the heat produced was **inexhaustible, 'caloric' could not be a material substance.**

Rumford **challenged** the caloric theory and laid the foundation for an understanding of the **science of thermodynamics** and the **law of conservation of energy** established later in the 19th century. The caloric theory of heat soon became **obsolete** and gave way to the **kinetic theory.**

Kinetic theory of heat

All matter is made of particles in constant motion and therefore these particles possess kinetic energy.

*The **thermal energy** of a body is the energy it possesses due to the **kinetic energy** of its particles.*

*Heat is the flow of **thermal energy** from a body of **higher temperature** to one of **lower temperature** due to a temperature difference between them.*

A body possesses **thermal** energy, **not heat** energy. The thermal energy of a body can be increased or decreased due to the absorption or emission of heat.

The higher the **temperature** of a body, the greater is the quantity of **thermal energy** it contains. At the extremely cold temperature of **absolute zero**, i.e. **0 K** (see Chapter 10), particles of matter would have no kinetic energy and would therefore **be at rest.**

Joule's role in establishing the principle of conservation of energy

James Joule proved experimentally that **energy can be transformed** from one type to another but is **always conserved**. His experiment is outlined below.

Two bodies, each of mass m, and attached to the ends of a string, were made to fall through a height, Δh, of about 2 m, as shown in Figure 7.1. As they descended, a pulley mechanism turned paddles in a mass, m_w, of water. The masses were quickly wound up using a slip ratchet winder to disengage the paddles and the process was repeated N times until the temperature of the water rose by a few degrees.

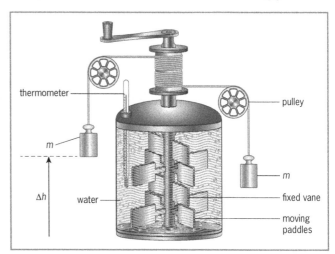

Observation: The **loss in gravitational potential energy** of the masses as they fell was equal to the **gain in thermal energy** of the water (see Chapter 12 for calculations).

Figure 7.1 *Joule's experiment, establishing the principle of conservation of energy*

$$N(2mg\Delta h) = m_w c\Delta T$$

Conclusion: As the masses fell, their **gravitational potential energy** was transformed into **kinetic energy** of the paddles which then transformed into **thermal energy** of the water, causing the temperature of the water to increase by an amount, ΔT.

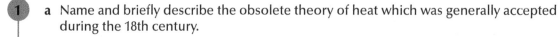

gravitational potential energy ⟶ kinetic energy ⟶ thermal energy

Put in another way, the potential **energy transformed** as the masses fell, did an equal amount of **work** in turning the paddles, and resulted in a transformation to thermal energy of the water.

work done = energy transformed

Recalling facts

1 **a** Name and briefly describe the obsolete theory of heat which was generally accepted during the 18th century.

  b State TWO phenomena which support the theory mentioned in **a** above and TWO phenomena which cannot be explained by it.

2 **a** Name the British physicist who challenged the caloric theory.

  b State the observations and conclusions he made which supported his challenge.

3 **a** Briefly describe the theory which replaced the obsolete theory of heat of the 18th century.

  b What is meant by the term 'thermal energy'?

  c At what temperature would the particles of a body have no kinetic energy?

4 **a** Name the scientist who established the principle of conservation of energy.

  b In the famous experiment which established the principle, what work was done, and what transformation of energy took place?

Applying facts

5 State TWO reasons why the loss in energy of the falling masses may **not** be equal to the gain in energy of the water, using the apparatus shown in Figure 7.1.

8 Phases of matter and the kinetic theory

Solids, liquids and gases are composed of tiny particles – **atoms, molecules** or **ions**. The study which explains the behaviour of all matter in terms of these particles and their energies is known as the **kinetic theory of matter.** Phenomena such as expansion and contraction, temperature change, evaporation, osmosis, diffusion, condensation, melting and freezing, can all be explained by the kinetic theory.

Solids, liquids and gases explained by kinetic theory

Learning objectives

- Recall that the particles of matter have **internal energy**.
- Distinguish among the **properties** of **solids, liquids** and **gases**.
- Use the **kinetic theory** to explain the properties of solids, liquids and gases.
- Give **examples** and perform **experiments** to demonstrate the **particulate nature** of matter.
- Demonstrate the existence of **intermolecular forces**.

Energy of the particles of matter

The internal energy of a substance is the sum of the potential and kinetic energies of the particles of the substance.

Kinetic energy

Particles of matter possess **kinetic energy** due to their **constant motion** of **vibration, translation** or **rotation**.

Potential energy

Particles of matter possess **potential energy** due to the **attractive** and **repulsive forces** existing between them when they are **close together**.

There is a space between the particles. When the space becomes very small (approximately the diameter of the particle), **strong forces** exist between them. The particles of **solids and liquids** therefore have potential energy.

When the particles are very **far apart**, there are **no forces** existing between them, and so they have **no potential energy**. Since this is the case for **gases**, except for the very brief time of collision, we consider the potential energy of a gas to be **zero**.

Physical properties of matter in terms of the forces and distances between its particles

Figure 8.1 shows the relative positioning of the particles of solids, liquids and gases, and Table 8.1 describes how the **forces** and **distances** between these particles explain their **physical properties.**

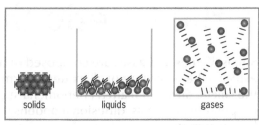

Solids

The particles of a solid are **very close together**. When they are closer than their mean separation, repulsive forces exist between them and when they are slightly further apart than their mean separation, the forces become attractive. The particles therefore constantly **vibrate** about their mean positions while being bonded in a **fixed lattice**.

Figure 8.1 *Particle proximity in solids, liquids and gases*

Liquids

The particles of a liquid have more energy than those of a solid. They are therefore **slightly farther apart** and overcome the very strong attractive forces that would bind them in a fixed lattice. This causes the **forces** between the particles to be **weaker** than those in solids and allows them to **translate** by sliding past each other. The weaker forces of attraction are still sufficient however, to keep the particles **together** in a certain **volume**, taking the shape of the part of the container with which they are in contact.

Gases

Except at the time of collision, the particles of a gas are **far apart,** and therefore the **forces** between them are **negligible**. They therefore **translate** freely, filling the container in which they are enclosed.

Table 8.1 *Physical properties of matter in terms of forces and distances between particles*

	Solids	Liquids	Gases
Density	**High** since the particles are **tightly packed.**	**High** since the particles are **tightly packed** – almost as close as in solids.	**Very low** since the **particles** are **far apart.**
Shape and volume	**Fixed shape and volume** due to **rigid bonds** resulting from **strong forces** between the particles.	Take the **shape of the part of the container** with which they are in contact, since the **bonds** between particles **are not rigid**. These **weaker forces**, however, still cause liquids to have a **fixed volume.**	They have **no fixed shape or volume. Forces** exist between the particles only at the **time of collision** and therefore gases spread out to fill the container which encloses them.
Ability to flow	**Cannot flow** since **strong forces** produce **rigid bonds** between the particles.	**Can flow** since the **weaker forces** between the particles **cannot form rigid bonds.**	**Can flow** since the **very weak forces** between the particles **cannot form rigid bonds.**
Ability to be compressed	**Not easily compressed** since the particles are **tightly packed**, making it difficult to push them closer.	**Not easily compressed** since the particles are **tightly packed**, making it difficult to push them closer.	**Easily compressed** since there is **much space** between the particles.

Evidence for the particulate nature of matter

Brownian motion

*Brownian motion is the **vigorous, haphazard** motion of small particles **suspended** in a liquid or gas due to the **bombardment** by molecules of the liquid or gas.*

Figure 8.2 shows an apparatus used to observe the motion of smoke particles suspended in air. **Smoke** is placed in the glass cell by inserting a piece of **smouldering string** below its lid. A cylindrical lens is used to concentrate light from a source into the cell. By means of the microscope, the smoke particles are observed as **bright specks** moving in a **jerky**, **haphazard** manner. Although the molecules of air are too small to see, it can be concluded that the jerky, haphazard motion of the smoke particles is due to **bombardment** by the **irregular motion** of the smaller **air molecules.**

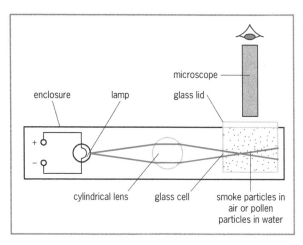

Figure 8.2 *Brownian motion*

By replacing the smoke suspended in air with **pollen** grains suspended in **water**, a jerky, haphazard motion of the grains is also observed, indicating the irregular motion of the bombarding water molecules.

Diffusion

*Diffusion is the movement of particles from a region of **higher concentration** to a region of **lower** concentration.*

To demonstrate **diffusion through gases** we can open a bottle of cologne in a room. The scent quickly spreads through the air as particles of the cologne, too small to see, leave the bottle and travel **randomly** in **all directions**. Similarly, when we smell the scent of our tasty meal, it is due to small particles from the food diffusing through the air and reaching our noses.

Diffusion in liquids can be demonstrated by dropping a crystal of **copper sulfate (blue)** or **potassium permanganate (purple)** through a funnel to the bottom of a beaker of water (see Figure 8.3). The particles of the crystal **dissolve** and **diffuse** to regions where they are **less concentrated**. The coloured solution formed is seen to spread out **randomly in all directions**. After some time, it acquires a **uniform colour**, indicating that the particle **concentration** of the dissolved crystal is **consistent**.

If the experiment is carried out in **warmer** water, the **rate of diffusion is increased**, since the **kinetic energy** of the particles **increases with temperature.**

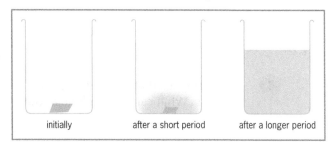

initially   after a short period   after a longer period

Figure 8.3 *Crystal of copper sulfate diffusing in water over time*

Evidence for intermolecular forces

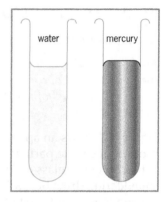

Figure 8.4 *Different menisci*

- When the nozzle of a syringe containing water is sealed, it is extremely difficult to compress the liquid by pushing on the piston. The molecules of a liquid are **very close** and so repelling **forces** are produced when any attempt is made to bring them closer.

- Figure 8.4 shows that water in a vertical test tube has a **concave meniscus.** This is so because the attractive forces between water molecules and glass molecules (**adhesive forces**) are stronger than the attractive forces between neighbouring water molecules (**cohesive forces**).

- Figure 8.4 shows that mercury in a vertical test tube has a **convex meniscus.** The cohesive forces between mercury atoms are stronger than the adhesive forces between mercury atoms and glass molecules.

Recalling facts

1 a What is responsible for the particles of a substance having

   i kinetic energy?

   ii potential energy?

  b What is meant by the 'internal energy' of a substance?

2 Use the kinetic theory to explain the following statements.

  a Particles of a solid vibrate.

  b Liquids do not necessarily fill the container which holds them, unlike gases, which always fill their container.

3 Define each of the following: a Brownian motion   b diffusion.

4 Describe a simple experiment to demonstrate diffusion through a liquid.

Applying facts

5 Janelle notices that the meniscus of a certain liquid curves upward at the sides of a test tube. What can she deduce about the forces between molecules that cause this occurrence?

6 Joshua accidentally breaks a mercury thermometer in his school laboratory. Explain, in terms of the mercury particles, why the liquid breaks into small globules on the floor.

7 Demario and Jaren are each given a syringe which has its spout sealed. Demario's syringe contains air and Jaren's contains water. Describe and explain what happens as they each try to reduce the volume of the contents of their syringes by pressing on the pistons.

8 a Draw a well-labelled diagram to show how Brownian motion can be observed in water.

   b Maya is observing Brownian motion in water. What can she do to increase the activity of the observed particles?

9 Explain the following in terms of the kinetic theory of matter.

   a Lime juice has a much higher density than air.

   b Water and air can flow, but solid rock cannot.

   c Falling onto a solid floor leaves it undisturbed but falling into a swimming pool creates a splash.

   d Particles of air are considered not to have potential energy.

9 Thermal expansion

Solids, liquids and gases are constantly **expanding** and **contracting** as temperatures change.

Changes in volume with temperature can **present problems**. Roofs shrink at night as the temperature falls, producing creaking sounds as joints move relative to each other. Some structures must be carefully designed to accommodate the very strong forces associated with expansion or contraction.

Changes in volume can also be **useful**. Thermometers (see Chapter 10), components such as bimetallic strips (see page 107), and many other applications, depend on changes in volume as temperature changes.

Thermal expansion and its related applications

Learning objectives

- Describe simple experiments to **demonstrate expansion** in solids, liquids and gases.
- Describe an experiment to demonstrate the **forces** associated with expansion and contraction.
- Explain **expansion** in terms of the **kinetic theory** of matter.
- Describe how some **problems of expansion** can be overcome.
- Describe the **anomalous expansion of water** and explain **phenomena** which arise from it.
- Explain the function of the **bimetallic strip** and describe some of its **applications**.

Demonstrating expansion

Simple experiments to demonstrate expansions in solids, liquids and gases are illustrated in Figure 9.1.

Solid: At room temperature, the hammer can just fit into the space and the ball can just fit through the hole. However, the ball and hammer expand when heated, causing the fit to be no longer possible.

Liquid: The flask expands slightly when heated, causing the liquid level to initially drop. With continued heating however, the heat energy reaches the liquid, causing it to expand up the tube.

Gas: The air in the glass bulb expands when heated and forces its way downward through the stem to produce bubbles in the water.

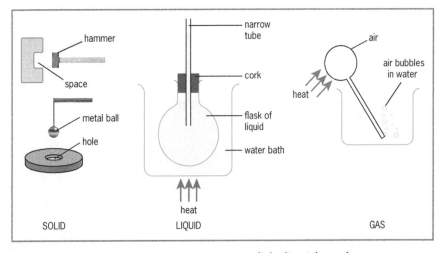

Figure 9.1 *Demonstrating expansion in solids, liquids and gases*

Demonstrating that strong forces are produced by expansion and contraction

Figure 9.2 shows an apparatus used to demonstrate the strength of the **forces** associated with expansion and contraction. The **steel bar** rests in **slots** within the pillars, A, B and C. The bar has a **hole** to accommodate a small **cast iron rod** placed **perpendicularly** to it. A **flame torch**, not shown in the diagram, is used to heat the bar.

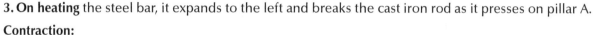

Figure 9.2 *Breaking rod experiment*

Expansion:

1. The cast iron rod is inserted through the hole in the steel bar.

2. The nuts are tightened with the cast iron rod just touching pillar A keeping the right end of the bar fixed.

3. **On heating** the steel bar, it expands to the left and breaks the cast iron rod as it presses on pillar A.

Contraction:

1. The cast iron rod is inserted through the hole in the steel bar and the nuts tightened so that rod just touches pillar B.

2. The steel bar is heated and expands a small amount to the left.

3. The nuts are then quickly adjusted, pulling the steel bar to the right until the cast iron rod once more just touches pillar B.

4. **On cooling**, contraction of the steel breaks the cast iron rod as it presses against pillar B.

Expansion in terms of the kinetic theory of matter

Solids: When a solid is heated, the **thermal energy** supplied converts into **kinetic energy** of its particles. The molecules of the solid **vibrate faster** and with **greater amplitude**, and so **occupy more space**.

Liquids and gases: When a liquid or gas is heated, the thermal energy supplied converts into kinetic energy of its particles. The molecules of the liquid or gas **translate faster** and so **occupy more space**.

Avoiding problems due to expansion and contraction

- **Concrete surfaces** are laid in **slabs** with **spaces** between them (Figure 9.3a). The spaces may be left unfilled, or they may be filled with pitch or other materials which can be **easily compressed**. During expansion of the concrete, destructive forces between adjacent slabs are avoided.

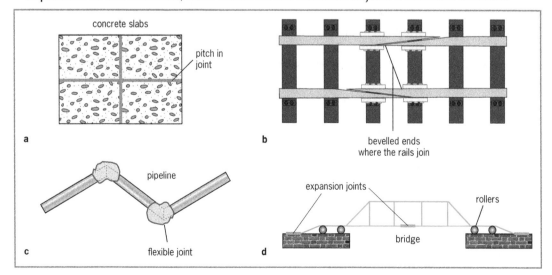

Figure 9.3 *Dealing with expansion*

- **Railway lines** are laid in **short lengths** with their **ends bevelled** and **overlapping** (Figure 9.3b). This prevents warping since they can then freely **slide over each other** as they expand and contract.

- **Oil pipelines** in deserts, and those which carry steam, are laid in **zig-zag** formation with **flexible joints** (Figure 9.3c). On expansion and contraction, they flex without being subjected to strong forces.

- **Large structures** such as bridges can be built with **rollers** and suitably placed **expansion joints** (Figure 9.3d) so that they can expand and contract freely, without creating destructive forces. Details of an **expansion joint** are shown in Figure 9.4.

- **Baking dishes** which undergo large temperature differences are made of materials which **expand** and **contract very little** with temperature change. This prevents them from cracking as they contract when removed from the oven into the much cooler environment. A common brand name of such a material is **Pyrex**.

Figure 9.4 *Expansion joint*

- **Power lines** strung between poles must be laid **slack in summer** so that strong forces of tension are not produced when they **contract** during the cooler **winter**.

- A tight metal lid on a glass jar is easily removed by running hot water onto it. Since the lid has a **higher expansivity** than the glass, it expands more, and is then not bound as tightly to it. The metal lid also has a **higher conductivity** than glass and so responds more rapidly to temperature changes.

The irregular expansion of water

Most substances contract when they change from solid to liquid. Water, however, is an exception. As water cools, it contracts as does other liquids, but only until its temperature is **4 °C**. It then **expands as it cools to 0 °C**, and further **expands**, by about 9%, **as it freezes**!

- A can of carbonated drink (beverage such as coke in which carbon dioxide is dissolved) will **burst when frozen.** As the water in the can freezes, it expands, increasing the pressure of the carbon dioxide present above the liquid to such an extent, that it breaks the can.

- Pipes which carry water can burst in temperatures below 0 °C as the water they contain expands on freezing.

- Figure 9.5 shows that as water **freezes** to form an **iceberg**, the **reduced density** causes the iceberg to float.

- A glass of water **cools rapidly** when ice floats in it. Water in contact with the ice cools, contracts and sinks, forming a **convection current** with warmer water rising and taking its place to be cooled.

Figure 9.5 *Icebergs float due to reduced density on freezing*

- In cold regions, the air above a pond can be below 0 °C, and this causes the temperature of the pond to decrease as heat leaves it. When the temperature decreases to **below 4 °C,** water becomes **less dense** and rises as shown in Figure 9.6. As the water freezes at the surface, the **ice formed** is even **less dense** than the **cold water**, and so forms a **floating cap.**

Since **ice is a poor thermal conductor,** it significantly **reduces** further **heat loss,** preventing the water below from freezing, and therefore **sustaining aquatic life** in the pond.

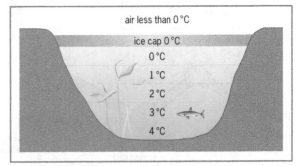

Figure 9.6 *Aquatic life preserved due to irregular expansion of water*

The bimetallic strip

A **bimetallic strip** consists of **two metal strips riveted together**. The strips bend on heating so that the metal which **expands more** is on the **outer side of the curve**. Brass expands more than invar when heated and this combination forms a good bimetallic strip. A material which expands more than another when heated will also contract more on cooling. Figure 9.7 shows what happens to a bimetallic strip which is initially straight at 50 °C, if it is cooled to 20 °C, and if it is warmed to 80 °C.

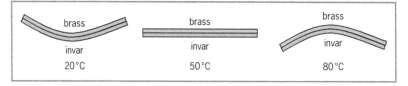

Figure 9.7 *The bimetallic strip*

Applications of the bimetallic strip

Simple fire alarm

A simple fire alarm is shown in Figure 9.8 (see circuit symbols, Chapter 26). Heat from the fire causes the **brass** to expand more than the **invar**, and so the right end of the bimetallic strip bends upward and closes the contacts. This completes the circuit and sounds the bell.

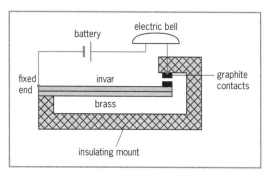

Figure 9.8 *Simple fire alarm*

Electric thermostat

An electric thermostat connected to the heater of an oven is shown in Figure 9.9. With the switch on, the heater **warms** the oven. The bimetallic strip then bends, causing the graphite **sliding contacts** to **separate** and the circuit to break. With the heater disconnected, the oven **cools**, the bimetallic strip straightens, the contacts **reconnect**, and the process repeats. The temperature **control knob** is an adjusting screw that can be advanced so that it forces the sliding contacts further over each other. The bimetallic strip must then bend more in order to break the circuit.

By connecting the thermostat to the electrical pump of a refrigerator with **the positions of the brass and invar** switched, the thermostat can be used to prevent it from becoming too cold. As

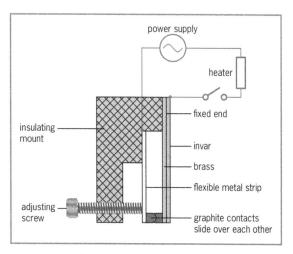

Figure 9.9 *Electric thermostat connected to heater*

the temperature in the refrigerator reduces, the bimetallic strip will bend outwards. At a certain set temperature, the contacts will separate, breaking the circuit and switching off the refrigerator.

Recalling facts

1. Describe an experiment to demonstrate that a metal sphere expands when heated.

2. Use the kinetic theory to explain why a liquid expands when heated.

3. a Why are concrete slabs laid with small spaces between them?

   b Name a suitable material which can be placed between the spaces.

4. Explain why 'Pyrex' dishes are used as baking ware.

5. How can Adrianna remove a tight metal lid from a glass bottle of jam using thermal expansion?

6. Describe, with the aid of a diagram, how aquatic life in lakes is preserved when the temperature of the air above the water is below the freezing point of water.

Applying facts

7. Explain why Akeem's teeth hurt when he eats ice cream after a bowl of warm soup.

8. Why do water pipes burst during cold winter months in regions of high latitude?

9. M has a higher expansivity than N.

   a Draw a diagram of a device which uses a bimetallic strip comprised of metals M and N to switch on a light bulb when the temperature falls below a certain value.

   b Explain how the device works.

10. A bimetallic strip consisting of X and Y at 60 °C is shown in Figure 9.10. Given that X expands more than Y when heated through the same temperature range, redraw the strip when cooled to –5 °C.

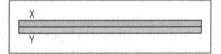

Figure 9.10 *Question 10*

11 Figure 9.11 shows the change in density which occurs as ice at −10 °C is warmed and melts, and the water formed is further warmed until its temperature reaches 100 °C. Use the graph to answer the following questions.

a At what temperature is water most dense?

b Why does ice float on water?

c What is the density of water at 100 °C?

d State, with reason, whether the volume of the water will increase or decrease as it is heated from 20 °C to 50 °C.

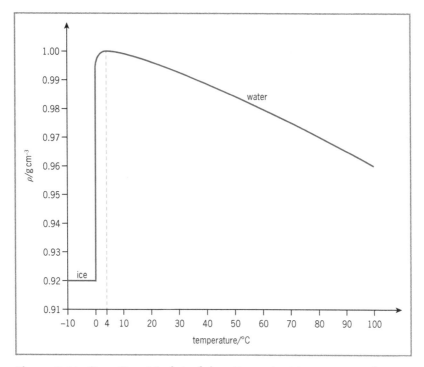

Figure 9.11 *Question 11 plot of density against temperature for ice and water*

10 Temperature and thermometers

When a 'red hot' pin at 900 °C is dropped into a bucket of water at room **temperature**, a quantity of **heat** leaves the pin and enters the water. However, heat and temperature are clearly **different** quantities, since as the temperature of the pin rapidly and significantly decreases, there is minimal increase in the temperature of the water. This chapter deals with temperature, and Chapter 12 discusses how heat energy can affect temperature change.

Temperature, temperature scales and thermometers

Learning objectives

- Distinguish between **heat** and **temperature.**
- Describe the **Celsius** temperature scale and the **Kelvin** temperature scale.
- Define the **fixed points** on the **Celsius** scale.
- Convert between the Celsius and Kelvin temperature scales.
- Identify **thermometric properties** and describe how they may be used to measure temperature.
- Relate thermometric properties to the **design of thermometers.**
- List the various **qualities** to be considered when selecting a suitable thermometer.
- Describe several **types of thermometer** and explain their function.

Temperature and temperature scales

Temperature is the degree of hotness of a body measured on a chosen scale.

As discussed in Chapter 7, the **temperature of a body** depends on the **kinetic energy of its particles;** the **faster** the particles, the **higher** is the **temperature** of the body.

Heat is thermal energy in the process of transfer from a region of higher temperature to a region of lower temperature, due to a temperature difference between them.

A thermometer is an instrument used to measure temperature and so must have an attached temperature scale. Two important temperature scales are the **Celsius** temperature scale and the **Kelvin** temperature scale (also known as the **absolute** or **thermodynamic** temperature scale).

The SI unit of temperature is the **kelvin (K)**, but it is also common in science to express temperature in **degrees Celsius (°C).**

Celsius temperature scale

Water is a very common substance, and the temperatures at its freezing and boiling points can be determined with minimum effort. On the Celsius scale the numbers 0 and 100 have been assigned to represent the temperatures at these fixed points. Since there are 100 intervals between the fixed points, we refer to the Celsius scale as a **centigrade scale.**

The lower fixed point (0 °C) is the temperature of pure melting ice at standard atmospheric pressure.

The upper fixed point (100 °C) is the temperature of steam from pure boiling water at standard atmospheric pressure.

To determine the lower fixed point, we can immerse an ungraduated thermometer into chips of **pure melting ice** as shown in Figure 10.1a. The lower fixed point is the point to which the mercury thread falls when the pressure is 1.0 atmosphere.

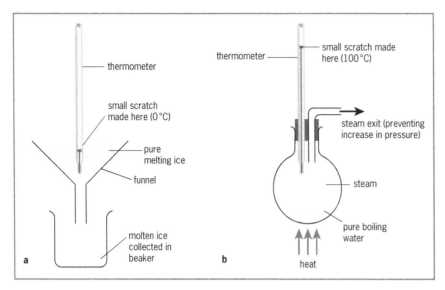

Figure 10.1 *Determining the lower and upper fixed points on the Celsius scale*

To determine the upper fixed point, we can place an ungraduated thermometer in the vessel shown in Figure 10.1b. The bulb of the thermometer is placed in the **steam – not in the water**. The **steam exit** ensures that the **pressure** in the vessel **does not increase** as the steam is produced. The upper fixed point is the point to which the mercury thread rises when the pressure is 1.0 atmosphere.

Kelvin temperature scale

Miles, kilometres, feet and inches are all absolute scales of measurement since 0 miles, 0 km, 0 feet and 0 inches, are all the same; they all mean that the length is minimum (ZERO). The Celsius scale is **not an absolute scale** since 0 °C is not the lowest temperature which may be attained by a body.

According to the kinetic theory of matter there is an **absolute zero of temperature.** This is the temperature at which the particles of matter will have **no energy** and therefore **no motion.** On the Kelvin scale this is **0 K**, and on the Celsius scale it is **–273 °C.** See Figure 10.2.

Note: Intervals of 1 K and 1 °C are **equal in size.**

Converting between degrees Celsius and kelvins

If θ is the temperature measured in °C and T is the temperature measured in K, then

$$T = \theta + 273$$

Examples of converting between °C and K are shown in Table 10.1.

Table 10.1 *Converting °C to K*

°C	K
200	200 + 273 = 473
27	27 + 273 = 300
0	0 + 273 = 273
−20	−20 + 273 = 253

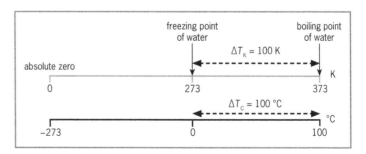

Figure 10.2 *Comparing temperature scales*

Thermometric properties of thermometers

A thermometric property of a material is one which varies with temperature.

Since thermometers may be used to measure temperatures under various circumstances, their design for a particular application will depend on the thermometric property most suitable to it.

Table 10.2 shows some thermometers and their thermometric properties.

Table 10.2 *Some thermometers and their thermometric properties*

Type of thermometer	Effect on thermometric property as temperature increases
Liquid-in-glass (Figures 10.3 and 10.5)	volume of liquid increases
Constant-volume gas thermometer (Figure 10.7)	pressure of gas increases
Thermoelectric thermometer (Figure 10.6)	electromotive force (emf) produced which varies non-linearly
Resistance thermometer (metallic)	resistance increases
Thermistor	resistance usually decreases but increases for some types

Selecting a suitable thermometer

To select a suitable thermometer, we may need to consider the following:

- cost
- accuracy
- response time
- sensitivity (ability to detect slight changes in temperature)
- temperature range
- ability to interface electronically with a digital device which can be accessed remotely, and which can manipulate data and produce graphs, etc.
- portability and ease with which it can be placed in the necessary location.

Liquid-in-glass thermometers

The liquid used in these thermometers is generally **mercury** or ethyl **alcohol**.

- Glass is a poor thermal conductor and so the bulb is made with a **thin wall** to facilitate the easy **transfer of heat** through it.
- The **bore** is very **narrow** so that any change in volume of the liquid will result in a noticeable change in length of the liquid thread (see Example 1).
- The **bulb** is relatively **large** so that the corresponding expansion or contraction of the liquid it contains is noticeable for small changes in temperature. The **larger** the bulb, the **longer** can be the stem, and therefore the more **precise** will be the instrument in measuring temperature.
- The **scale** is positioned very close to the bore to **reduce parallax error** while taking readings.

Laboratory mercury thermometer

Figure 10.3 shows a liquid-in-glass thermometer which is generally designed to read temperatures between –10 °C and 110 °C. It is useful for measurements made in the school laboratory since experiments performed there are usually within this range.

Constant pressure if the vessel is freely expandable

- If the container is **freely expandable**, and the pressure instantaneously increases or decreases, it will expand or contract so that the gas contained once more has the same pressure as the environment (usually atmospheric pressure). Such is the case with a gas trapped in a syringe with a well lubricated piston. It can therefore be assumed that the **pressure** of the gas remains **constant**.

Constant volume if the container is strong and rigid

- If the container is strong and rigid, any increase or decrease in its volume will be negligible. It can therefore be assumed that the volume of the gas remains constant.

Combining the gas laws to a general gas law

For a fixed mass of an ideal gas changing from state 1 to state 2, the following relationship holds:

$$\frac{PV}{T} = \text{constant} \quad \therefore \frac{P_1 V_1}{T_1} = \frac{P_2 V_2}{T_2}$$

- P and V may be in **any unit** of pressure and volume respectively.
- T **must** be measured on the **Kelvin scale**.
- In problems where one of the variables is constant, it may be **omitted** from the equation.

Applying the gas laws

Example 1

5.0 cm³ of an ideal gas at a pressure of 4.0×10^5 Pa is slowly compressed to 2.5 cm³. The containing vessel is made of a very good conducting material. Determine the new pressure of the gas.

Note: Since the compression is done slowly, and since the vessel is a good conductor, any increase in thermal energy will quickly conduct through the walls of the container and then radiate to the environment. The temperature is therefore constant and can be omitted from the calculation.

$$P_1 V_1 = P_2 V_2$$
$$4.0 \times 10^5 \times 5.0 = P_2 \times 2.5$$
$$\frac{4.0 \times 10^5 \times 5.0}{2.5} = P_2$$
$$8.0 \times 10^5 \text{ Pa} = P_2$$

Example 2

4.0 gallons of an ideal gas at a temperature of 27 °C is slowly heated to 177 °C in a freely expandable vessel. Determine the new volume of the gas.

Note: Since the vessel is freely expandable, the increase in pressure is only temporary. The pressure returns to atmospheric pressure when the vessel and gas cease to expand. The pressure can therefore be considered as constant and can be omitted from the calculation.

$$\frac{V_1}{T_1} = \frac{V_2}{T_2}$$
$$\frac{4.0}{(273 + 27)} = \frac{V_2}{(273 + 177)}$$
$$\frac{4.0}{(300)} = \frac{V_2}{450}$$
$$\frac{4.0 \times 450}{300} = V_2$$
$$6.0 \text{ gallons} = V_2$$

Example 3

20 litres of an ideal gas at a pressure of 900 mmHg and temperature, 7.0 °C, is heated to 287 °C in a strong steel vessel. Determine the new pressure.

Note: A 'strong steel vessel' implies that the volume may be considered as constant. This assumption is valid since solids expand only very slightly when heated whereas gases expand substantially. The strong steel vessel prevents the gas from expanding by any significant amount. The volume can therefore be considered as constant and can be omitted from the calculation.

$$\frac{P_1}{T_1} = \frac{P_2}{T_2}$$

$$\frac{900}{(273 + 7.0)} = \frac{P_2}{(273 + 287)}$$

$$\frac{900}{280} = \frac{P_2}{560}$$

$$\frac{900 \times 560}{280} = P_2$$

$$1800 \text{ mmHg} = P_2$$

Example 4

7.2 m³ of an ideal gas at a temperature of 400 K and a pressure of 1.0×10^5 Pa is cooled to −73 °C as its volume is decreased to 1.8 m³. Determine the final pressure.

$$\frac{P_1 V_1}{T_1} = \frac{P_2 V_2}{T_2}$$

$$\frac{1.0 \times 10^5 \times 7.2}{400} = \frac{P_2 \times 1.8}{(273 - 73)}$$

$$\frac{1.0 \times 10^5 \times 7.2}{400} = \frac{P_2 \times 1.8}{200}$$

$$\frac{1.0 \times 10^5 \times 7.2 \times 200}{400 \times 1.8} = P_2$$

$$2.0 \times 10^5 \text{ Pa} = P_2$$

Verifying the gas laws

Experimental verification of Boyle's law

Boyle's law can be verified using the apparatus shown in Figure 11.1. The pressure, *P*, and volume, *V*, of the trapped air are measured and recorded. The pressure is increased by use of the pump and the new readings of *P* and *V* are taken. This is repeated until a total of 6 pairs of readings are tabulated. $\frac{1}{V}$ is calculated and recorded for each *V* and a graph of *P* vs $\frac{1}{V}$ is plotted.

- The scale used to measure the gas could be of **length instead of volume**. Since the tube is of **uniform cross-sectional area**, any change in length will produce a proportional change in volume.

- Increasing the pressure will also increase the temperature. Before taking readings, a **short period** should be allowed after increasing the pressure for the air to **return to room temperature**.

- The **straight line through the origin** of the graph **verifies the law**.

See question 14 at the end of this chapter for a simpler apparatus to carry out the same experiment.

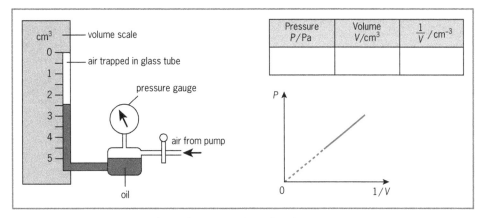

Pressure P/Pa	Volume V/cm³	$\frac{1}{V}$ / cm⁻³

Figure 11.1 *Experimental verification of Boyle's law*

Experimental verification of Charles' law

Charles' law can be verified using the apparatus shown in Fig. 11.2. The volume, V, and Celsius temperature, T_C, of the trapped air is measured and recorded. By heating the water bath, the temperature is then increased several times by about 10 °C, each time taking a new pair of readings of V and T_C. The temperatures are converted to the Kelvin scale and a graph is plotted of V against T_K.

- The scale behind the capillary tube could be of **length instead of volume**. Since the tube is of **uniform cross-sectional area**, any change in length will produce a proportional change in volume.

- Readings are only taken when the **bead** of sulfuric acid is **steady**, indicating that the pressure of the gas has returned to its **initial value.**

- The **straight line through the origin** of the graph **verifies the law.**

- **Zero** on the Kelvin scale is the temperature at which an ideal gas will **occupy no space**. This point of **absolute zero** is used in the **establishment** of the **Kelvin temperature scale.**

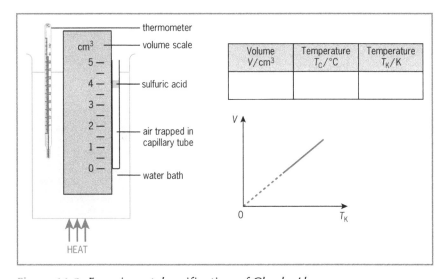

Volume V/cm³	Temperature T_C/°C	Temperature T_K/K

Figure 11.2 *Experimental verification of Charles' law*

Experimental verification of the pressure law

The pressure law can be verified using the apparatus shown in Figure 11.3. The pressure, P, and Celsius temperature, T_C, of the gas trapped in the spherical flask is measured and recorded. By heating the water bath, the temperature of the trapped air is increased several times by intervals of about 10 °C, each time taking a new pair of readings of P and T_C. The temperatures are converted to the Kelvin scale and a graph is plotted of P against T_K.

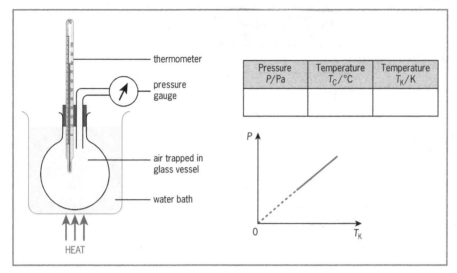

Pressure P/Pa	Temperature T_C/°C	Temperature T_K/K

Figure 11.3 *Experimental verification of the pressure law*

- The volume of the gas is considered **constant** since the **expansion** of the vessel is **negligible.**
- The **straight line through the origin** of the graph **verifies the law.**
- **Zero** on the Kelvin scale is the temperature at which an ideal gas will **exert no pressure** since its molecules would have **no motion.** This point of absolute zero is used in the **establishment of the Kelvin temperature scale.**

Other important graphs

In addition to the graphs used to verify the gas laws experimentally, Figure 11.4 shows other important graphs illustrating gas law relationships.

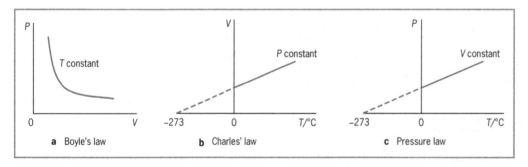

Figure 11.4 *Other important graphs illustrating gas law relationships*

Gas pressure in terms of kinetic theory

The particles of a gas (atoms or molecules) move **randomly**, bombarding each other and the walls of their container. As a particle of mass, m, and velocity, u, collides with the walls of its container, it imparts a force, f_1, onto it for a brief impact time, t. See Figure 11.5.

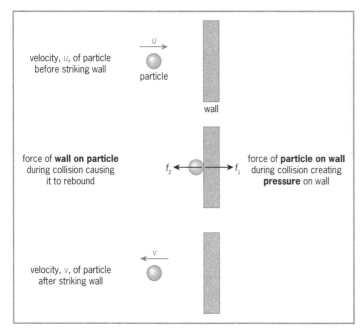

Figure 11.5 *Pressure caused by the force of gas particles on the wall of a container*

- In accordance with **Newton's third law** of motion (see Chapter 4), the container provides a **reaction force, f_2, equal in magnitude** but **opposite in direction,** onto the particle that causes it to rebound with a new velocity, **v.**

- This force produces a **rate of change of momentum** in accordance with **Newton's second law** of motion.

$$f_2 = \frac{mv - mu}{t}$$

At any instant in time, the **gas pressure, P,** is caused by the **total force, F,** exerted by all the gas particles on the **total area, A,** of the walls of the container.

$$P = \frac{F}{A}$$

At higher temperatures, the **increased thermal energy** causes the particles to move at **higher velocities.** This causes the rate of change of momentum (force, F) on collision to be greater and therefore causes the pressure to increase.

Kinetic theory and Boyle's law

As a gas is **compressed** at **constant temperature**, the average **speed** of its particles is **unchanged** and therefore the **force** exerted on the walls of the container **remains constant**. However, since the **volume decreases**, the collisions are on a **smaller area ($A\downarrow$)** and therefore the **pressure** (force per unit area) **increases ($P\uparrow$).** See Figure 11.6.

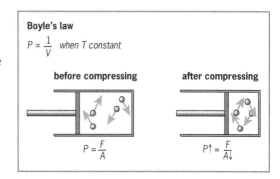

Figure 11.6 *Kinetic theory and Boyle's law*

Kinetic theory and Charles' law

As the **temperature** of a gas **rises**, the average **speed** of its particles **increases** and therefore the average **force** it exerts on the walls of its container **increases** (**F↑**). If the vessel is **freely expandable**, its **volume** will **increase**, and therefore the **area** of the walls will also **increase** (**A↑**). However, since the force and the area **increase by the same factor**, the **pressure remains constant**. See Figure 11.7.

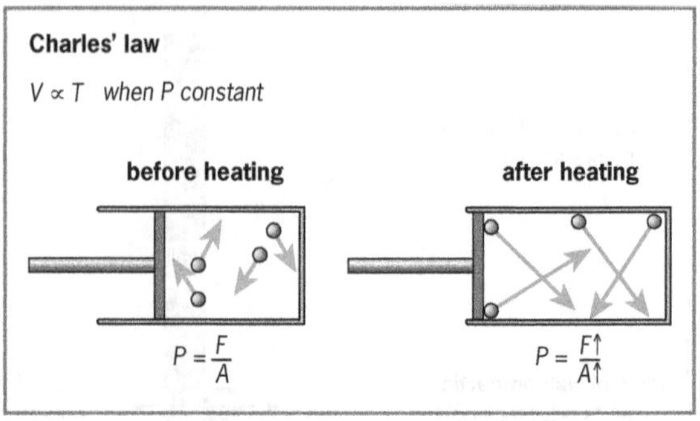

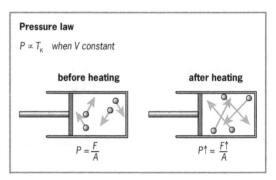

Charles' law

$V \propto T$ when P constant

before heating

after heating

$P = \dfrac{F}{A}$

$P = \dfrac{F\uparrow}{A\uparrow}$

Figure 11.7 *Kinetic theory and Charles' law*

Kinetic theory and the pressure law

As the **temperature** of a gas **rises**, the average **speed** of its particles **increases** and therefore the average **force** it exerts on the walls of its container **increases** (**F↑**). If the **volume** remains **constant**, the **area** of the walls remains **constant**. Since the force increases as the area of the walls remain constant, the **pressure** on the walls **increases** (**P↑**). See Figure 11.8.

Pressure law

$P \propto T_K$ when V constant

before heating

after heating

$P = \dfrac{F}{A}$

$P\uparrow = \dfrac{F\uparrow}{A}$

Figure 11.8 *Kinetic theory and the pressure law*

Recalling facts

1 State each of the THREE gas laws.

2 During experiments to verify the gas laws, how can we ensure that:

  a volume is constant?

  b pressure is constant?

  c temperature is constant?

3 Write an equation expressing the general gas law in terms of the pressure, P, volume, V, and absolute temperature, T, of a gas.

4 Describe an experiment to verify Charles' law. You should include:

  a a suitable method

  b a well-labelled diagram of the apparatus

  c a sketch of any necessary graph from the observed data

  d the precaution taken to ensure that one of the variables remains constant

  e how you conclude that the law is verified.

5 Sketch the following graphs showing the relationships between pressure, P, volume, V, absolute temperature, T, and Celsius temperature, θ.

  a P against V

  b P against $\dfrac{1}{V}$

  c P against T

  d P against θ

  e V against θ

Example 3

Calculate the thermal energy released when ethanol, of mass 200 g, cools from 30 °C to 10 °C.

$E_H = mc\Delta T$

$E_H = 0.200 \times 2400 \times (30 - 10)$

$E_H = 9600$ J

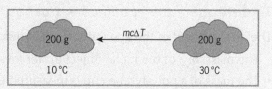

Figure 12.2 *Example 3*

Example 4

A copper block of mass 2.0 kg cools from 150 °C as it is immersed in water of mass 4.0 kg, initially at 30 °C. Calculate the final temperature, *X*.

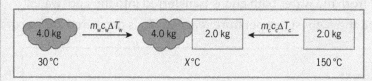

Figure 12.3 *Example 4*

heat gain of water = heat loss of copper

$$m_w c_w \Delta T_w = m_c c_c \Delta T_c$$

$$4.0 \times 4200 \times (X - 30) = 2.0 \times 390 \times (150 - X)$$

$$16\,800(X - 30) = 780(150 - X)$$

$$16\,800X - 504\,000 = 117\,000 - 780X$$

$$16\,800X + 780X = 504\,000 + 117\,000$$

$$17\,580X = 621\,000$$

$$X = \frac{621\,000}{17\,580} = 35\ °C$$

Final temperature = 35 °C

Example 5

A vessel of **heat capacity** 40 J K⁻¹ contains 400 g of water at 10 °C. When heated for 3.0 minutes, the temperature rises to 90 °C. Calculate the power of the heater.

Since the temperature of the water is initially 10 °C, the temperature of the containing vessel is also 10 °C. The heater raises the temperature of both the **vessel** and the **water** to 90 °C.

power = $\dfrac{\text{energy}}{\text{time}}$

$$P = \frac{m_v c_v \Delta T_v + m_w c_w \Delta T_w}{t}$$

$$P = \frac{C_v \Delta T_v + m_w c_w \Delta T_w}{t} \ \dots \text{ since } m_v c_v = C_v$$

$$P = \frac{(40 \times 80) + (0.400 \times 4200 \times 80)}{3 \times 60}$$

$$P = \frac{3200 + 134\,400}{180}$$

$$P = 760 \text{ W to 2 sig. fig.}$$

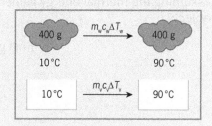

Figure 12.4 *Example 5*

Experiments to determine specific heat capacity

To determine the specific heat capacity of a metal by the method of mixtures

The arrangement of the apparatus used in this experiment is shown in Figure 12.5.

- The mass, m_m, of the metal object under investigation is measured and recorded using a balance.
- The mass of a polystyrene cup is measured empty, m_c, and then when half filled with water, m_{cw}. The mass of water, m_w, is then calculated as follows: $m_w = m_{cw} - m_c$
- The temperature, T_2, of the cool water in the cup is measured and recorded.
- The metal object is heated in boiling water by hanging it from a string so that it is submerged in the water bath. The temperature, T_1, is measured and recorded.
- The hot metal object is removed from the water bath and is transferred into the cool water.
- The water in the cup is stirred and the highest temperature reached, T_3, is measured and recorded.

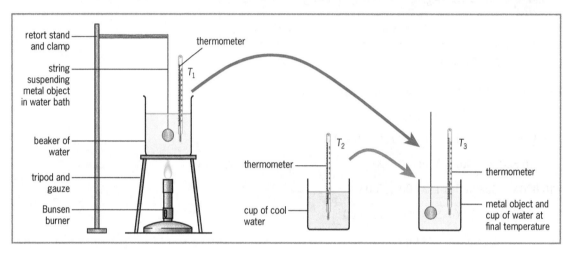

Figure 12.5 *Determining the specific heat capacity of a metal by the method of mixtures*

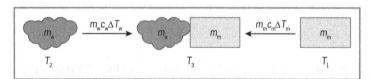

Figure 12.6 *Calculating the specific heat capacity of a metal by the method of mixtures*

Assuming the thermal energy gained by the water is equal to the thermal energy lost by the metal object:

$$\text{heat gained by water} = \text{heat lost by metal}$$

$$m_w c_w (T_3 - T_2) = m_m c_m (T_1 - T_3)$$

$$\frac{m_w c_w (T_3 - T_2)}{m_m (T_1 - T_3)} = c_m$$

Errors and precautions

- **Error:** As the hot metal object is transferred to the cool water, heat is lost to the surrounding air. **Precaution:** To reduce this error, the hot metal is **quickly** transferred.
- **Error:** Thermal energy from the hot metal is also transferred to the polystyrene cup and to the bench top. This is not considered in the calculation.

- **Error:** Evaporation of water from the surface of the metal object as it is transferred to the cooler water removes **latent heat** of vaporisation (see Chapter 13). The temperature of the metal on reaching the cool water is therefore **less** than T_1.

- **Precaution:** The metal object is left in the boiling water for a **few minutes** to ensure that it acquires the temperature of the hot water.

- **Precaution:** The metal object should be suspended so that it is **not in contact** with the walls of the hot water bath or polystyrene cup since the temperature of the walls may be different than the temperature of the water. It is the temperature of the water that is being measured.

- **Precaution:** The water in the polystyrene cup is **stirred** with the thermometer or a stirrer to ensure that the temperature recorded, T_3, is the **mean** temperature reached.

To determine the specific heat capacity of a liquid by the method of mixtures

The experimental arrangement of the apparatus used is shown in Figure 12.7.

- The mass, m_b, of the beaker and the mass, m_c, of the polystyrene cup are measured and recorded using a balance.

- The liquid under investigation is poured into the beaker and water is poured into the polystyrene cup, and the new masses, m_{bL} and m_{cw}, respectively, are measured and recorded.

- The mass, m_L, of the liquid, and the mass, m_w, of the water in the cup, are calculated as follows:
  $m_L = m_{bL} - m_b \quad m_w = m_{cw} - m_c$

- The temperature, T_2, of the cool water in the cup is measured and recorded.

- The liquid is heated for a few minutes and its temperature, T_1, is measured and recorded.

- The hot liquid is poured into the cool water.

- The highest temperature reached, T_3, is measured and recorded.

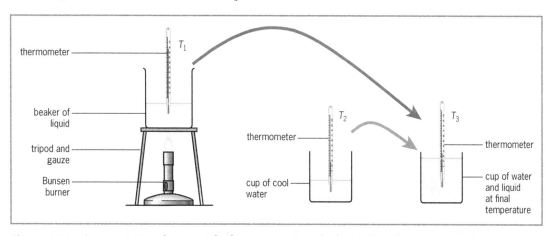

Figure 12.7 *Determining the specific heat capacity of a liquid by the method of mixtures*

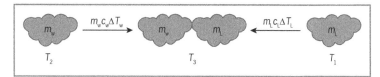

Figure 12.8 *Calculating the specific heat capacity of a liquid by the method of mixtures*

Assuming the thermal energy gained by the water is equal to the thermal energy lost by the liquid:

heat gained by water = heat lost by liquid

$$m_w c_w (T_3 - T_2) = m_L c_L (T_1 - T_3)$$

$$\frac{m_w c_w (T_3 - T_2)}{m_L (T_1 - T_3)} = c_L$$

Errors and precautions

- **Error:** As the hot liquid is transferred to the cool water, heat is lost to the air.
  Precaution: To reduce this error, the hot liquid is **quickly** transferred.

- **Error:** Thermal energy is transferred to the cup and to the bench. This is not considered in the calculation.

- **Precaution:** The liquids are **stirred** with the thermometer or a stirrer to ensure that the temperatures recorded are the **mean** temperatures reached.

To determine the specific heat capacity of a metal by an electrical method

- The metal to be investigated is designed as a block with **slots** to fit a heater and a thermometer. The mass, **m**, of the block is measured by means of a balance and is recorded.

- The apparatus is set up as shown in Figure 12.9a and the heater is switched on.

- After a short while, the temperature, T_1, of the block is measured and a stopwatch is started simultaneously.

- When the temperature has risen by about 10 °C, the new temperature, T_2, is measured and recorded and the heater is switched off.

- The electrical energy consumed, E_H, is the reading on the **joulemeter**. If we assume that it is all used in raising the temperature of the block, then $E_H = mc(T_2 - T_1)$.

If a joulemeter is unavailable, E_H can be obtained as in Figure 12.9b by using the readings V and I from a voltmeter and an ammeter respectively, together with the time, t, of heating, obtained from a stopwatch.

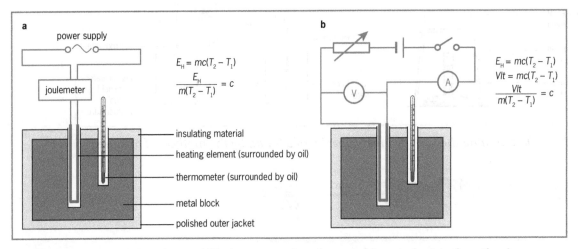

Figure 12.9 *Determining the specific heat capacity of a metal by an electrical method*

Errors and precautions

- **Error:** Some of the thermal energy of the block is transferred to the surroundings.
  Precaution: An insulating material such as **cotton wool** can be placed around the block to **reduce conduction** to the surroundings and a sheet of **aluminium foil** can be placed around this lagging to **reduce radiation** to the air (see Chapter 14).

- **Precaution:** A small quantity of oil should be placed around the bulb of the thermometer and around the heater element to improve conduction with the block.

To determine the specific heat capacity of a liquid by an electrical method

- The mass, m, of the liquid is measured by a balance and is recorded.

- The apparatus is set up as shown in Figure 12.10a and the heater is switched on.

- After a short while, the temperature, T_1, of the liquid is measured and a stopwatch is started simultaneously.

- When the temperature has risen by about 10 °C, the new temperature, T_2, is measured and recorded, and the heater is switched off.

- The electrical energy used, E_H, is the reading on the **joulemeter**. If we assume that all this energy is used in raising the temperature of the liquid, then $E_H = mc(T_2 - T_1)$.

If a joulemeter is unavailable, E_H can be obtained as in Figure 12.10b by using the readings, V and I, from a voltmeter and an ammeter respectively, together with the time, t, of heating, obtained from a stopwatch.

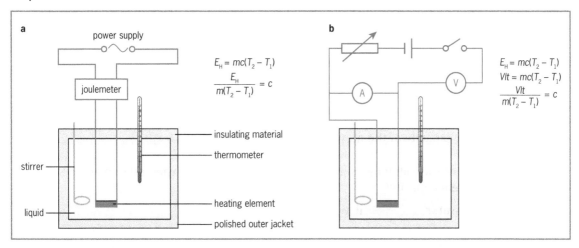

Figure 12.10 *Determining the specific heat capacity of a liquid by an electrical method*

Errors and precautions

- **Error:** Some of the thermal energy gained by the liquid is transferred to the surroundings.
  Precaution: An insulating material such as **cotton wool** can be placed around the container to **reduce conduction** to the surroundings, and a sheet of **aluminium foil** can be placed around this lagging to **reduce radiation** to the air.

- **Precaution:** The liquid should be **stirred** to ensure that the mean temperature is recorded.

- **Precaution:** The heater element should be placed near the **bottom** of the liquid to allow effective heating by **convection** (see Chapter 14).

Recalling facts

1

  a Define the following quantities and state the SI unit of each:

    i the specific heat capacity of a substance

    ii the heat capacity of a body.

  b Write an equation relating *heat capacity* to *specific heat capacity*, stating what each symbol in your equation represents.

2 Describe an experiment to determine the specific heat capacity of aluminium by an electrical method. You should include:

  a a well-labelled diagram of the apparatus

  b a suitable method

  c two precautions taken to ensure that sources of error are reduced

  d the equation on which your calculation is based.

Applying facts

Table 12.1 on page 127 should be referred to when necessary.

3 Janelle found that it takes the same amount of heat energy to warm 500 g of liquid X through 30 °C than it does to warm 250 g of liquid Y through the same temperature range. State what can be deduced about each of the following:

  a the *specific heat capacity* of X relative to that of Y

  b the *heat capacity* of 500 g of X relative to that of 250 g of Y.

4 Andrea is told that the heat capacity of a metal garden fork is 1800 J K^{-1}. Determine the specific heat capacity of the metal if the mass of the fork is 4.0 kg.

5

  a For **EACH** of the following, state, with a reason, which is expected to have the higher specific heat capacity:

    i soil or water   **ii** wood or iron.

  b Liquid **A** has a higher specific heat capacity than liquid **B**. Based on this information, which would be more suitable for use as a coolant for the radiator of a car?

  c **X** and **Y** are materials used in the construction of a saucepan. **X** is a poor conductor of specific heat capacity 2000 J kg^{-1} K^{-1} and **Y** is a good conductor of specific heat capacity of 350 J kg^{-1} K^{-1}. State, with a reason, which part of the saucepan each should be used for.

6 A block of iron and a block of copper of the same mass are heated through the same temperature range. State, with a reason, which block will absorb the most heat.

7 Calculate the heat needed to raise the temperature of 500 g of lead from 20 °C to 50 °C.

8 Calculate the heat released when 2.0 kg of water cools from 80 °C to 30 °C.

9 A block of copper of mass 2.0 kg and temperature 80 °C is placed into 2.5 kg of water initially at 20 °C. Determine the final temperature of the mixture.

10 An iron pulley wheel of mass 1.2 kg is placed into water of mass 2.0 kg and temperature 20 °C. The final temperature is 24 °C. Determine the initial temperature of the iron.

11 A hot piece of copper at 120 °C is placed into 2.0 kg of water at 0 °C which is in a pot of *heat capacity* 200 J °C⁻¹. The final temperature is 10 °C. Determine the mass of the copper.

12 A heater supplies energy at a rate of 350 J s⁻¹ to water of mass 500 g, initially at 10 °C. Determine the temperature after 2.5 minutes, stating TWO assumptions made for your calculation.

13 Water at 2.0 °C flows over the edge of a waterfall of height 210 m. Given that the acceleration due to gravity is 10 m s⁻², calculate the temperature of the water at the bottom of the falls if 80% of its gravitational potential energy is converted to thermal energy and evaporation is negligible.

14 2.0 kg of water at 10 °C is heated by a 1.2 kW heater to a temperature of 80 °C. Determine the time taken.

15 A vessel of *heat capacity* 40 J °C⁻¹ contains 0.30 kg of water at 10 °C. When heated for 4.0 minutes and 20 seconds, the temperature rises to 90 °C. What is the power of the heater?

Analysing data

16 **a** An immersion heater of power 250 W is used to warm 2.0 kg of liquid A from 20 °C to 50 °C. Figure 12.11 shows how temperature varies with time for the process.

   i Determine the slope of the graph for liquid A.

   ii How long does it take for the temperature of liquid A to reach 45 °C?

   iii Write an equation relating the power of the heater to the change in temperature and time of the process.

   iv From your equation, calculate the specific heat capacity of liquid A.

   b The experiment is repeated with a second liquid B of the same mass; the graph illustrates the process. What can be said about the specific heat capacity of this liquid compared to the first?

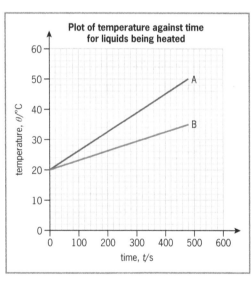

Figure 12.11 *Question 16 Plot of temperature against time for liquids being heated*

13 Heat and phase change

When a substance is heated at its melting point or boiling point, it changes from one state to a next (it changes phase) **without a change of temperature**. The quantity of heat energy, E_H, required to change the state of the substance is dependent on the mass, m, of the substance and a property of the substance known as its **specific latent heat, l**.

Since a substance can change from the solid phase to the liquid phase, or from the liquid phase to the gaseous phase, there are two specific latent heats; the **specific latent heat of fusion, l_f** and the **specific latent heat of vaporisation, l_v**.

These variables are related in the equations $E_H = ml_f$ and $E_H = ml_v$

Latent heat and specific latent heat

Learning objectives

- Demonstrate that **temperature** remains **constant** during a **change of phase**.
- Define **latent heat, specific latent heat of fusion** and **specific latent heat of vaporisation**.
- Understand and use information from **heating curves** and **cooling curves**.
- Derive the **units** of **specific latent heat** of fusion and specific latent heat of vaporisation.
- Apply the relation $E_H = ml_f$ and $E_H = ml_v$ to solve problems.
- **Experimentally determine** the **specific latent heat of fusion** of ice and the **specific latent heat of vaporisation** of water by an **electrical method**.
- Distinguish between **evaporation** and **boiling**.
- Use the **kinetic theory** to explain evaporation and boiling.
- Use the **kinetic theory** to explain how various **factors** can affect the **rate of evaporation**.
- Give examples and explanations of **applications** utilising the **cooling effect** of evaporation.

Experimentally demonstrating a change of phase

Liquid to gas

- A beaker of water, initially at room temperature, is heated until about one quarter of it boils away (Figure 13.1). During the heating process, the temperature of the water is measured and recorded at regular intervals of **one minute**.

- The time at which boiling begins is noted; it is indicated by the production of a **constant flow of bubbles** rising through the body of the fluid.

- A graph known as a **heating curve** is plotted of temperature against time. The graph illustrates that from the time boiling begins, the **temperature remains constant**.

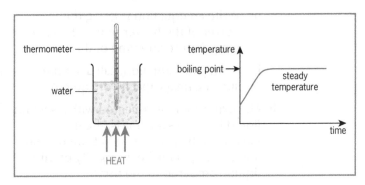

Figure 13.1 *Temperature is constant during boiling*

Liquid to solid

- Figure 13.2 shows chips of **acetamide** added to a boiling tube which is submerged in a water bath. (Flakes of candle wax may be used instead of the acetamide.)
- The water bath is heated until the temperature increases to about **90 °C**. At this temperature the acetamide will be in the liquid phase.
- The stand and clamp, together with the boiling tube and its contents, are removed from the water bath, and a thermometer is placed in the liquid to monitor its temperature as it cools.
- Temperature readings are taken **every minute** until the temperature reaches about **70 °C**.
- A graph, known as a **cooling curve**, is plotted of temperature against time. The graph illustrates that during the period when the acetamide is **solidifying**, the **temperature remains constant**. This temperature is the **melting point** or **freezing point** of acetamide, and is seen to be **80 °C**.

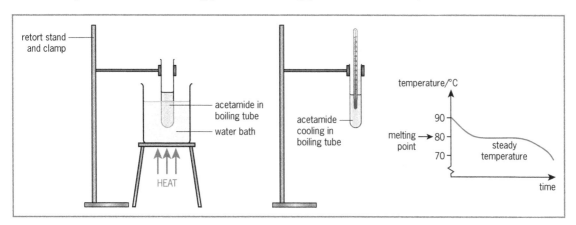

Figure 13.2 *Temperature is constant during solidification*

Defining latent heat and specific latent heat

Latent heat is the heat needed to change a solid to a liquid or a liquid to a gas without a change of temperature.

The specific latent heat of fusion of a substance is the heat needed to change a unit mass of the substance from solid to liquid without a change of temperature.

The specific latent heat of vaporisation of a substance is the heat needed to change a unit mass of the substance from liquid to gas without a change of temperature.

Unit of specific latent heat

$$E_H = ml \quad \therefore l = \frac{E_H}{m} \qquad \qquad \therefore \text{the SI unit of } l = \frac{J}{kg}, \text{ i.e. J kg}^{-1}$$

Latent heat and the kinetic theory

Latent heat of fusion is necessary to provide potential energy to:

a) do **internal work** in **breaking the bonds** between the particles of a solid

b) do **external work** as the particles' spacing expands *slightly* against the **atmospheric pressure**.

Latent heat of vaporisation is necessary to provide potential energy to:

a) do **internal work** in **overcoming attractive forces** between the particles of the liquid

b) do **external work** as the particles' spacing expands *significantly* against the **atmospheric pressure.**

The particles of a gas have more energy than the particles of a liquid, and the particles of a liquid have more energy than the particles of a solid, as shown in Figure 13.3.

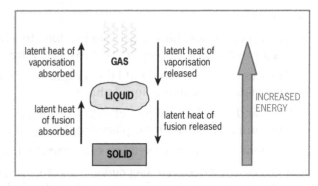

Figure 13.3 *Latent heat and phase changes*

When a substance changes phase to a higher energy state, it must absorb latent heat, and when it changes phase to a lower energy state, it must release latent heat.

Since the external work done against the atmospheric pressure when a liquid changes to a gas is much greater than that done when a solid changes to a liquid, the specific latent heat of vaporisation of a substance is much greater than its specific latent heat of fusion.

Table 13.1 below demonstrates this for the phase changes of water.

Table 13.1 *Specific latent heat of ice and water*

Specific latent heat of fusion of ice	3.3×10^5 J kg^{-1}	3.3×10^2 J g^{-1}
Specific latent heat of vaporisation of water	2.3×10^6 J kg^{-1}	2.3×10^3 J g^{-1}

Imagine being burnt by a kilogram of steam. As the steam condenses at 100 °C, it first releases **2 300 000 J** of energy onto your body, and the water formed then releases a further **4 200 J** for every degree fall in its temperature!

Heating curves and cooling curves

Consider a block of ice of mass m at −5 °C being heated until it changes to steam at 110 °C (assuming that the steam produced at 100 °C has been captured and is further heated). Figure 13.4 indicates the changes of temperature and phase which occur during the process.

On heating, the temperature of the ice increases from −5 °C to 0 °C. The ice then begins to melt and continues doing so at **constant temperature**. When all the ice has melted, further supply of heat increases the temperature to 100 °C. The water then begins to boil and continues doing so at **constant temperature**. When all the water has boiled, further supply of heat to the captured steam raises its temperature once more.

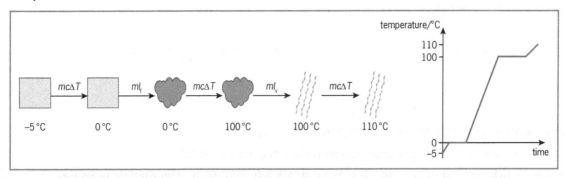

Figure 13.4 *Temperature diagram and corresponding heating curve as ice converts to steam*

Examine Figures 13.5a and 13.5b to obtain a better understanding of how to interpret and use heating curves.

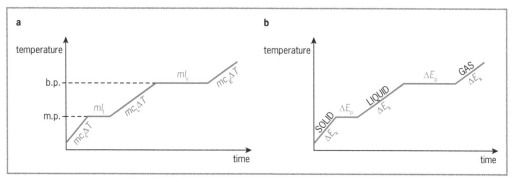

Figure 13.5 *Understanding heating curves*

Note the following with reference to figure 13.5:

- At the **melting point** and the **boiling point**, the **slope** of the graph is **zero**, as there is **no change in temperature.**
- As the **temperature rises**, the heat supplied **increases** the **speed** and **kinetic energy** of the particles of the substance.
- When the **temperature remains constant**, the heat supplied increases the **potential energy** of the particles of the substance. The particles obtain enough potential energy to overcome the rigid bonds of the solid to become a liquid, or to overcome the attractive forces of neighbouring particles of the liquid to become a gas.
- Heat energy associated with the changes in temperature is calculated from $E_H = mc\Delta T$.
- Heat energy associated with changes of phase is calculated from $E_H = ml$.

Figure 13.6 shows a cooling curve of a body of lead as it changes from liquid to solid.

- The melting point is seen to be 328 °C.
- As the **temperature decreases**, the **speed** and **kinetic energy** of the particles of the substance **decrease.**
- When the **temperature remains constant**, the **potential energy** of the particles of the substance **decreases.**

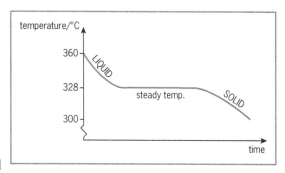

Figure 13.6 *A cooling curve of lead*

Applying the relationship $E_H = ml$

Tackling problems involving heat and state change

- Always draw a temperature diagram to assist in formulating your equation. See examples below.
- If there is a temperature difference, place cooler bodies on the left and warmer bodies on the right. Temperature change, ΔT, is then calculated as '**warmer subtract cooler**', i.e. '**right subtract left**'.
- Changes of **temperature and phase** must be represented as **separate changes** in the diagram.
- Indicate above each arrow whether the heat involved represents a **temperature change** ($mc\Delta T$) or **a phase change** (ml). These will be individual terms in your equation.
- Show the **same mass** at **each end** of each arrow.
- For situations where **heat gain is equal to heat loss**, your diagram should show each chain of arrows having **all bodies meeting at the same temperature.**

Table 13.1 together with the data below may be used in the following examples

specific heat capacity of ice = 2100 J kg⁻¹ K⁻¹

specific heat capacity of water = 4200 J kg⁻¹ K⁻¹

Example 1

Determine the power of a heater which takes 5 minutes to convert 600 g of water at 100 °C to steam at 100 °C.

$$P = \frac{E}{t} = \frac{ml_v}{t}$$

$$P = \frac{0.600 \times 2.3 \times 10^6}{5 \times 60}$$

$$P = 4600 \text{ W}$$

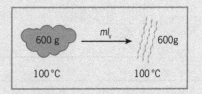

Figure 13.7 *Example 1*

Example 2

Determine the heat released when 4.0 kg of water at 0 °C solidifies to ice at 0 °C.

$$E = ml_f = 4.0 \times 3.3 \times 10^5 = 1.3 \times 10^6 \text{ J}$$

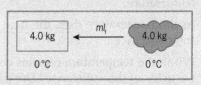

Figure 13.8 *Example 2*

Example 3

Determine the final temperature reached as 1.2 kg of ice at −4 °C is placed into 3.0 kg of water at 40 °C, given that the ice is completely melted in the process.

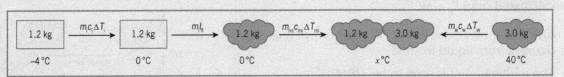

Figure 13.9 *Example 3*

heat gained by ice and molten ice = heat lost by warm water

$$m_i c_i \Delta T_i + m_i l_{fi} + m_{mi} c_{mi} \Delta T_{mi} = m_w c_w \Delta T_w$$

$$(1.2 \times 2100 \times 4) + (1.2 \times 3.3 \times 10^5) + (1.2 \times 4200 \times x) = 3.0 \times 4200 \times (40 - x)$$

$$10\,080 + 396\,000 + 5040x = 504\,000 - 12\,600x$$

$$5040x + 12\,600x = 504\,000 - 396\,000 - 10\,080$$

$$17\,640x = 97\,920$$

$$x = \frac{97\,920}{17\,640}$$

$$x = 5.6 \text{ °C}$$

Example 4

A vessel of *heat capacity* 5000 J °C⁻¹ contains 2.0 kg of ice chips at 0 °C. Water at 80 °C is poured onto the ice and the final temperature becomes 20 °C. Determine the mass of water added to the ice.

Note:

- There are three bodies involved in this heat exchange: the ice, the vessel and the warm water. They must each have their own arrow chain in the diagram representing the heat change to which they contribute.

- Since the heat capacity ($C = mc$) of the vessel is given, its mass, m, or specific heat capacity, c, is not required.

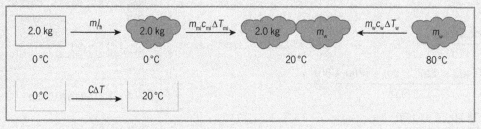

Figure 13.10 *Example 4*

heat gained by ice, molten ice and vessel = heat lost by warm water

$$m_i l_{fi} + m_{mi} c_{mi} \Delta T_{mi} + C_v \Delta T_v = m_w c_w \Delta T_w$$

$$2.0 \times 3.3 \times 10^5 + 2.0 \times 4200 \times 20 + 5000 \times 20 = m_w \times 4200 \times 60$$

$$6.6 \times 10^5 + 168\ 000 + 100\ 000 = 252\ 000 m_w$$

$$9.28 \times 10^5 = 252\ 000 m_w$$

$$\frac{9.28 \times 10^5}{2.52 \times 10^5} = m_w$$

mass of warm water = 3.7 kg

Example 5

Water at 60 °C is added to a mixture of 400 g of melting ice chips and 800 g of water, both at 0 °C, until all the ice *just melts*. Calculate the mass of water added.

Note: Since the process is stopped when the ice *just melts*, the heat from the warm water is used only to change the phase of the ice and not to raise the temperature of the molten ice. Therefore:

- the final temperature of the mixture remains at 0 °C

- the 800 g of water, initially mixed with the ice at 0 °C, has no part in the calculation.

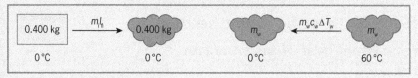

Figure 13.11 *Example 5*

heat gained by ice = heat lost by water

$$m_i l_{fi} = m_w c_w \Delta T_w$$

$$0.400 \times 3.3 \times 10^5 = m_w \times 4200 \times 60$$

$$m_w = \frac{0.400 \times 3.3 \times 10^5}{4200 \times 60}$$

$$m_w = 0.52 \text{ kg}$$

Example 6

A vessel of heat capacity 200 J K⁻¹ contains 400 g of water at 20 °C.

a Find the time taken to raise the temperature to 100 °C using a heater of power 500 W.

b Find the extra time required to boil the water until just 140 g remains.

a Note: Since the water, initially at 20 °C, is in the vessel, they are both at the same temperature. The temperature of both (the vessel and the water) will increase to 100 °C.

$$P = \frac{E}{t}$$

$$\therefore t = \frac{E}{P}$$

$$t = \frac{mc\Delta T + C\Delta T}{P}$$

$$t = \frac{(0.400 \times 4200 \times 80) + (200 \times 80)}{500}$$

$$t = \frac{134\,400 + 16\,000}{500}$$

$$t = 300 \text{ s (to 2 sig. fig.)}$$

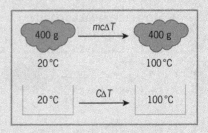

Figure 13.12 *Example 6a*

b Note:

- Since 140 g of water will remain, only 260 g will boil.

- The **vessel will receive no additional heat** since the temperature is not going to increase above 100 °C. Only the water will absorb heat from the heater as it changes from the liquid phase to the gaseous phase **without a change of temperature.**

$$P = \frac{E}{t}$$

$$\therefore t = \frac{E}{P}$$

$$t = \frac{ml_v}{P}$$

$$t = \frac{0.260 \times 2.3 \times 10^6}{500} = 1200 \text{ s (or } \frac{1200}{60} = 20 \text{ minutes) (to 2 sig. fig.)}$$

Figure 13.13 *Example 6b*

Experiments to determine specific latent heats

To determine the specific latent heat of fusion of ice by an electrical method

- The mass, m_b, of an empty beaker is measured.
- The apparatus is set up as shown in Figure 13.14a with the heater submerged in chips of melting ice and the circuit switched on.
- As the ice melts and drips into the beaker, more is added, so that the heater is always submerged.
- When the beaker is about half filled, the funnel is removed and the mass, m_{bw}, of the beaker and water is measured. The mass of melted ice, m_i, is calculated from $m_i = m_{bw} - m_b$.
- The electrical energy used, E_H, is the reading on the **joulemeter**. If we assume that it is all used in melting the ice, then $E_H = m_i l_f$.

If a joulemeter is unavailable, E_H can be obtained as in Figure 13.14b by using the readings of V and I from a voltmeter and ammeter respectively, together with the time, t, of heating, obtained from a stopwatch.

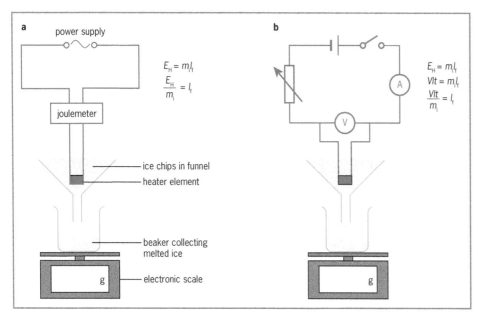

Figure 13.14 *Determining the specific latent heat of fusion of ice by the electrical method*

Errors and precautions

- **Error:** The ice transferred to the heater will always have a layer of water with it which has melted without energy from the heater. Since its mass is included in the calculation, the result will be slightly in error.
  Precaution: To reduce this error, the ice should be dabbed in a paper towel, just before being placed around the heater.

- **Error:** The temperature near the centre of each ice chip may be below 0 °C. Since the calculation considers the temperature of the ice to be zero, the result will be slightly in error.
  Precaution: The ice chips should be small so that their mean temperature will be approximately 0 °C.

- **Error:** Thermal energy from the environment is absorbed by the ice. This energy is not considered in the calculation.
  Precaution: To reduce this problem, the funnel should be made of a material such as styrofoam, or a lagging can be placed around the funnel, to reduce heat flow from the environment. *Alternatively*, the experiment may be modified as in question 21 at the end of this chapter.

- **Precaution:** Ensure that the heater is always submerged in ice.

To determine the specific latent heat of vaporisation of water by the electrical method

- The apparatus is set up as shown in Figure 13.15a and the heater is switched on.
- When the water is steadily boiling, the initial mass, m_1, is measured.
- After about 5 minutes, the new mass, m_2, is measured.
- The mass of water boiled, m_w, is calculated as $m_1 - m_2$.
- The electrical energy used, E_H, is the reading on the **joulemeter**. If we assume that it is all used in boiling the water, then $E_H = m_w l_v$.

If a joulemeter is unavailable, E_H can be obtained as in Figure 13.15b by using the readings of V and I from a voltmeter and an ammeter respectively, together with the time, t, of heating, obtained from a stopwatch.

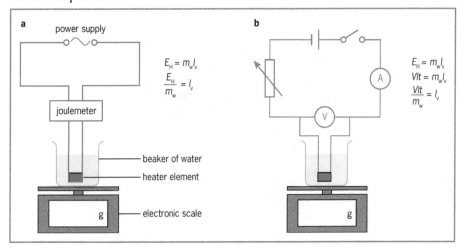

Figure 13.15 *Determining the specific latent heat of vaporisation of water by the electrical method*

Errors and precautions

- **Error:** Some heat is lost to the surroundings.
  Precaution: To help prevent heat loss to the environment, the beaker can be **lagged** with a good insulator to reduce **conduction**, and **aluminium foil** can be wrapped around it to reduce **radiation**.
- **Precaution:** The water is boiled for a few minutes before any readings are taken, to ensure that it is all at the boiling point.
- **Precaution:** The heater is placed near the **bottom** of the water to produce **convection currents**, which transfer thermal energy to the entire mass.

Evaporation and boiling

We in the Caribbean know only too well how small puddles can quickly disappear on hot days, and how wet clothes from the wash can soon become dry when hung onto a line in the yard. We are also familiar with the cooling effect on our bodies when perspiration on our skin vaporises due to a welcoming breeze. These occurrences, where a liquid **vaporises over a range of temperature**, are due to a process known as **evaporation**.

When we cook food in a pot of water and see bubbles of steam rising to the surface, we observe a second way a liquid can vaporise. We notice that the bubbles of steam are produced only at a **specific temperature**. This process is known as **boiling**.

*Evaporation is the process whereby a liquid vaporises to a gas, over a **range of temperature**.*

*Boiling is the process whereby a liquid vaporises to a gas, at a **given temperature** dependent on the surrounding **pressure**.*

Evaporation and boiling explained by kinetic theory

During the process of evaporation, the **more energetic molecules overcome the attractive forces** of their neighbours at the surface and escape to the gaseous phase. The remaining molecules have **less kinetic energy** and therefore the **temperature** of the liquid **decreases**.

Evaporation occurs at a **higher rate** as the **temperature increases**. At a certain critical temperature (the boiling point), bubbles of **steam** begin to rise **through the liquid** by the process known as **boiling**. This occurs when the **vapour pressure** of the liquid is equal to the **atmospheric pressure**. At this temperature, molecules transfer from the liquid phase to the gaseous phase **at all points** within the liquid. The process immediately stops if the heat supply is removed, indicating that for a liquid **to remain boiling**, an **external source** must be **supplying heat to it**.

Table 13.2 *Summary of differences between evaporation and boiling*

Evaporation	Boiling
Occurs only at the surface of the liquid	Occurs throughout the body of the liquid
Occurs over a range of temperatures	Occurs at one temperature for a given pressure
Does not require an external heat source	Requires an external heat source

Factors affecting the rate of evaporation explained by kinetic theory

1. **Temperature:** Molecules **move faster** at **higher temperature** and therefore possess **more kinetic energy**. They have a better chance of overcoming the attractive forces of the neighbouring molecules and escaping as a gas.

2. **Humidity:** If the humidity is high, molecules escaping from the surface of the liquid are more likely to crash into other molecules and rebound to the liquid, thereby reducing the amount of evaporation. If the air becomes saturated with water vapour, the net rate of evaporation is zero since the rate of molecules leaving the liquid is equal to the rate of molecules returning to it. This explains why the perspiration on our bodies is unable to adequately cool us by evaporation when the humidity is high.

3. **Wind:** Moving air removes evaporated molecules from above the surface of a liquid, allowing other evaporating molecules to have a better chance of escaping completely without colliding and rebounding to it.

4. **Surface area:** Evaporation is a surface phenomenon and therefore the larger the surface area, the greater is the chance for molecules of the liquid to escape.

Cooling due to evaporation explained by kinetic theory

During evaporation, the **more energetic** molecules have **higher velocities** and can **overcome the attractive forces** of the other molecules of the liquid. In so doing, they escape from the surface of the liquid at a rate dependent on the temperature. Some may return to the liquid after rebounding from molecules above the surface, but most escape completely. The remaining liquid becomes cooler since the more energetic molecules **absorb latent heat of vaporisation** from the less energetic ones as they evaporate.

Demonstrating cooling due to evaporation

A beaker of ether is placed in a small pool of water, as shown in Figure 13.16a. Ether is a **volatile** liquid, that is, it **readily evaporates**. When the air above the liquid is saturated with vapour from the ether, there is less chance of further evaporation occurring. However, by pumping air through the body of the ether, the following conditions for evaporation become favourable:

- Due to the **bubbles of air** produced, there is **increased surface area** between the liquid and the air, and therefore there is increased evaporation.

- **Since unsaturated air** is passing through the ether, evaporation can readily occur.

As the ether evaporates, it draws latent heat of vaporisation from the beaker and surrounding water, causing the pool of water to **freeze**.

Another way to demonstrate evaporative cooling is shown in Figure 13.16b.

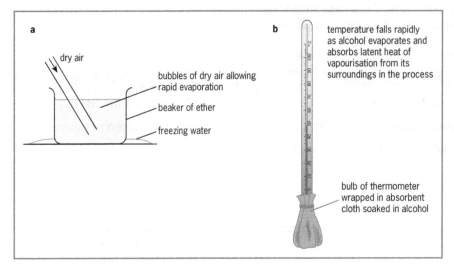

Figure 13.16 *Demonstrating cooling by evaporation*

Applications using evaporative cooling

Refrigerator

The refrigerator is a device which utilises **evaporative cooling** of a liquid. The liquid is one which evaporates **easily** and is known as a refrigerant. It is circulated through a narrow tube by an electrical **pump**, as shown in Figure 13.17.

At the evaporator: The **liquid** refrigerant is pumped to an **expansion valve**, where it **evaporates** as it is sprayed through a fine nozzle. **Latent heat is absorbed** through the tubes by **conduction** from the contents of the fridge, decreasing the temperature of the air, food and drink it contains.

At the condenser: The **gaseous** refrigerant is then pumped to the condenser where it is **compressed** and **condenses** back to a liquid. This **releases the latent heat** energy previously absorbed. A grill made of **copper** or **aluminium** is attached to the condenser so that heat can quickly be **conducted** from the refrigerant to the grill and then **radiated** to the surrounding air.

The condenser should be in a **ventilated area** so that the thermal radiation emitted by the grill can readily dissipate to the environment. The grill is painted black since **black is a good emitter** of radiation and allows the heat to be easily transferred to the surrounding air (see Chapter 14).

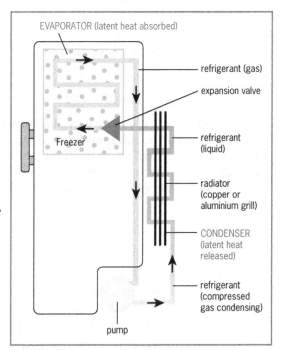

Figure 13.17 *Refrigerator*

Air conditioner

As shown in Figure 13.18, this is much like a refrigerator.

At the evaporator: Air from the room being cooled is sucked into the evaporator section by means of a fan. As it passes through the aluminium grill, heat is conducted from it to the evaporator tubes (coils), providing the necessary latent heat of vaporisation for the **liquid** refrigerant to change to **gas** as it is sprayed through the expansion valve. The cooled air then returns to the room from which it was drawn.

At the condenser: The refrigerant **gas** is pumped to the **condenser** where it is **compressed** and condenses back to a **liquid**, releasing the latent heat energy previously absorbed. The heat energy is **conducted** to the **aluminium grill** and is radiated to the surroundings. A fan produces **forced convection** of the hot air away from the unit, to the outside.

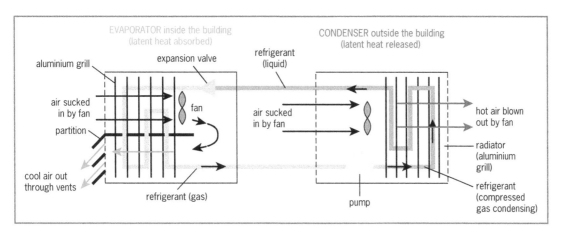

Figure 13.18 *Air conditioner*

Earthenware pots

Pots made of baked clay are called earthenware pots. These vessels are used for potting plants and for keeping water cool. The material of the pot has **small holes** through which water in the pot can slowly pass to its outer surface where it evaporates. The **latent heat energy** necessary for the evaporation is absorbed from the pot and from the water in the pot, and therefore the water is kept cool.

Hurricanes

Solar energy incident on the ocean causes the evaporation of water and the production of clouds. The extent of evaporation in the summer months can be very high, and the upward rush of the vapour can cause **low-pressure** regions that may develop into storms. As the vapour condenses in the clouds, **latent heat of vaporisation** is released, and this energy **drives the winds faster**, increasing the rate of evaporation and the power of the storm. The region can become covered with storm clouds, producing much condensation and the release of a tremendous amount of energy evidenced by hurricane-force winds.

Perspiration

When we use excessive power, for example by exercising or lifting heavy objects, our body temperature rises due to the higher rate of conversion of chemical energy. Our temperature may also increase due to the surroundings on a hot day. To achieve a cooling effect, our bodies secrete perspiration, which on evaporation extracts **latent heat of vaporisation** from our skin. Cats and dogs also cool by evaporation, but since their bodies are totally covered by hair, they depend on evaporation from their wet noses and thick wet tongues.

Spraying aerosols

When we press on the button at the top of a can of deodorant an expansion valve opens, and the liquid contents of the can spray through a fine nozzle. On expanding into the atmosphere under reduced pressure, the liquid changes to a gas, absorbing **latent heat of vaporisation** from the can and its remaining contents in the process.

Storing meats in a cooler with ice

Meats and other foodstuffs can be kept fresh when stored on ice in a cooler. The **specific latent heat of fusion** of ice is **large** (3.3×10^5 J kg^{-1}), and so as the ice melts, it absorbs a large amount of heat from the food, keeping its temperature low and preventing rapid decay.

Steam cooking

The **specific latent heat of vaporisation** of water is **large** (2.3×10^6 J kg^{-1}). When food is cooked by a steaming process, a large amount of heat is released onto the food as the steam condenses, reducing the time necessary for cooking.

Recalling facts

1. **a** Define the following, stating the associated SI unit in each case.

    i latent heat

    ii specific latent heat of fusion

    iii specific latent heat of vaporisation.

    b Write an equation which can be used to determine the latent heat of a body. Each symbol in your equation should be identified.

2. **a** Latent heat provides energy to the particles of a body as it changes phase. Does it increase the potential energy or the kinetic energy of those particles?

    b Give TWO reasons that this energy is needed as a liquid changes phase to a gas.

3. Why is a burn from steam at 100 °C much more severe than a burn from hot water at the same temperature?

4. Distinguish between evaporation and boiling.

5. Why does the temperature of the can of an aerosol air freshener decrease as its contents are sprayed into the air?

6. Scott returns home from spear fishing at mid-day and hangs his wet towel on the clothesline. List FOUR factors which will affect the rate of evaporation of water from the towel and explain the effect of each of the factors in terms of the kinetic theory.

7. Ethanol has a melting point of −114 °C and a boiling point of 79 °C. Sketch a graph to show how its temperature changes with time as it is heated from −130 °C to 85 °C. No values are required for the x-axis. Your graph should indicate the following:

    a the solid, liquid and gaseous states

    b when the heat supplied is providing kinetic energy, and when it is providing potential energy

    c the melting point and the boiling point of ethanol.

8   a Use kinetic theory to explain why liquids cool due to evaporation.

   b Describe how you can demonstrate that liquids cool due to evaporation.

9 Describe an experiment to determine the specific latent heat of vaporisation of water using an *electrical method.*

Your experiment should include:

   i a diagram of the apparatus as it is being used

   ii a suitable method

   iii **TWO** precautions taken to ensure that sources of error are reduced

   iv the equation on which your calculation is based.

10 Explain how excessive evaporation can lead to the formation of a storm.

11 Explain how water stored in earthenware pots is kept cool.

12 Figure 13.19 shows a refrigerator.

   a Describe the role of latent heat in the function of the following:

      i the evaporator

      ii the condenser.

   b Why must the refrigerant be a volatile liquid?

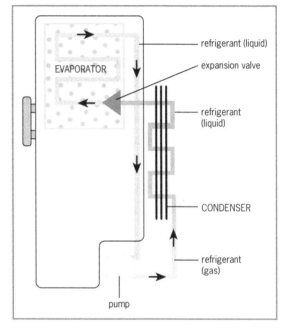

Figure 13.19 *Question 12*

Applying facts

13 Alvin's fishing boat has an *ice box* in which he stores his daily catch. His wife, Brenda, often steam-cooks the fish when he returns at night. How are the specific latent heat of fusion of ice and the specific latent heat of vaporisation of water significant in:

   a storing the fish?

   b steaming the fish?

   Use the data listed in Table 13.1 and immediately before Example 1 (page 140) to solve the following questions.

14 Determine the heat required to convert 8.0 kg of ice at 0 °C to water at 5.0 °C.

15 Determine the heat released when 400 g of steam at 100 °C is cooled to water at 80 °C.

16 Determine the final temperature when 2.0 kg of ice at −5.0 °C is placed into 4.0 kg of water at 80 °C, given that all the ice melts.

17 A vessel of heat capacity 2000 J K^{-1} contains ice chips at 0 °C. Water of mass 1.8 kg and temperature, 60 °C is poured onto the ice until the final temperature becomes 25 °C. Determine the mass of ice initially in the vessel.

18 Calculate the mass of water at 50 °C which can **just melt** 20 g of ice at 0 °C when poured onto it.

19 A heater of power 630 W is used to raise the temperature of 500 g of water, initially at 10 °C. Determine:

a the time taken to bring the water to its boiling point of 100 °C

b the additional time required to continue boiling until only 300 g of water remain.

Analysing data

20 Figure 13.20 shows a heating curve for a solid of mass 200 g being heated by a heater of power 100 W until it changes to a liquid at 100 °C.

a Which section of the graph represents:

   i the solid state?   **ii** the liquid state?

b What is the melting point of the substance?

c How long did the melting process take?

d Calculate:

   i the specific heat capacity of the solid

   ii the heat capacity of the solid

   iii the specific latent heat of fusion of the solid.

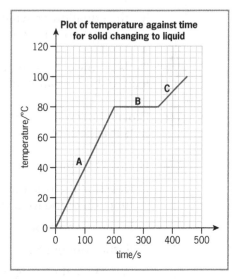

Figure 13.20 *Question 20*

21 Ice chips at 0 °C were placed in a funnel and left to melt. The ice which melted in a period of 105 s was collected and its mass was measured.
A heater was then placed in the ice and the mass of ice melted during the **same** period was again measured. Data for the experiment is given below.

mass of ice melted by thermal energy from the environment = 3.0 g
mass of ice melted by thermal energy from the environment and from the heater = 18.0 g
power of heater = 50 W

a What mass of ice was melted by energy from the heater?

b Calculate the specific latent heat of fusion of ice from the data given.

14 Thermal energy transfer

Walking bare footed on a hot tar road can become unbearable due to thermal **conduction** from the road. Looking down from a roof into an active chimney is extremely uncomfortable due to heat transferred by **convection** currents rising within the chimney. Standing near to a bonfire warms us due to outward **radiation** of electromagnetic waves from the burning twigs. These examples demonstrate the **three processes** by which heat can be transferred from one point to the next.

Conduction, convection and radiation

Learning objectives

- Define the three processes of thermal energy transfer and distinguish among their transfer capabilities through solids, liquids, gases and a vacuum.
- Explain **conduction** and **convection** in terms of the **kinetic theory**.
- Describe simple experiments to **demonstrate conduction** and **convection.**
- Relate the **conductivity** of **materials** to their use.
- Relate **convection** to **common phenomena** including land and sea breezes.
- Explain the transfer of thermal energy by **radiation.**
- Describe an experiment to show that **radiation does not require a medium** for transmission.
- Distinguish among the terms **emission**, **reflection** and **absorption**, and describe experiments to demonstrate the factors on which absorption and emission depend.
- Relate **absorption** and **emission** of **radiation** to common situations, including **global warming**.
- Describe and explain the **design of devices** utilising different methods of heat transfer.

Processes of heat transfer

Conduction: The transfer of thermal energy between two points in a medium by the relaying of energy between adjacent particles of the medium.

Convection: The transfer of thermal energy between two points in a medium by movement of the particles of the medium due to regions of different density.

Radiation: The transfer of thermal energy between two points by means of electromagnetic waves.

Thermal transfer capability through various media

Table 14.1 *Thermal energy transfer through solids, liquids, gases and a vacuum*

Conduction	Occurs significantly in solids (to a lesser extent in non-metals than in metals), less in liquids, and very little in gases. Cannot occur in a vacuum.
Convection	Occurs in liquids and gases. Cannot occur in a vacuum.
Radiation	Occurs readily through gases and through a vacuum.

Thermal transfer by conduction explained by kinetic theory

Non-metals

Figure 14.1a shows what happens when one end of a non-metallic bar is warmed. The heat energy supplied converts to kinetic energy of the particles (**atoms or molecules**), which causes them to **vibrate faster** and with **greater amplitude**. They **bombard** their neighbours with **greater force** than before, passing on the increased vibration. In this manner, thermal energy is **relayed** between adjacent particles to the other end of the bar. The **temperature** of a substance is **proportional to the kinetic energy** of its particles, and therefore the temperature at the cooler end of the bar increases.

Metals

A similar process occurs in metals as shown in Figure 14.1b. Metals contain vibrating **cations** (positively charged ions). When a metal bar is heated, the heat energy supplied is converted to kinetic energy of the cations, causing them to vibrate faster and with greater amplitude. They then bombard their neighbours with greater force, relaying the **kinetic energy** from **cation to cation** along the bar.

Metals also contain a **'sea' of free electrons** which **translate** between the cations. When these electrons are supplied with heat energy, their kinetic energy increases, causing them to **translate faster**. On **collision** with a vibrating cation, the **kinetic energy** is relayed from **electron to cation**. Since the temperature of a substance is proportional to the kinetic energy of its particles, the temperature at the cooler end of the bar increases.

Since **metals** have **two modes** of thermal conduction, whereas **non-metals** have only **one**, metals are better conductors of heat.

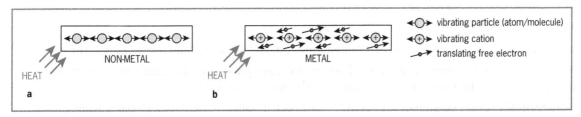

Figure 14.1 *Conduction in non-metals and metals*

Experiments involving thermal conduction

Comparing thermal conduction through solids

Figure 14.2a shows three rods of equal length being heated by the **same** flame. A bit of **wax** is placed at the end of each rod. It is noticed that the wax melts first from the copper, then from the iron and finally from the wood, showing that copper is the best conductor of the three materials and wood is the worst.

Figure 14.2b shows a wooden rod joined to a copper rod. Paper is **tightly** wrapped around the region where the rods are joined. When the combined rod is lowered over a flame, it is noticed that the paper **chars over the wood** but not over the copper. Heat conducted through the paper to the copper quickly spreads through the copper, since **copper is a good thermal conductor**. However, since **wood is a poor thermal conductor**, heat conducted to it through the paper collects at the interface with the paper, causing it to char.

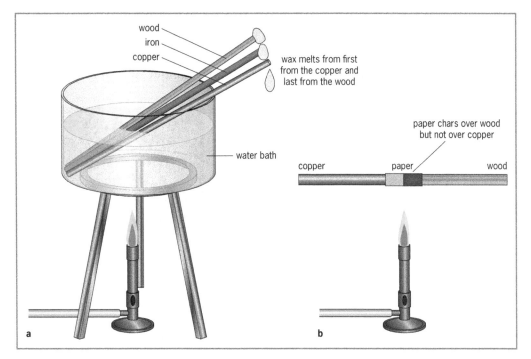

Figure 14.2 *Demonstrating conduction in solids*

Demonstrating that water is a poor thermal conductor

Figure 14.3 shows a piece of ice wrapped in copper mesh and submerged in a test tube of water. The weight of the mesh keeps the ice at the bottom of the tube. The water, **heated at the top** to **avoid convection**, rapidly comes to a boil, although the ice remains solid for quite some time. Since the only process by which thermal energy can transfer from water at the top of the tube to the ice below is by conduction, water must be a **poor thermal conductor**.

Good and poor thermal conductors

Metals are **good thermal conductors** and are useful in the manufacture of important devices.

- **The bases and sides of kettles, pots and pans** are usually made of **copper, aluminium** or **stainless steel** so that heat can readily transfer through them to the food by conduction.
- **Boilers** that are heated externally are generally made of alloys such as **steel**. They need to be strong and to pass the heat by conduction from the burning fuel readily to the liquids they contain.
- **Heat exchangers** such as radiators are generally made of **aluminium**. They transfer heat by conduction from the fluid in the radiator to their outer surface which then passes the heat to the air that is blown through it.

Non-metals are **poor thermal conductors** and are used daily for many purposes.

- The **handles** of **cooking utensils** and **garden tools** are generally made of **plastic** or **wood** to reduce the conduction of heat to those who hold them when they are hot. Thick **leather gloves** are also useful for this purpose.
- **Water heater storage tanks** are lagged with **polyurethane** to reduce outward thermal conduction.

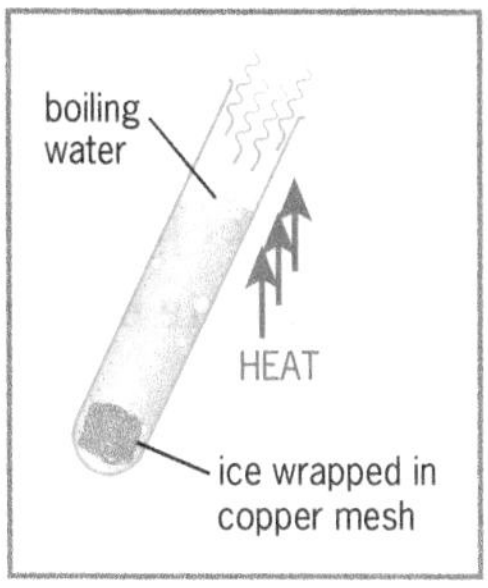

Figure 14.3
Demonstrating that water is a poor conductor

- **Bricks**, **clay** and materials such as **grasses** are used to build places of shelter that reduce conduction of heat, allowing homes to stay cooler during hot periods and warmer during cool periods.
- **Igloos** reduce outward conduction of heat since ice is a poor thermal conductor (Figure 14.4a).

a   b   c

Figure 14.4 *Applications of poor thermal conduction*

- **Air is a very poor thermal conductor.**
  - **Woollen jackets** keep us warm since they have **air pockets** trapped between fibres that reduce heat from conducting outward from our bodies.
  - **Expanded polystyrene** (XPS) contains much **air** and is used as a thermal insulator, reducing the flow of heat through the walls of **buildings, refrigerators, ovens** and **water heater storage tanks**.
  - **Roofs** made of **leaves**, such as those from palm trees, trap air and therefore insulate shelters.
  - **Feathers**, **hair** and **fur** insulate animals from the cold. Air trapped by these covers offers further protection. Birds **fluff their feathers** to trap more air when necessary (Figure 14.4b).
  - **Concrete blocks** and **clay blocks** containing air pockets act as thermal insulators (Figure 14.4c).

Why a stone floor seems cooler than a rug

Figure 14.5 shows a person of **body temperature 37 °C** standing with one foot on a rug and the other on a stone floor. The stone and the rug are initially **at the same temperature** of 25 °C. To the person, it seems that stone floors are cooler than rugs. This misconception occurs because the stone is a much better thermal conductor than the rug. Heat flowing from the person to the stone can **quickly spread** out within the **stone**, but most of the thermal energy flowing from the foot to the rug remains in the region under the foot. The temperature of the rug under the foot therefore soon becomes 37 °C, giving the impression that the rug was always warmer.

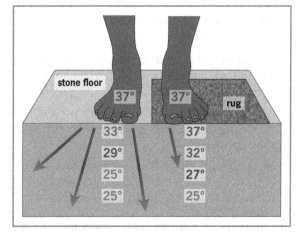

Figure 14.5 *Conduction of stone floor and rug*

Thermal transfer by convection explained by kinetic theory

Figure 14.6 illustrates the process of convection. On warming a liquid or gas **from below**, the increased thermal energy in its lower region gives the particles there more **kinetic energy**, causing them to **translate** more vigorously and to occupy **more space**. The region therefore becomes **less dense**, resulting in its particles **rising** and allowing **cooler particles** from **denser regions** to **fall** and take their place.

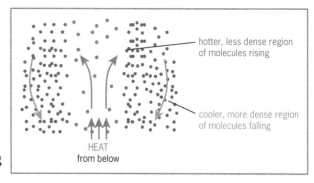

Figure 14.6 *Convection in liquids and gases*

Demonstrating convection

In liquids

A crystal of **potassium permanganate** (purple in colour) is added to a beaker of water and is heated from below, as shown in Figure 14.7a. As the crystal dissolves it forms a purple solution, which rises from the region being heated and displays the path of the **convection current**.

Alternatively, by placing **tiny aluminium flakes** in the heated region of the water, the glittering flakes will be seen in the path of the convection current.

In gases

A small hand-held fan is placed in the air just over a flame, as shown in Figure 14.7b. The upward air draft of the convection current will create a force on the fan, which causes it to spin.

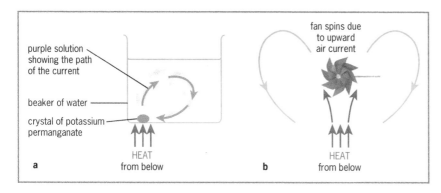

Figure 14.7 *Demonstrating convection in liquids and gases*

Phenomena involving convection

Land and sea breezes

During the day: The Sun's **radiation** warms the land more than it warms the sea. As the air in contact with the land is heated by **conduction**, it becomes less dense and rises, allowing denser air from over the sea to take its place in the form of a **cool onshore sea breeze**. These **convection currents** mean that coastal regions do not experience the extremely high day temperatures that can be found further inland.

The air in contact with the ground further inland also becomes hot during the day **but cannot rise** since there is no cooler, denser surrounding air to take its place (Figure 14.8a).

During the night: The land loses heat **radiation** at a higher rate than the sea and therefore the surface of the land becomes cooler than that of the sea. Air in contact with the sea is warmed by **conduction**, becomes less dense and rises, allowing denser air from over the land to take its place in the form of a **cool offshore land breeze**. These **convection currents** mean that coastal regions do not experience the extremely low night temperatures that can be found further inland (Figure 14.8b).

The air in contact with the ground further inland also becomes cool at night but cannot flow from the region since there is no nearby rising warm air for it to take the place of.

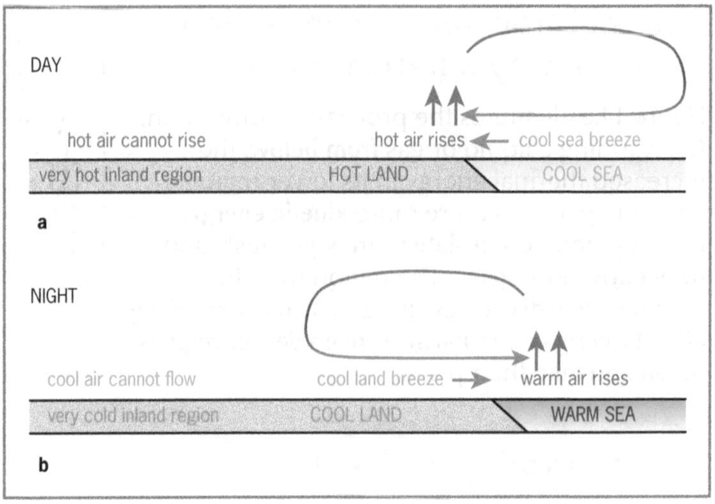

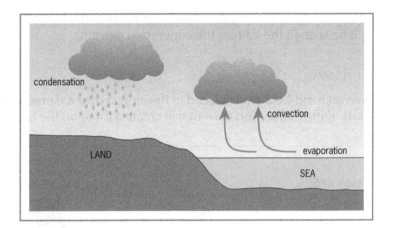

Figure 14.8 *Land and sea breezes*

Convectional rainfall

During the day, the Sun's radiation produces **evaporation** of water from lakes and oceans. Air in contact with the water is warmed by conduction, becomes less dense and rises, carrying the moisture upward in a **convection current** (Figure 14.9). As it rises, it cools, and the water vapour condenses to produce clouds and rainfall. This type of rainfall is particularly dominant in equatorial regions such as Guyana.

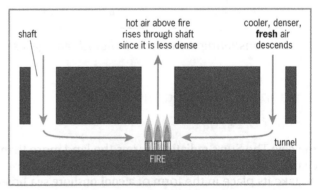

Figure 14.9 *Convectional rainfall*

Using convection currents to obtain fresh air

The air around miners in underground tunnels can become very poor due to the surrounding materials and the depletion of oxygen as the workers breathe. By boring vertical shafts, as shown in Figure 14.10, and by creating a small fire at the base of one of the shafts, a **convection current** can be set up, which brings fresh air to the tunnel.

Figure 14.10 *Using convection currents to obtain fresh air*

Thermal transfer by radiation

All matter is composed of **charged particles** that oscillate. As they do so, they are constantly **accelerating**, and this results in the emission of **electromagnetic waves**.

Thermal radiation is electromagnetic radiation produced as a result of the thermal motion (and hence temperature) of the particles of a body.

All matter emits a type of electromagnetic radiation known as **infrared radiation** (see Chapter 22). Due to the temperature of matter commonly encountered, a high proportion of the radiation it emits is **infrared radiation**. As our nerve endings absorb this radiation, we detect a sensation of **warmth**.

- The **energy** of an electromagnetic wave is **proportional to its frequency**.
- **Hotter bodies** emit **more** radiation and **higher frequency** radiation than do cooler bodies.
- Electromagnetic waves are the **only means by which thermal energy can be transported through a vacuum.** This is how we receive energy from the **Sun**.

Demonstrating that radiant energy can propagate through a vacuum

Figure 14.11 shows an arrangement used to demonstrate that radiation does not require a medium for its transmission. The jar is first **evacuated** by means of a **pump** and the heater is then switched on. The glass jar becoming warm is evidence that radiation is transferred to it across the vacuum from the **glowing heater** coil.

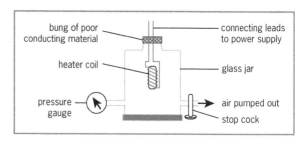

Figure 14.11 *Radiant energy can propagate through a vacuum*

Emitters and absorbers of thermal radiation

All bodies emit and absorb infrared radiation. Figure 14.12 shows how the temperature of a body relative to its surroundings determines whether it is a net emitter or net absorber at any time.

- Bodies that are **hotter than their surroundings** are **net emitters** of thermal radiation – they emit more energy than they absorb.
- Bodies that are **cooler than their surroundings** are **net absorbers** of thermal radiation – they absorb more energy than they emit.
- The surface of a body will usually reflect some radiation. A good absorber is a poor reflector and a poor absorber is a good reflector.

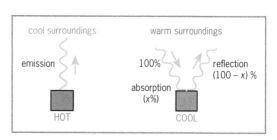

Figure 14.12 *Emitters and absorbers of radiation*

Factors affecting the absorption and emission of radiation

1. Colour of surface
2. Texture of surface
3. Area of surface

Table 14.2 *Effect of colour and texture on absorption and emission of thermal radiation*

Physical properties of surface	Absorber/reflector	Emitter
Black – matt, dull, rough	Good absorber/poor reflector	Good emitter
White or silver – gloss, shiny, smooth, polished	Poor absorber/good reflector	Poor emitter

The colour and texture of a body **does not determine** whether a body is a net absorber or a net emitter. This depends on its **temperature relative to the surroundings** at the time.

To determine if a body is a good absorber or good emitter at any given time

Use the following **two-step** process:

1. Consider its **temperature relative to the surroundings** to determine if the body is **an absorber** or **an emitter** at the given time.

2. Consider the **physical properties** of the surface to determine **how good** an **absorber** or **emitter** it is.

a) To be most efficient, which of gloss black, matt black, gloss white or matt white is best to paint the outside of a refrigerator?

   1. **Temperature relative to surroundings:** When keeping foodstuffs cool, the contents of a refrigerator are cooler than the surroundings. It is therefore a net **absorber** of radiation.

   2. **Physical properties of surface:** Since an efficient refrigerator should prevent heat from entering, it should be a **poor absorber** (good reflector). It should therefore be painted **gloss white.**

b) To be most efficient, which of gloss black, matt black, gloss white or matt white is best to paint the outside of an oven?

   1. **Temperature relative to surroundings:** When cooking, the contents of the oven are very hot relative to the surroundings. It is therefore a net **emitter** of radiation.

   2. **Physical properties of surface:** Since an efficient oven will prevent heat from leaving, it should be a **poor emitter.** It should therefore be painted **gloss white.**

c) Akil works in the freezer room of an ice company. State, with a reason, whether he should wear a black or white suit, and whether it should have a rough or glossy texture.

   1. **Temperature relative to surroundings:** Akil's body temperature is about 37 °C, but the freezer room is below 0 °C. He is therefore a net **emitter** of radiation.

   2. **Physical properties of surface:** In order to remain warm, he must be a **poor emitter**, and therefore should wear a **glossy, white** suit.

Experiment to investigate how the properties of a surface affect its rate of emission of radiation

Figure 14.13 shows a **Leslie's cube** containing **hot water.** The outer faces of such a cube each have a **different texture** and are painted either black, white or a random colour. They are all at the **common temperature** of the hot water. This cube has one of its vertical faces painted **shiny white** and another painted **matt black.** A **radiometer** (radiation detector) placed at a fixed distance from the surfaces detects maximum radiation emitted from the matt black surface and minimum radiation from the shiny white surface.

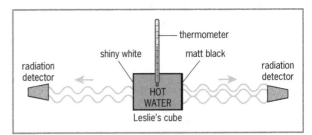

Figure 14.13 *Comparing radiation emitted from different types of surface*

By altering the temperatures of the water in the cube, it can be shown that **more radiation** is emitted with **increased temperature.**

Experiment to investigate how the properties of a surface effects its rate of absorption or reflection of radiation

Figure 14.14 shows a plan view of an arrangement where radiant energy from a candle is incident on several plates made of the same metal, each having a **cork** stuck to its outer surface by **wax**.

The **distance** between the candle and each plate is the **same** and therefore the radiant energy incident on each plate is the same.

The **inner** surface of each plate is **painted differently** but the outer surfaces are the same.

Radiation from the candle, incident on the plates, is **absorbed** by their inner surfaces and is then **conducted** to the wax on their outer surfaces. The cork attached to the plate that has its inner surface painted matt black is the first to fall, and the one painted shiny silver is the last to fall. This indicates that the **matt black** surface is the **best**, and the **shiny silver** surface is the **worst**, at **absorbing** the radiant energy.

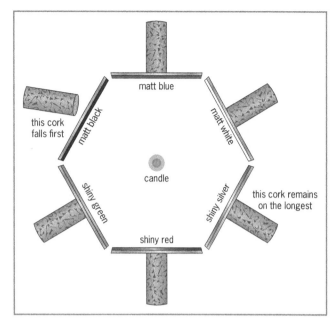

Figure 14.14 *Comparing radiation absorbed by different types of surface*

The experiment also shows that the **shiny silver** surface is the **best**, and the **matt black** surface is the **worst**, at **reflecting** the radiant energy.

Applications that utilise principles of thermal energy transfer

The vacuum flask

The vacuum flask shown in Figure 14.15 has several features that can reduce thermal energy transfer and so maintain its contents at an almost constant temperature.

- The **vacuum** within the double walls **prevents** energy transfer by **conduction and convection**.

- The **double walls** are made of **glass**, a poor conductor, to **reduce conduction** to or from the contents of the flask.

- The **inner, facing walls** of the vacuum region are **silver coated**. If the contents of the flask are hot, thermal energy will be conducted from it through the first wall of the vacuum and then radiated though the vacuum to the facing wall. Since silver is a **poor emitter**, the energy radiated is minimised,

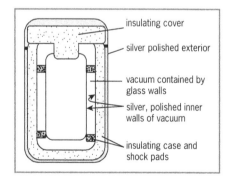

Figure 14.15 *The vacuum flask*

and since it is also a **poor absorber**, very little energy is absorbed (a large amount is reflected) by the facing wall. If the contents of the flask are cold, heat trying to enter from the outside will be similarly minimised.

- The **cover, case** and **shock pads** are made of **insulating** material to **reduce conduction**.

- The **cover** also **prevents convection**, as it prevents hot air from rising and leaving the flask.

- The **outer wall of the case** is **polished silver**. If the contents are hot, radiation emitted is minimised since **shiny silver surfaces are poor emitters** of radiation. If the contents are cold, radiation absorbed is also minimized, since **shiny silver surfaces are also poor absorbers** (good reflectors) of radiation.

The glass greenhouse

- The Sun's surface temperature is approximately 6000 K and therefore it emits a large amount of **high frequency** (short wavelength) electromagnetic waves, including **ultraviolet**, **visible light** and **short wavelength infrared (IR)** radiation.

- These waves **easily enter** the greenhouse through its **glass** walls (Figure 14.16).

- The contents of the green house **absorb** the radiation, become warm, and then **emit** their own radiation. However, since their **temperature** is much **lower** than that of the Sun, the waves emitted are mainly of **longer wavelength** (lower frequency) **infrared** radiation.

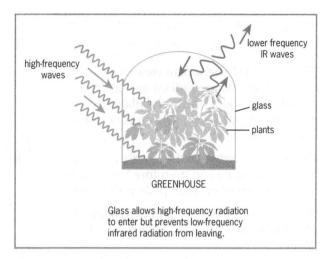

Figure 14.16 *The glass greenhouse*

- This longer wavelength infrared radiation **cannot transmit through the glass.** Some is **reflected**, but the rest is **absorbed** and then in turn emitted by the glass, much of it returning to the plants within the chamber. The greenhouse therefore acts as a **heat trap.**

- The sealed **roof** and **sides** of the greenhouse **prevent convection**, as they prevent the hot air from leaving the system.

The greenhouse effect and global warming

The Earth's atmosphere behaves like the glass of the greenhouse (Figure 14.17).

- **High frequency** (short wavelength) electromagnetic waves emitted by the Sun – mainly **ultraviolet, visible light** and **high frequency infrared** radiation – easily penetrate our atmosphere.

- Some of this energy is reflected into space, but much of it is **absorbed by the planet.**

- As the planet becomes warm, it **emits its own radiation**, but since its temperature is much **lower** than that of the Sun, the waves emitted are instead mainly of **lower frequency** (longer wavelength) **infrared** radiation.

- These longer wavelength **IR** waves are **absorbed** by **greenhouse gases** in the atmosphere, mainly **carbon dioxide, water vapour, methane, nitrous oxide** and **ozone.** The heated gases then **re-emit IR** radiation in all directions, much of it returning to further warm the planet.

Note: The greenhouse effect **prevents** our seas and lakes from **freezing.** However, with the increased **burning of trees** and the **extensive use of fossil fuels**, the levels of greenhouse gases have risen to the extent that the planet is experiencing **global warming.** This is causing our weather to become more severe, the polar ice caps to melt, and the ecology of the planet to be negatively affected.

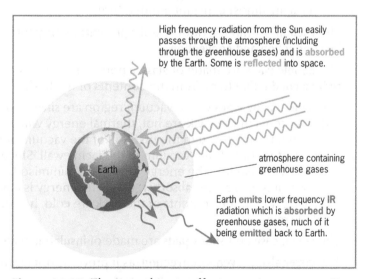

Figure 14.17 *The greenhouse effect*

The features of the solar water heater are illustrated in Figure 14.18.

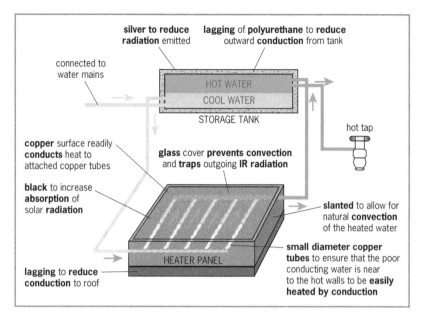

Figure 14.18 *The solar water heater*

- **Flow between the tank and the heater panel**: During the day, as the water is warmed in the heater panel, it becomes less dense and rises by **natural convection** to the **storage tank**. At the same time, cooler, denser water falls to the **heater panel**, where it is in turn heated.

- **Flow between the water mains supply, through the tank, to the user's hot water tap**: When the hot tap is opened, cool water from the mains supply is able to enter the bottom of the storage tank and force hot water from the top outward to the hot tap.

- All hot water entries and exits are at the top and all cool water entries and exits are at the bottom of the panel and tank.

- **Water** is a **poor thermal conductor**, so that the less dense hot water in the tank remains hot and rests on the denser, cooler water below it.

- If the tank is too heavy for the roof, it may be placed on the ground. An electric **pump**, however, will then be required to **force** the **hot, less dense** water **downward** to the storage tank, and the **cooler, denser** water **upward** to the heater panel on the roof.

Recalling facts

1 **a** Define each of the THREE processes by which heat energy can be transferred.

  b Which type of heat transfer is responsible for the propagation of energy through outer space?

2 **a** In terms of particles and their energies, explain the process of heat transfer through a metal bar.

  b How is this process different through non-metals?

3 Describe a simple experiment to demonstrate each of the following:

  a water is a poor thermal conductor

  b copper is a better thermal conductor than wood.

4 Give ONE example of how each of the following may be utilised:

  a the poor thermal conductivity of air

  b the good thermal conductivity of metals.

5 **a** In terms of particles and their energies, explain the process of convection in a liquid.

  b Describe how convection currents may be demonstrated in liquids.

6 **a** What behavior of electric charges causes the emission of electromagnetic radiation?

  b What type of electromagnetic radiation do our bodies detect as warmth?

  c How does the frequency of emitted radiation depend on the temperature of the emitting body?

7 Redraw and complete Table 14.3 to indicate if the properties of emission, absorption and reflection of thermal energy are **good** or **poor** for a dull black surface and a shiny white surface.

Table 14.3 *Question 7*

Surface properties	Emission	Absorption	Reflection
Dull black			
Shiny white			

Applying facts

8 Sasha notices that there is an offshore breeze blowing during the night in her coastal neighbourhood. Explain

  a how convection may be the reason for this breeze

  b why the temperature in her neighbourhood does not decrease by as much as that further inland.

9 Dylan is going for a run dressed in black when the outdoor temperature is 40 °C. Comment on his choice of clothing.

10 Figure 14.19 shows the design of a 'vacuum flask' which Anita would like to construct. The double walls C have a space D between them.

  a i Suggest a suitable material for the cover A.

   ii What processes of heat transfer would be reduced by A?

  b i What type of coating is B?

   ii What process of heat transfer would B reduce?

  c i State the material from which the walls of C should be made.

   ii Give the reason for your choice.

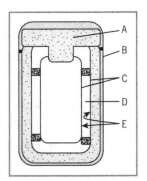

Figure 14.19
Question 10

d i Describe the region, D.

   ii What processes of heat transfer are prevented by D?

e i What should the inner facing walls of C be coated with?

   ii What process of heat transfer would this coating reduce?

11 Denzil lives in a cold region but wants to grow plants that need warm temperatures.

a Draw a diagram of a suitable greenhouse he can build to trap the heat entering.

b How is heat prevented from leaving the greenhouse by convection?

c Describe how the frequency of the radiation from the Sun differs from that emitted by the plants.

d What causes the frequencies mentioned in **c** above to be different?

e Briefly explain how the walls of the greenhouse act as a heat trap.

f Name THREE gases in the atmosphere that play a similar role to the walls of a greenhouse.

g What is the atmospheric phenomenon related to the greenhouse known as?

h Briefly describe the phenomenon referred to in **g** above.

12 Figure 14.20 shows the design of Sammy's solar water heater.

a Describe the features required of component A.

b State the material from which component B should be made.

c State a suitable material from which C could be made.

d Explain the function of EACH of the following:

   i A   **ii** B   **iii** C   **iv** the pump.

e Sammy finds that the he can never get hot water early in the morning.

   i State TWO adjustments he can make to the storage tank to overcome this problem.

   ii Explain why these adjustments should remedy the problem.

f How can Sammy rearrange the heater and storage tank so that he would no longer have to use the electric pump?

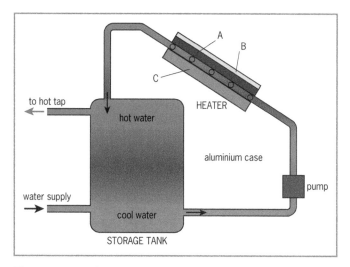

Figure 14.20 *Question 12*

Section C
Waves and optics

15 Light rays and rectilinear propagation

Luminous objects **emit** light. When we shine a powerful torch into the night sky, its direction of travel is clearly visible. The path in which the light energy propagates is known as a **ray**, and rays will always travel in straight lines **through a given medium** unless they pass through an extremely small gap (as discussed in Chapter 19).

Non-luminous objects, for example the Moon and trees, **reflect** light emitted by luminous objects.

Learning objectives

- Define a **ray** and a **beam** of light.
- Demonstrate the **rectilinear propagation** of light within a given medium.
- Demonstrate the formation of **shadows** produced from **point sources** and from **extended sources** of light.
- Calculate the **magnification** of **shadows**.
- Describe and explain the formation of **lunar eclipses** and **solar eclipses**.
- Describe and explain the function of the **pinhole camera**.
- Calculate the **magnification** produced by a **pinhole camera**.

*A **ray** of light is the direction in which light propagates.*

*A **beam** of light is a stream of light energy.*

Experiment to demonstrate that light travels in a straight line

The three sheets of card shown in Figure 15.1 are mounted vertically by small chunks of plasticine so that the **holes** at their centres are **aligned**. Light from a source placed at one end of the arrangement can be seen from the other end by an observer. If any of the three sheets of card are **slightly shifted**, the observer will no longer see light through the holes. This proves that light travels in a straight line within a given medium; a phenomenon known as **rectilinear propagation**.

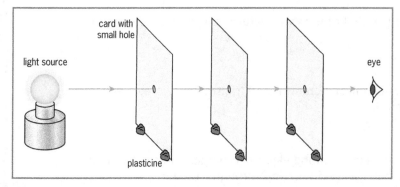

Figure 15.1 *Demonstrating the rectilinear propagation of light*

Shadows

The rectilinear propagation of light is responsible for the formation of shadows, including eclipses of the Moon and of the Sun.

Figure 15.2 shows a shadow cast by a cat. Note that the light reflects from and is absorbed by the cat, and there is no straight path for it to reach the shaded region.

The section that follows shows how light sources of different sizes placed in front of the same object produce shadows with different characteristics.

Figure 15.2 Shadow produced by rectilinear propagation of light

Shadow produced by a point source of light

A close approximation to a point source of light can be obtained by enclosing a lamp within an opaque box which has a **small hole** in one of its faces. The small hole acts as the point source. Figure 15.3 shows that a **distinct** shadow is produced known as the **umbra**. It is of **uniform obscurity** and has a **sharp edge**.

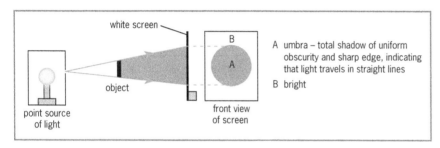

Figure 15.3 Shadow produced by a point source of light

Magnification of shadow produced by a point source of light

$$\text{magnification} = \frac{\text{size of shadow}}{\text{size of object}} = \frac{\text{distance of shadow from light source}}{\text{distance of object from light source}}$$

Shadow produced by an extended source of light source of light

An extended source of light can be obtained by enclosing a lamp within an opaque box with a **large hole** of approximate diameter 2 cm in one of its faces. The large hole acts as the extended source and can be interpreted as several point sources, each casting a shadow that overlaps with the shadows from the other neighbouring 'point' sources. Figure 15.4 shows the shadows produced by an extended source. Surrounding the umbra is a region known as the **penumbra**, where the shadow is partial and **gradually** changes from **total obscurity** to **total brightness**.

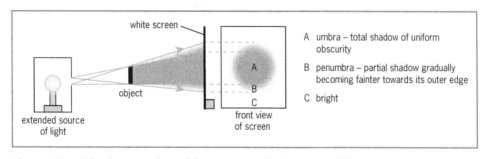

Figure 15.4 Shadow produced by an extended source of light

Eclipse of the Moon (lunar eclipse)

The Moon is a **non-luminous** body and is therefore seen from the Earth by **reflection** of light **from the Sun**. The orbit of the Moon generally passes outside the Earth's shadow, but at times it passes through the cone of the shadow, as shown in Figure 15.5. An eclipse of the Moon occurs only when there is a **full moon**, that is, when the Sun and Moon are on opposite sides of the Earth. As the Moon enters the Earth's umbra, light from the Sun can no longer reach it to be reflected to Earth, and a lunar eclipse occurs. It can take more than 1½ hours before it emerges from the other side of the umbra.

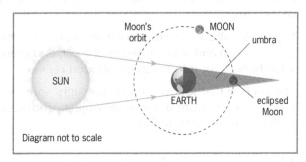

Figure 15.5 *Eclipse of the Moon*

Eclipse of the Sun (solar eclipse)

At times the orbit of the Moon can pass directly through the line joining the Sun to the Earth as in Figure 15.6. When this occurs, the Moon's umbra reaches the Earth's surface over a small region from which a **total solar eclipse** can be observed. People in the penumbra will observe a **partial eclipse**.

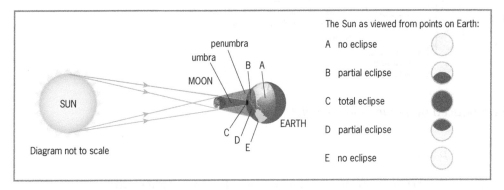

Figure 15.6 *Eclipse of the Sun*

Figure 15.7 illustrates that at times the Moon's **umbra does not reach the Earth**. An annular eclipse is then observed when viewed from the location A on Earth. A **bright ring** is seen surrounding the obscure shadow of the Sun.

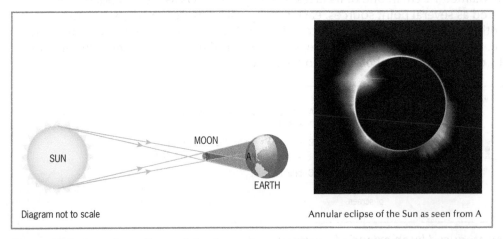

Figure 15.7 *Annular eclipse of the Sun*

The pinhole camera

Construction

A pinhole camera is shown in Figure 15.8. It may be constructed as follows.

- The **inner walls** of a cardboard box are painted **black** or lined with black paper.
- The back face of the box is removed and a sheet of **tracing paper** is stuck over the opening to act as a screen for the image.
- To obtain a **permanent image**, the back of the box should instead be fitted with a sheet of **photographic paper (film)** to its inner wall.
- A small opening is made near the middle of the front face of the box and a piece of black paper with a **small hole** at its centre is stuck over the opening.
- All openings, excluding the small hole at the front of the box, are sealed with opaque tape.

Figure 15.8 *A pinhole camera*

Function

Since light travels in a straight line, rays from any point on the object have only one possible path into the box, and consequently only one point at which they meet the screen. Figure 15.9 shows that a **focused**, **real**, **inverted** image is therefore **always** produced.

Decreasing the **size** of the **hole increases** the **sharpness** of the **image** up to a certain point. If reduced further, the image becomes less focused due to the phenomenon of diffraction (explained in Chapter 19).

An observer can view the image from behind the translucent screen. The image is **faint** due to the **low intensity** of light entering the camera through the **small opening**.

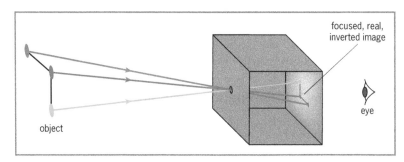

Figure 15.9 *Formation of image by pinhole camera*

Disadvantages

- A pinhole camera fitted with film **cannot take instant photos** (snapshots) since the intensity of light entering through the small hole is not sufficient. It can however be used for photographing **stationary scenes** such as **landscapes**.
- Due to the **low rate of light energy entering** the camera, it can take several minutes or **even hours** to produce a suitable image on the film.

*Effect of changing the depth of the box, the distance of the object
from the box and the size of the hole*

Figure 15.10a shows the formation of an image in a pinhole camera and Figures 15.10b, c and d show the effect on the image produced by making different changes to the camera or object. Note that the front face of each box is aligned so that you may relate each of diagrams b, c and d to a.

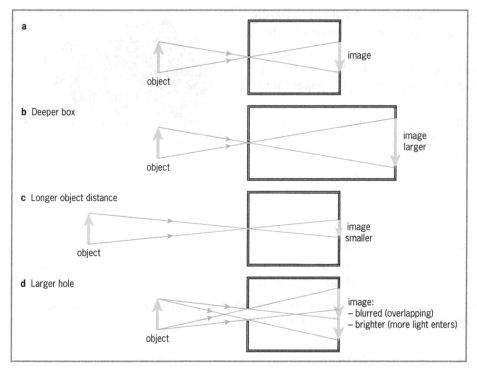

Figure 15.10 a *Formation of the image in a pinhole camera* **b** *The effect of changing the depth of the box* **c** *The effect of changing the distance of the object from the box* **d** *The effect of changing the size of the hole*

- On increasing the depth of the box, the image becomes larger because the rays spread more before reaching the screen (see Figure 15.10b).
- On increasing the distance of the object from the box, the image becomes smaller because the rays enter at a smaller angle (see Figure 15.10c).
- On increasing the size of the hole in the front of the box, the image becomes:
  ○ blurred due to the formation of overlapping images
  ○ brighter due to the increased intensity of the light entering the box (see Figure 15.10d).

Magnification of image produced in a pinhole camera

$$\text{magnification} = \frac{\text{size of image}}{\text{size of object}} = \frac{\text{distance of image from pinhole}}{\text{distance of object from pinhole}}$$

Recalling facts

1 a Distinguish between a ray of light and a beam of light.

   b What is meant by the phrase *rectilinear propagation of light*?

2 a Describe how an eclipse of the Moon occurs.

   b Why can the duration of an eclipse of the Moon be as much as 1½ hours?

   c Does an eclipse of the Moon occur during a new moon, a first quarter moon or a full moon?

3 a What are the relative positions of the Earth, Moon and Sun during a solar eclipse?

   b Why is it that as certain regions on Earth experience a total solar eclipse, other regions experience a partial solar eclipse and some regions experience no eclipse?

4 a Sketch a labelled diagram of a pinhole camera which produces an image on photographic paper.

   b Why is it that the pinhole camera can only be used to take photographs of stationary objects?

   c Explain why a pinhole camera always produces a real, focused image of a stationary object.

Applying facts

5 Draw a well-labelled ray diagram to show the formation of the shadow cast on the wall by the object shown in Figure 15.11. The light source is a long tube.

6 An opaque disc of diameter 5.0 cm is placed 40 cm to the right of a point source of light. Light from the source reaches the centre of the disc perpendicularly.

   a Draw a diagram to show the shadow produced on a vertical white screen placed 20 cm to the right of the disc.

   b Calculate the diameter of the shadow.

   c Describe the obscurity of the shadow.

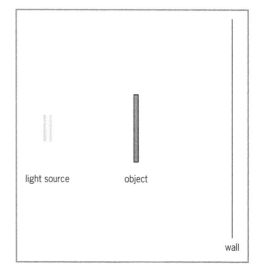

light source   object

wall

Figure 15.11 *Question 5*

7 a Jenna makes a pinhole camera at school. She is told that she has made the hole in the front of the camera too large. With the aid of a diagram, explain what effect/s this will have on the image.

   b Zoe helps her to make another pinhole camera, this time with a small 'pinhole' opening. She sets up the camera with a sheet of photographic paper on its back inner wall and they use the camera to take a photograph of a beautiful countryside landscape. What can be said about the time necessary to print an adequate image? Give a reason for your answer.

8 A pinhole camera is constructed with its back wall made of tracing paper. A pencil is held vertically in front of the camera and the image on its back wall is observed from behind the camera.

  a Describe the image.

  b Draw a diagram to show how the image changes if the pencil is moved closer to the box.

9 The distance between the screen and pinhole of a pinhole camera is 30 cm. A tree 4.0 m tall stands 15 m from the front face of the camera. Determine:

  a the magnification of the image

  b the height of the image.

16 Reflection and refraction

In the last chapter we learnt that we see the Moon because light from the Sun reflects from its surface to Earth. Indeed, that is how it is possible to see the trees, buildings, our books and any other **non-luminous** object around us. During **reflection**, light changes direction as it rebounds from a surface.

White light is comprised of **several colours**. When a non-luminous object appears to be a particular colour, it is because it is reflecting that colour and absorbing all the other colours. A green leaf reflects green light and a red rose reflects red light. A truly black object absorbs all the colours and therefore reflects no light. An excellent reflector, such as a mirror, reflects all the colours.

Light can also change direction as it enters a second medium. The process is known as **refraction** and is why a pencil, partly submerged in a glass of water, appears crooked.

Reflection

Learning objectives

- State and apply the **laws of reflection**.
- Experimentally **verify** that the **angle of incidence** is **equal** to the **angle of reflection**.
- State the **characteristics of images** formed in **plane mirrors**.
- Construct **ray diagrams** illustrating **reflection**.
- Locate **virtual** images experimentally using a method of **no-parallax**.

Laws of reflection

Law 1 – The incident ray, the reflected ray and the normal, at the point of incidence, are on the same plane.

Law 2 – The angle of incidence is equal to the angle of reflection (Figure 16.1).

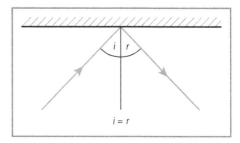

Figure 16.1 *Laws of reflection*

Experiment using a ray box to verify that the angle of incidence is equal to the angle of reflection

A **ray box** is a small light source which can provide various types of light beams. Figure 16.2 shows the path of a **narrow beam** from such a source as it reflects from a plane mirror.

An experiment to verify the second law of reflection using a narrow beam from a ray box is outlined below.

- Secure a large sheet of paper to the desktop using adhesive tape.
- Rule a line across its centre and construct a line NO to act as a normal as shown in Figure 16.3.
- Construct six lines on the paper from point O to act as rays with incident angles ranging from about 10° to about 60°. Mark each of the lines sequentially, 1 through 6.

Figure 16.2 *Light beam reflecting from plane mirror*

- Place a plane mirror to stand with its reflecting surface on the line drawn across the centre of the page.
- Dim the lighting in the room and direct a ray from the ray box along one of the lines.
- Mark the position of the reflected ray by two small crosses far apart on the paper, labelling each cross with the number of the corresponding incident ray.
- Repeat the procedure for the other incident lines.
- Remove the mirror and draw straight lines from O through each pair of crosses labelled with the same number. These lines represent the reflected rays.

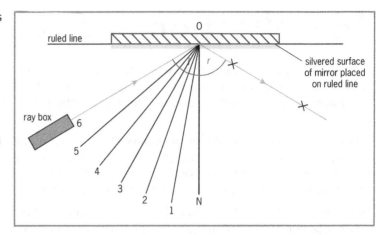

Figure 16.3 *Experiment using a ray box to verify that the angle of incidence is equal to the angle of reflection*

- Measure and record the angles of incidence, *i*, and reflection, *r*, for each pair of rays with the same number.

It will be observed that, within the limits of experimental error, the angle of incidence is always equal to the angle of reflection.

Experiment using pins to verify that the angle of incidence is equal to the angle of reflection

- Rule a line across the centre of a large sheet of paper and construct a line NO to act as a normal.
- Construct six lines on the paper from point O to represent rays with incident angles ranging from about 10° to about 60°. Mark each of the lines sequentially, 1 through 6.
- Secure the paper to a **pinboard** using thumb tacks and place a plane mirror to stand with its reflecting surface on the line drawn across the centre of the paper.
- Insert two large pins into the pinboard, **vertically** and **far apart** on one of the lines to act as objects.
- By viewing from a position as shown in Figure 16.4, insert two **search pins**, vertically and far apart so that they are aligned with the **images** of the two **object pins**.
- Mark the positions of the search pins by **small crosses** on the paper, labelling each cross by the number of the corresponding incident ray.
- Repeat the procedure for the other incident lines.
- Remove the mirror and draw straight lines from O through each pair of crosses labelled with the same number. These lines represent the reflected rays.
- Measure and record the **angles** of **incidence**, *i*, and **reflection**, *r*, for each pair of rays with the same number.

It will be observed that, within the limits of experimental error, the angle of incidence is always equal to the angle of reflection.

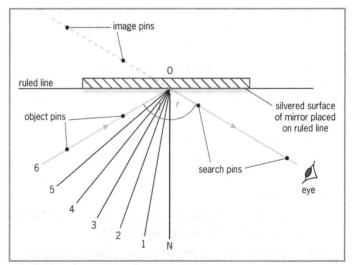

Figure 16.4 *Experiment using pins to verify that the angle of incidence is equal to the angle of reflection*

Characteristics of the image formed in a plane mirror

The image is:

- the **same size** as the object
- the **same distance** perpendicularly behind the mirror as the object is in front of it
- **virtual**
- **erect**
- **laterally inverted**.

Virtual – not real/apparent/cannot be formed on a screen/light is not where the image appears to be.

Laterally inverted – the left and right sides of the image are reversed relative to the object. Figure 16.5 shows that if you raise your right hand in front of a mirror, your image will appear to be raising its left hand. Similarly, the image of the word PHYSICS placed in front of the mirror is reversed.

Figure 16.5 *Lateral inversion in a plane mirror*

Reflection ray diagrams using point objects

The following method describes how these diagrams (see Figure 16.6) may be drawn.

- Draw a line to represent the mirror and place a small dot to represent the point object in its correct position to scale on the diagram.
- Place the point image on the diagram the same distance perpendicularly behind the mirror as the object is in front. A faint construction line, originating at the object and passing perpendicularly through the mirror, is useful here.
- Draw an eye at the location from which the image is viewed.
- Draw two rays from the **image to the edges of the eye** (the lines behind the mirror should be broken since they represent virtual rays).
- Draw two more rays, this time **from the object, to where the first two rays intersect the mirror.**

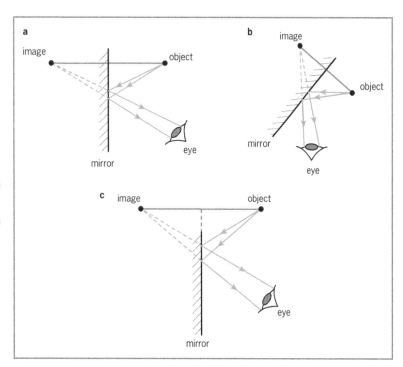

Figure 16.6 *Reflection ray diagrams*

Notes:

- The construction line between the object and image is particularly useful in Figures 16.6b and c.
- Figure 16.6c shows that the object does not have to be perpendicularly in front of the mirror to create an image. In this case the mirror line is extended upward so that the construction line between the object and image passes perpendicular to it.
- Figure 16.7 shows that if **straight lines** from the image to the eye do not intersect the mirror, then the eye cannot view the image from that position.
- A simpler reflection ray diagram using only one ray from the point object can be drawn as shown in Figure 16.8. In this case, the eye can be drawn much smaller.

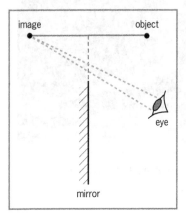

Figure 16.7 *Image not seen from this position*

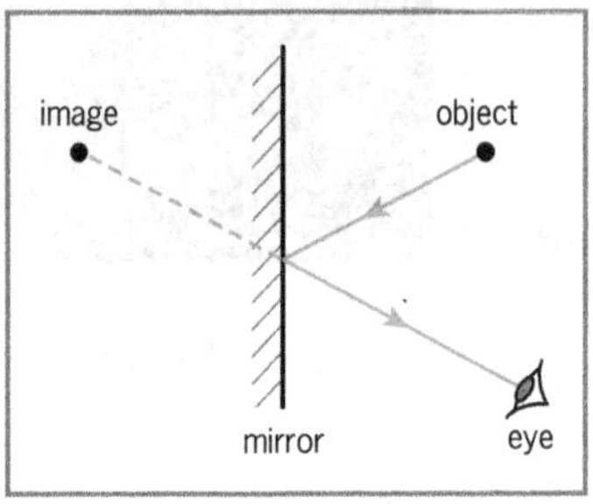

Figure 16.8 *Single ray reflection*

Reflection from inclined mirrors

Figure 16.9 illustrates that an object can produce multiple images in mirrors inclined to each other.

Figure 16.10 shows an eye viewing an image I_{AB} caused by reflection from two mirrors. Object O is in front of mirror A and mirror B and therefore forms **primary images** I_A and I_B, respectively. However, I_A is in front of mirror B and I_B is in front of mirror A, and they form overlapping **secondary images**, I_{AB} and I_{BA}. The eye sees these overlapping images after two reflections.

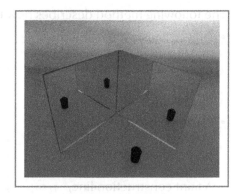

Figure 16.9 *Multiple reflections caused by inclined mirrors*

Drawing the rays:

- Start by drawing two rays **from the secondary image being viewed (I_{AB}) to the eye**.
- This reflection is in mirror B, so the object creating it must be in front of mirror B. This is the virtual object I_A. Two rays should then be drawn from I_A to where the first two rays intersected mirror B.
- From the direction of the rays leaving A we see that I_A is the image of the real object O. Two rays should be drawn from O to where the rays leave mirror A.

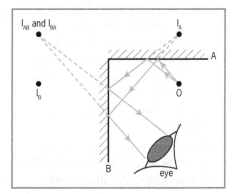

Figure 16.10 *Reflection from inclined mirrors*

- Join each incident ray to its corresponding emergent ray through the outline of the block.
- Measure and record the angles of incidence, i, and the angles of refraction, r, at the point where the rays enter the block.
- Record the corresponding values of sin i and sin r.
- Plot a graph of sin i against sin r. The **gradient** (slope) of the graph is the **refractive index** of the material of the block.

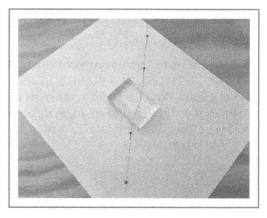

Figure 16.18 *Aligning pins as seen through a glass block*

Deviation of light on passing from one medium to another

When light enters a second medium perpendicular to its interface (Figure 16.19a):

- it does not deviate – the angles of incidence and refraction are both zero
- it travels slower in the more optically dense medium.

When light enters a more optically dense medium other than perpendicularly (Figure 16.19b):

- it refracts towards the normal – the angle between the ray and the normal decreases
- its speed decreases.

When light enters a less optically dense medium other than perpendicularly (Figure 16.19c):

- it refracts away from the normal – the angle between the ray and the normal increases
- its speed increases.

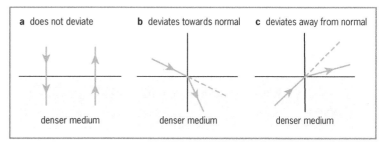

a does not deviate **b** deviates towards normal **c** deviates away from normal

denser medium denser medium denser medium

Figure 16.19 *Deviation of light on passing between media of different optical densities*

Light passing through a rectangular glass block

As light passes through the glass block shown in Figure 16.20 it deviates through an angle, d, in a **clockwise** direction, and then through the same angle in an **anticlockwise** direction. The **total deviation**, D, is therefore zero. There is, however, a **net lateral displacement**, s; the greater the width, w, of the block, the greater is the lateral displacement.

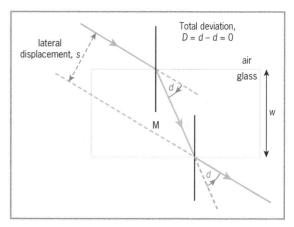

Figure 16.20 *Light through a rectangular glass block*

Total deviation, $D = d - d = 0$

lateral displacement, s

air
glass

M

w

Light passing through a triangular glass prism

As light enters the triangular glass prism shown in Figure 16.21 it deviates through an angle d_1 in a **clockwise** direction, and as it leaves the prism it deviates through an angle d_2, again in a **clockwise** direction. The total deviation is therefore $d_1 + d_2 = D$ (clockwise).

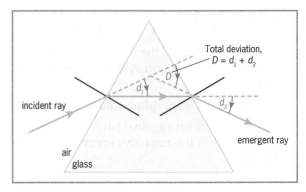

Figure 16.21 *Light through triangular glass prism*

Refraction ray diagrams for light travelling from a denser medium to a less dense medium

Figure 16.22 shows that the refraction of light can cause the object viewed to appear distorted. This can be explained by drawing diagrams illustrating the refraction of rays, as follows (see Figure 16.23).

- Draw a line to represent the interface between the media and place a small dot to represent the point object in its correct position.

- Place the point image in the diagram. An estimate will be sufficient here unless the exact position is given or is required by calculation. The image is closer to the surface than is the object.

- Draw an eye at the location from which the image is viewed.

Figure 16.22 *Straight rod appears crooked*

- Draw two rays from the **image to** the edges of the **eye** (using broken lines before they reach the interface).

- Draw two more rays, this time from the **object to** where the **first two rays intersect the interface**.

- Figure 16.23c shows how the diagram may be altered to show why a straight stick, partially submerged in water, appears crooked when viewed from above the surface.

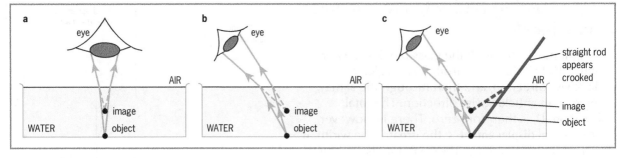

Figure 16.23 *Refraction ray diagrams*

Apparent depth of an object viewed from a less dense medium

Figure 16.23 illustrates that the image is closer to the water surface than is the object. When viewed from the air **vertically above** the object as in Figure 16.23a, the ratio of the real depth to the apparent depth gives the refractive index of the denser medium (water) relative to the less dense medium (air):

$$\frac{\eta_w}{\eta_a} = \frac{\text{real depth}}{\text{apparent depth}}$$

Since the refractive index of air is 1:

$$\eta_w = \frac{\text{real depth}}{\text{apparent depth}}$$

Dispersion of white light

*The **dispersion** of white light is the separation of white light into its constituent colours.*

Figure 16.24 shows how **rainbows** are formed due to the dispersion occurring in suspended **water droplets** in the sky. The light is first **refracted on entering** the droplet, then **internally reflected** (see Chapter 17) as it meets its edge, and finally **refracted again on leaving** the droplet.

Figure 16.25 shows the dispersion of white light by a glass prism.

The phenomenon of the dispersion of white light was first explained by Sir Isaac Newton.

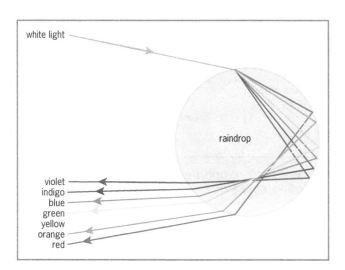

Figure 16.24 *Dispersion of white light by a raindrop*

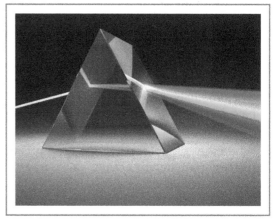

Figure 16.25 *Dispersion of white light produced by a triangular glass prism*

Dispersion and Newton's associated experiments

Sir Isaac Newton used the arrangement shown in Figure 16.26 to demonstrate the dispersion of white light into its constituent colours. A **narrow** beam of **white** light emerging from a **double slit collimator** is incident on a glass prism. On entering the prism, the beam separates into beams of different colours. The speed of the light of each colour is reduced by a different extent, and each colour deviates through a different angle. The **speed** of **red** light is **reduced least**, and that of **violet, most,** and this results in **red** light **deviating least**, and **violet most**. Light emerging from the prism is incident on a white screen where it forms a visible spectrum of the sequenced colours: red, orange, yellow, green, blue, indigo, violet (**ROYGBIV**).

Note: The beam of white light incident on the prism **must be narrow**. A **wider beam** can be considered as several narrow beams, each producing their own **overlapping spectra, recombining the colours.**

Newton's approach is to be commended since he **experimented** rather than simply **theorised.**

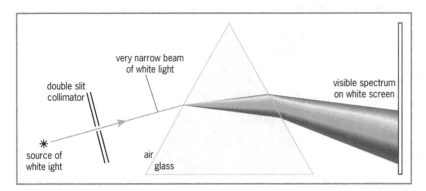

Figure 16.26 *Dispersion of white light*

He made several adjustments to the experiment to prove that the **origin** of the coloured light was **not within the prism.** Some of these are outlined below.

- When prisms of different transparent materials were used, the same colours were incident on the screen. This proved that **the colours were not of the material of the prism.**

- When light of one colour was incident on a prism (Figure 16.27) there was no change in the colour incident on the screen, indicating that properties of the glass were not being transferred to the light. This also shows that white light is a **physical mixture** of several different colours, and once those colours are separated, they cannot be further subdivided. (For simplicity, only the red and violet rays have been shown refracting through the first prism.)

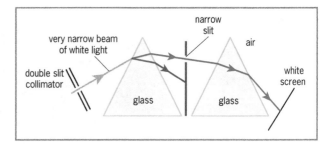

Figure 16.27 *Properties of glass not affecting colour of light*

- When the colours separated by one prism were recombined by adding a similar but inverted prism, as shown in Figure 16.28, white light emerged from the second prism. (For simplicity, only the red and violet rays have been shown). Since white light can be separated into various colours without chemical reaction, and reproduced by simple combination of those colours, it must be a physical mixture.

Note: the incident beam used here is not narrow as were the beams in Figures 16.26 and 16.27. The emergent beam is comprised of several overlapping spectra of different colours which are laterally displaced and parallel, and which superpose on each other to produce white light.

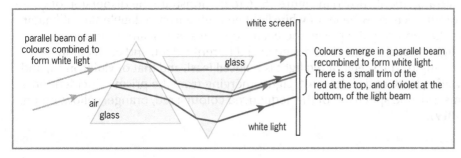

Figure 16.28 *Recombining the colours*

Recalling facts

1   **a**   Sketch a diagram to illustrate what is meant by the angle of incidence and the angle of reflection.

     b   State the laws of reflection.

2   A book is held in front of a plane mirror. Describe FIVE features of the image of the book.

3   Explain how the roof of a house can appear red when white light from the Sun is incident on it.

4   **a**   How do the following change when light travels to a medium of greater refractive index?

       i   The speed.

       ii   The angle between the ray and the normal.

     b   Under what condition can light travel to a different medium and not be deviated?

5   Sketch a diagram to show a ray of monochromatic light obliquely entering a triangular transparent prism and passing through the prism. Mark on the diagram:

     a   the angle D representing the total deviation

     b   the angles of deviation d_1 and d_2 at the interfaces, as the ray enters and leaves the prism.

6   **a**   Sketch a ray diagram illustrating how a visible spectrum can be produced.

     b   Name the scientist who first carried out this experiment.

     c   Sketch a diagram to show how the apparatus can be modified to reproduce white light.

     d   What is the splitting of white light into its constituent colours known as?

Applying facts

7   **a**   A ray of light from a point object 5 cm in front of a plane mirror is incident on the mirror at an angle of 20°. Sketch a diagram showing how the incident ray is reflected. Include in your diagram the location of the object and image and the angles of incidence and reflection.

     b   What would be the angle between the incident ray and reflected ray if the angle of incidence is increased by 15°?

8   Redraw and complete Figure 16.29 using TWO rays from the point object to show how the eye views the image.

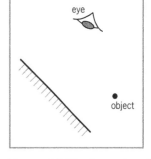

Figure 16.29 *Question 8*

9 Redraw Figure 16.30 (not to scale) and complete the path of the ray until it reflects from mirror M. Draw a normal at each point of reflection and indicate in the diagram the values of all angles of incidence and reflection.

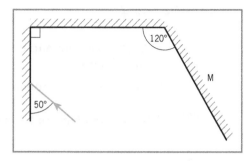

10 The word "**catch**" is printed on a piece of paper and held up in front of a mirror. Show how the image will appear in the mirror.

Figure 16.30 *Question 9*

11 a Jason views a marble which has fallen into a pond. Draw a ray diagram showing TWO rays from the marble to illustrate how he sees it from the air **vertically above**.

   b Given that the depth of the pond is 50 cm and that the refractive index of the water is 1.3, determine the distance between the image of the marble and the floor of the pond.

12 a Redraw and complete the path of the ray through the glass block shown in Figure 16.31, marking the angles of deviation at each surface with '*d*'.

   b What is the total deviation as it passes through the block if it deviates clockwise by 7° on entering its upper surface?

   c With the aid of construction lines, mark on the diagram the lateral displacement, *x*, of the ray.

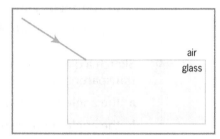

Figure 16.31 *Question 12*

13 A ray travelling in a transparent liquid approaches an interface with air at an angle of incidence of 45° and refracts into the air at an angle of 60°. Given that the speed of light in air is 3.0×10^8 m s^{-1}, calculate the speed of light in the liquid.

14 a Use Table 16.1 (page 178) to determine the value of *x* in each of the following situations shown in Figure 16.32.

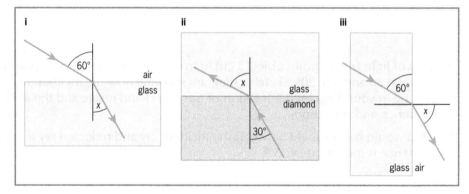

Figure 16.32 *Question 14*

   b Use Table 16.1 to determine the speed of light in diamond, if the speed of light in air is 3.0×10^8 m s^{-1}.

15 Figure 16.33 shows rays in air incident on three glass prisms. Redraw the diagrams and complete the rays through them given that total internal reflection does not occur. Construct a normal at each point of refraction except where the ray meets the glass perpendicularly.

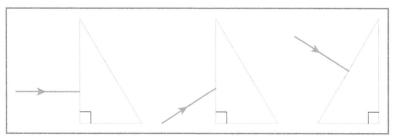

Figure 16.33 *Question 15*

Analysing data

16 Table 16.2 shows the results of an experiment carried out to determine the refractive index of peridot.

Table 16.2 *Angles of incidence and refraction for rays entering peridot*

Angle of incidence, i	60.0	50.0	40.0	30.0	20.0	10.0
Angle of refraction, r	30.5	28.0	22.0	16.0	12.0	6.0

a Redraw and complete the table by adding rows to include values for sin i and sin r for each value of i and r.

b Plot a graph of sin i against sin r.

c When sin i = 0.71, determine the corresponding values of:

  i sin r

  ii i

  iii r

d Determine the slope of the graph.

e Deduce the refractive index of peridot.

17 Critical angle and total internal reflection

Light travelling to a medium of **lesser optical density** will not always refract into it. If the angle of incidence is greater than a certain critical angle, the light energy will be totally reflected. It is the reason diamonds sparkle, and the phenomenon is utilised in devices such as periscopes, binoculars, telecommunications systems and in optical instruments used in medicine.

Learning objectives

- Define and explain **critical angle** and **total internal reflection**.
- Experimentally **determine** the **critical angle** of glass.
- Use **diagrams to illustrate** critical angle and total internal reflection.
- **Solve problems** involving critical angle and total internal reflection.
- Be familiar with the use of total internal reflection in **fibre optic cables** and **reflecting prisms**.

Critical angle and total internal reflection

Figure 17.1 shows that light directed towards a medium of lesser optical density (smaller refractive index) at small angles of incidence is **partly reflected** and **partly refracted**. The reflected ray is weak, and the refracted ray is strong. At larger angles of incidence, the angle of refraction increases and the reflected ray increases in strength. At a certain **critical angle, c**, the refracted ray grazes along the interface. This is the maximum angle of incidence for which there can be refraction into the medium of lesser density. At angles of incidence **greater than c**, there is **total internal reflection**.

Figure 17.2 shows a laser beam in water reaching the surface at an angle of incidence greater than the critical angle and being totally internally reflected.

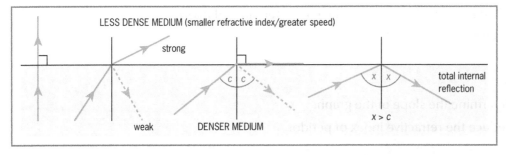

Figure 17.1 *Critical angle and total internal reflection for angles greater than c*

*The **critical angle** of a material is the **largest** angle at which a ray can approach an interface with a medium of lesser refractive index and be **refracted into it**.*

or

*The **critical angle** of a material is the **smallest** angle at which a ray can approach an interface with a medium of lesser refractive index and be **totally internally reflected by it**.*

Total internal reflection occurs at the interface between two transparent media when a wave travelling in the medium of greater refractive index approaches the other medium at an angle greater than the critical angle.

Figure 17.2 *Total internal reflection from a water surface*

- The ray must approach the second medium from one of greater refractive index, that is, from one in which its speed is less. For **light** waves the medium of greater refractive index (slower wave speed) is the **denser** medium.
- The angle of approach must be greater than the critical angle.

Experiment to determine the critical angle of glass

- Place a semicircular glass block on a piece of paper and outline its perimeter with a pencil.
- Remove the block and place a small dot at the centre of the diameter of the outline. Construct a normal though the dot as shown in Figure 17.3.
- With the lights in the room dimmed, direct a ray towards the dot, perpendicular to the curved surface of the glass block and at a large angle with the normal. The ray will be directed along the radius of the semicircle to the centre of curvature of the block and be totally internally reflected.
- Decrease the angle of incidence until a refracted ray just becomes visible along the straight edge of the block. The angle of incidence is then the critical angle.
- Mark the paths of the rays with small crosses, placed far apart.

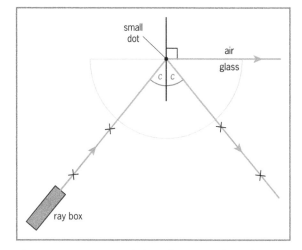

Figure 17.3 *Determining the critical angle of glass*

- Remove the block and draw the paths of the ray through the crosses to the dot on the normal.
- Measure the angle between the incident and reflected rays and divide the result by 2 to obtain the critical angle, c.

The mirage

Solid surfaces, such as the black tar of a road or the sand and stone in deserts, can reach high temperatures during the day as they are heated by the Sun's rays. Air in contact with the land becomes heated by conduction from it and becomes less dense as it expands. If the surrounding region is also hot, the less dense air has difficulty in rising since there is no nearby cooler (denser) air to take its place.

A ray of light from low in the sky will continuously refract away from the normal as it passes through layers of hotter, less dense air, of lesser refractive index. The refraction continues until the ray reflects upward just above the surface and then continuously refracts towards the normal as it enters the cooler, denser, air above. An observer receiving this ray will see a virtual image of the sky and may interpret it as a pool of water as shown in Figure 17.4.

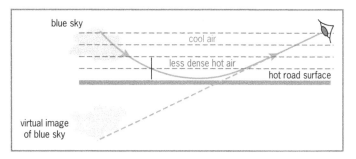

Figure 17.4 *The mirage*

View from above a water surface

A lamp is placed on the level floor of a pond at night as shown in Figure 17.5. An observer above the surface of the pond sees a glowing disc of diameter *d*, but beyond the edge of the disc there is darkness since rays from the lamp reach the water surface at angles greater than the critical angle (*x* > *c* and *y* > *c*) and are totally internally reflected. The reflected rays then continuously reach the floor and water surfaces with the same angle of incidence and are repeatedly reflected.

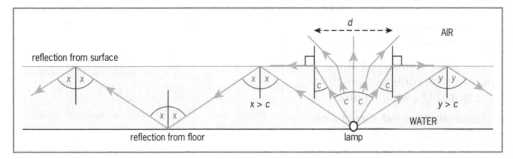

Figure 17.5 *View from above a water surface*

View from below a water surface – the fish's view

Figure 17.6 shows light from above a water surface entering the eye of a fish. As the rays enter the denser medium they refract towards the normal. For an angle of incidence of 90°, the angle of refraction is the critical angle. Light travelling in the water and approaching the surface at an angle greater than the critical angle undergoes total internal reflection. The fish:

- obtains a view of range **180° above the water** surface by rotating its eye through just **98° (twice the critical angle)**

- can view objects below its eye by looking upwards to receive light reflected from the water surface at angles **greater than the critical angle**.

Figure 17.7 shows the reflection seen in the water–air interface by an observer in front of an aquarium.

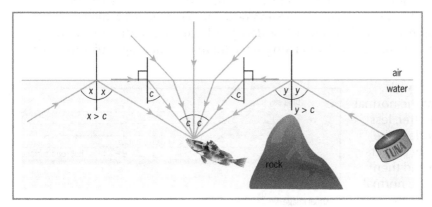

Figure 17.6 *View from below a water surface – the fish's view*

Figure 17.7 *Reflections viewed from below the water surface*

Calculations involving critical angle

In the previous chapter we learnt that the following relation can be applied when a ray travels from one medium to another:

$$_1\eta_2 = \frac{\eta_2}{\eta_1} = \frac{v_1}{v_2} = \frac{\sin\theta_1}{\sin\theta_2}$$

This can be applied to a situation where a ray approaches a second medium at the critical angle. The refracted angle is then 90° as shown in Figure 17.8, and therefore:

$$\frac{\eta_2}{\eta_1} = \frac{v_1}{v_2} = \frac{\sin c}{\sin 90°}$$

$$\frac{\eta_2}{\eta_1} = \frac{v_1}{v_2} = \sin c \ \dots \ (\text{since } \sin 90° = 1)$$

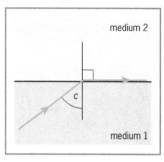

Figure 17.8 *Calculation involving the critical angle*

Notes for tackling problems

- In this last equation, since **sin c is always less than 1**, the **numerator is less than the denominator** of each ratio.

- It is useful to **sketch a diagram** showing the **medium that contains c**. The critical angle is always in the medium of **greater refractive index**. For light rays this is the **denser** medium (the medium in which the light travels **slower**).

- If the critical angle or refractive index of a medium is referred to without mention of a second medium, it can be assumed that the second medium is air or a vacuum.

Relation between critical angle and refractive index if one of the media is air or a vacuum

Consider the situation shown in Figure 17.9.

$$\frac{\eta_a}{\eta_x} = \frac{\sin\theta_x}{\sin\theta_a}$$

$$\frac{1}{\eta_x} = \frac{\sin c}{\sin 90°} \quad (\textit{refractive index of air is 1})$$

$$\frac{1}{\eta_x} = \sin c \quad \dots \ (\text{since } \sin 90° = 1)$$

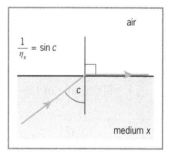

Figure 17.9 *Critical angle if one of the media is air or a vacuum*

Example 1

Given that the refractive index of ruby and water are 1.7 and 1.3 respectively, calculate the critical angle of ruby with respect to:

a air   **b** water.

a Since one of the media is air, the simpler relation can be used:

$$\sin c_r = \frac{1}{\eta_r}$$

$$\sin c_r = \frac{1}{1.7}$$

$$c_r = \sin^{-1}\left(\frac{1}{1.7}\right) = 36°$$

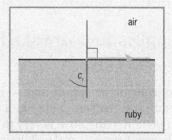

Figure 17.10 *Example 1a*

---

b **None of the media is air** so the general refraction formula may be used:

$$\frac{\sin \theta_r}{\sin \theta_w} = \frac{\eta_w}{\eta_r}$$

$$\frac{\sin c_r}{\sin 90°} = \frac{1.3}{1.7}$$

$$\frac{\sin c_r}{1} = \frac{1.3}{1.7} \quad \text{... } \sin c \text{ is found by placing the}$$
smaller value in the **numerator**

$$c_r = \sin^{-1}\left(\frac{1.3}{1.7}\right) = 50°$$

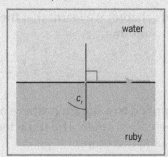

Figure 17.11 *Example 1b*

Example 2

Figure 17.12 shows light rays received by a fish that is 2.0 m below the surface of the sea. Calculate:

a the angle *x*

b the refractive index of the seawater

c the distance *y*.

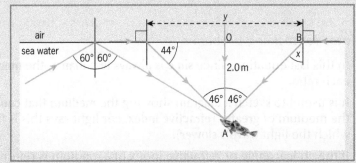

Figure 17.12 *Example 2*

a *x* is alternate to the angle marked 46°, therefore *x* is 46°.

b The refractive index of the seawater can be calculated using the critical angle. The angle, *x*, between the ray and the normal at the surface is the critical angle, i.e. the critical angle is 46°.

Since **one of the media is air**:

$$\eta = \frac{1}{\sin c} \qquad \eta = \frac{1}{\sin 46°} \qquad \eta = 1.4$$

Alternatively, we can apply the general equation:

$$\frac{\eta_w}{\eta_a} = \frac{\sin \theta_a}{\sin \theta_w} \qquad \frac{\eta_w}{1} = \frac{\sin 90°}{\sin 46°} \qquad \eta_w = \frac{1}{\sin 46°} \qquad \eta_w = 1.4$$

c Considering triangle AOB:

$$\tan 46° = \frac{OB}{2.0} \qquad \therefore OB = 2.0 \tan 46°$$

$$\therefore y = 2(2.0 \tan 46°) = 4.1 \text{ m}$$

Applications of total internal reflection

Optical fibres

Principle of the optical fibre – Light entering through one end of a thin, transparent, glass fibre is totally internally reflected as it reaches the inner walls of the fibre at an angle greater than its critical angle. Since the fibre is very thin, the reflected light travels only a very short distance before reaching the opposite wall and will be incident on it at approximately the same angle. It is then repeatedly reflected at that angle until it emerges from the other end of the fibre.

Since the fibre is very **thin** it is **flexible**.

Scratches on the outside of the fibre can cause light to exit through its sides. A **cladding** made of a **different type of glass** is placed around the core to protect it as shown in Figure 17.13. Since light travels through the core, scratches on the outside of the cladding have no effect on the reflections. For total internal reflection to occur, the **refractive index of the cladding must be less** than that of the core.

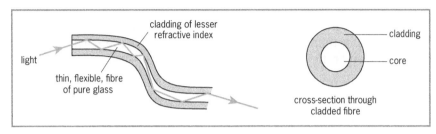

Figure 17.13 *An optical fibre*

Uses of optical fibres

- **Telecommunications** – Electronic communications for cable TV, telephone and the internet are mainly transmitted by pulses of electromagnetic radiation through fibre-optic cables. These cables can transfer information over **longer distances** and can provide **faster data transfer** (greater bandwidth) than can electrical cables made of copper.

- **Endoscopes**

    1. **Diagnostic imaging** – Light is transmitted to a cavity within the patient through an **incoherent bundle** of optical fibres in a medical device known as an **endoscope**. To prevent distortion of the image, the reflected light returns through a **coherent bundle** of optical fibres as shown in Figure 17.14. The image of the inside of the cavity can be viewed by the doctor through an eyepiece or can be collected by a digital camera which displays it on a screen, and which can save it for further enhancement by a computer.

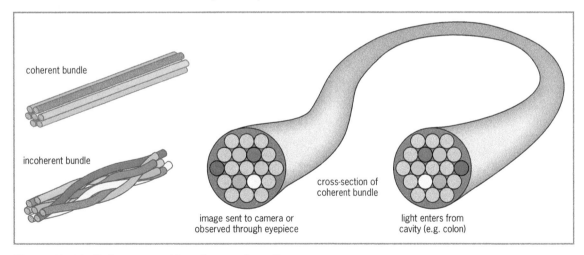

Figure 17.14 *Coherent and incoherent bundles*

    2. **Therapy** – Many types of surgery can be performed with the assistance of fibre-optic bundles, for example:
        - **polyps** may be removed from the colon during a **colonoscopy** by small cutting tools
        - **ligaments** can be reconstructed during **arthroscopy**
        - a **laser** beam may be directed along optical fibres to **destroy tumours** in solid organs, which are difficult to remove using a scalpel.

- **Fibrescopes**, like endoscopes, use fibre-optic cables for seeing into difficult places such as the inside of plumbing pipes or in the cavities of a car engine.

Reflecting light using right-angled isosceles glass prisms

Reflection of light using total internal reflection produces **stronger** and **better resolved** images than reflection of light using glass mirrors. Figure 17.15 shows how **right-angled isosceles glass prisms** can be used to reflect light.

Light entering such a prism perpendicularly through one of its sides meets the facing wall at an angle of incidence of **45°**. Since this is greater than the critical angle of glass (42°), there is **total internal reflection**. All angles of incidence and reflection are 45° in each of the prisms of Figure 17.15. The medium surrounding the prisms is air.

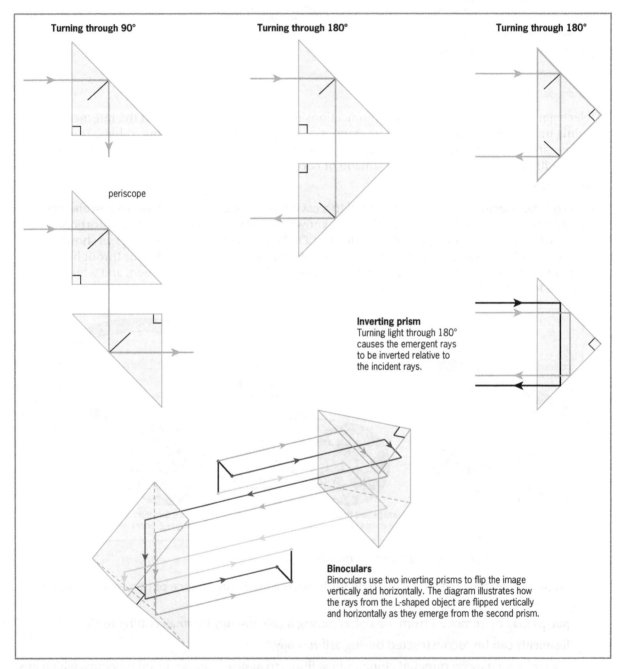

Figure 17.15 *Total internal reflection using right-angled isosceles glass prisms*

Sparkle of diamonds

Due to diamond's high refractive index of 2.42, it has a small critical angle of 24.5°. This increases the probability of total internal reflection of rays which refract into the gemstone. Specially cut diamonds seem to sparkle since light refracting into them can undergo more than one total internal reflection before it refracts back into the air in various directions from its differently angled faces.

The rays of Figure 17.16 each undergo two internal reflections within the diamond.

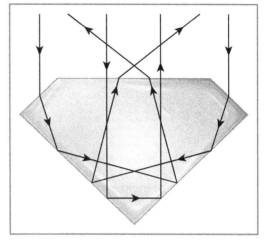

Figure 17.16 *Diamonds sparkle due to total internal reflection*

Reflecting light using glass mirrors

A glass mirror has a silver coating on its back surface. Figure 17.17 illustrates that when light from an object, **O**, is incident on such a mirror, a strong image, **I**, accompanied by multiple weak images are produced. Each time the light travelling in the glass reaches the interface with the air, there is a **partial refraction** and a **partial internal reflection**. The **thicker the mirror**, the **greater is the displacement** between the emergent rays, and so the image produced by a thick glass mirror is **not well resolved**.

Figure 17.17 *Reflecting light using glass mirrors*

Recalling facts

1　**a** Define critical angle.

　　b State the conditions necessary for total internal reflection to occur.

2　Sketch ray diagrams to show how each of the following may be achieved.

　　a TWO right-angled isosceles glass prisms being used to laterally displace a horizontal ray in a periscope.

　　b A single right-angled isosceles glass prism being used to invert TWO parallel rays as it turns them through 180°.

3　**a** Draw a labelled diagram of a ray passing through an optical fibre having a cladding.

　　b Why is the cladding important?

　　c Should the refractive index of the cladding be greater or lesser than that of the core?

---

4 **a** Explain how a fibre-optic bundle can carry light to a cavity in the body such as the stomach and produce an image for a doctor to view on a screen.

b Why have fibre-optic cables replaced the older copper electrical cables used in telecommunication networks such as the telephone, radio, television and the internet?

Applying facts

5 **a** Sketch a diagram of a ray in a solid transparent medium reaching an interface with the air at an angle less than its critical angle of 24°. Indicate a suitable value for the angle of incidence in your diagram.

b Add a ray to the diagram drawn in **a** which is totally internally reflected. Mark on the diagram a value for the incident angle that can cause such a reflection.

6 Calculate

a the critical angle of sapphire given that its refractive index is 1.8

b the refractive index of peridot given that its critical angle is 36°.

7 A ray of light travelling in glass towards ethanol refracts just about parallel to the interface. If the refractive index of glass and ethanol are 1.5 and 1.4 respectively, determine the critical angle of approach to the interface.

8 The speeds of light in diamond and air are 1.25×10^8 m s^{-1} and 3.0×10^8 m s^{-1}, respectively. Calculate the critical angle of diamond.

9 Figure 17.18 shows a ray incident on a transparent prism made of material of refractive index 1.7.

a Calculate the angle of refraction at AB.

b Calculate the critical angle of the material of the prism.

c Redraw and complete the path of the ray until it leaves the surface BC. Indicate the angle the ray then makes with the normal and give the reason for your answer.

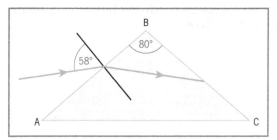

Figure 17.18 *Question 9*

10 A powerful light source is seated on a seabed at a depth of 10.0 m. Light from the source exits the water from within a cone of angle 96° as shown in Figure 17.19.

a Calculate:

  i the critical angle of the seawater

  ii the refractive index of the seawater

  iii the radius r of the cone at the surface.

b Explain what happens to rays of light from the source that approach the surface outside of the cone of rays shown in the diagram.

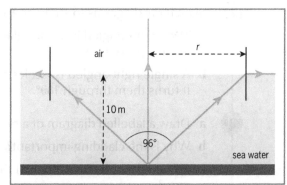

Figure 17.19 *Question 10*

11 Table 17.1 shows readings of the angles of incidence, i, and refraction, r, as light travels from a solid transparent block to the air above it.

a Redraw the table and add columns giving the values of sin i and sin r for each pair of readings.

b Plot a graph of sin i against sin r.

c Calculate the gradient of the graph.

d Determine the refractive index of the material of the block.

e Calculate the critical angle of the material of the block.

f Extend the graph to the point when sin r = 1.0 and record the corresponding value of sin i.

g Determine the values i and r relating to the values of sin i and sin r found in part f.

h Explain why the value found for i in part g is the critical angle.

Table 17.1

i	14	22	27	33	40	45
r	20	30	40	50	60	80

18 Lenses

A lens is a specially shaped transparent object, having one or more curved surfaces that can converge or diverge light to form focussed images. They are used in microscopes, telescopes, spectacles, cameras, the eye and other optical mechanisms. This chapter discusses the most common type of lens: the spherical lens, which has two outer surfaces shaped like part of the surface of a sphere.

Learning objectives

- Distinguish between **converging** and **diverging** lenses.
- Define and illustrate **important terms** associated with lenses.
- Apply the equations of **magnification** to solve problems.
- Use **scale diagrams** to illustrate and determine the **properties of images** produced by lenses.
- Use the **lens formula**, $\frac{1}{f} = \frac{1}{u} + \frac{1}{v}$.
- Perform experiments to **measure** the **focal length** of a converging lens.

Converging and diverging lenses

*A **converging lens** is one that is thicker at its centre. It can converge parallel rays of light to produce a **real** image.*

*A **diverging lens** is one that is thinner at its centre. It can diverge parallel rays of light to produce a **virtual** image.*

Figure 18.1 shows different types of converging and diverging lenses.

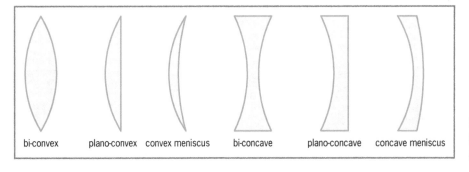

bi-convex   plano-convex   convex meniscus   bi-concave   plano-concave   concave meniscus

Figure 18.1 *Different types of converging and diverging lenses*

Lenses deviate light by the process of refraction

Figure 18.2 shows that a lens can be considered as a series of truncated prisms. Each section of the lens behaves as part of a prism with a slightly different angle.

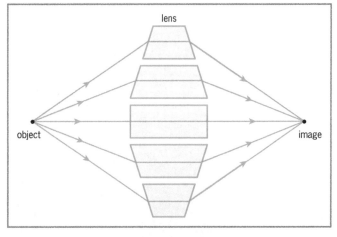

Figure 18.2 *Lens considered as a series of truncated prisms*

Important terms associated with lenses

The following terms relating to converging and diverging lenses are illustrated in Figure 18.3.

*The **optical centre**, O, of a lens is the point at the centre of the lens through which all rays pass without deviation.*

*The **principal axis** of a lens is the line that passes through the centres of curvature of its surfaces (or the line that passes through the optical centre and is perpendicular to the faces of the lens).*

*The **principal focus**, F, of a lens is the point on the principal axis through which all rays parallel and close to the axis converge, or from which they appear to diverge, after passing through the lens.*

*The **focal length**, f, of a lens is the distance between its optical centre and principal focus.*

*The **focal plane** of a lens is the surface perpendicular to the principal axis and containing the principal focus.*

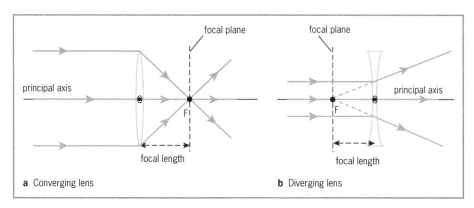

Figure 18.3 *Important terms associated with lenses*

Figure 18.4 shows that parallel rays will focus on the focal plane even if they are not parallel to the principal axis.

To determine where each ray focuses, draw an incident ray straight through the optical centre to meet the focal plane. Then connect the other parallel rays, after passing through the lens, to the same point.

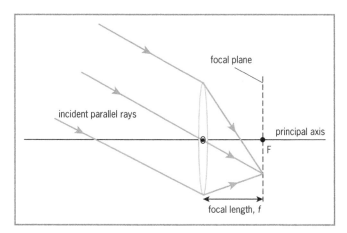

Figure 18.4 *Parallel rays focus on the focal plane*

Magnification

Magnification, m, is the ratio of the size of the image to the size of the object.

$$\text{magnification} = \frac{\text{image height}}{\text{object height}} \qquad m = \frac{I}{O}$$

Magnification has **no unit** since the units of the numerator and denominator of the ratio cancel.

Figure 18.5 shows an image of height **I** produced by a converging lens from an object of height **O**. The distances of the image and the object from the lens are respectively, **v** and **u**.

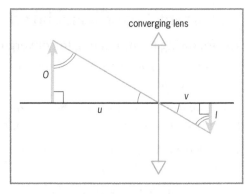

Figure 18.5 *Ratios for magnification*

Since the triangles are similar, the ratios of their corresponding sides are equal.

$$\frac{I}{O} = \frac{v}{u}$$

and therefore:

$$\text{magnification} = \frac{\text{distance of image from lens}}{\text{distance of object from lens}} \qquad m = \frac{v}{u}$$

A pencil is placed 20 cm in front of a converging lens so that its length is perpendicular to the principal axis. The image produced is of length 36 cm and is 60 cm from the lens. Determine:

a the magnification   **b** the length of the pencil.

a $m = \dfrac{v}{u} = \dfrac{60 \text{ cm}}{20 \text{ cm}} = 3.0$

b $m = \dfrac{I}{O}$ $3.0 = \dfrac{36 \text{ cm}}{O}$ $O = \dfrac{36 \text{ cm}}{3.0} = 12 \text{ cm}$

Real and virtual images

*Real images are those produced at a point to which light rays **converge**.*

*Virtual images are those produced at a point from which light rays appear to **diverge**.*

Tables 18.1 and 18.2 give examples and characteristics of real and virtual images produced by real objects.

Table 18.1 *Examples of real and virtual images*

Examples of real images	Examples of virtual images
Image produced on the retina	Image produced in a mirror
Image produced in a lens camera	Image produced by a converging lens acting as a magnifying glass
Image produced on the screen at the cinema	Image produced by a concave lens
Image produced in a pinhole camera	Image produced in water by an object submerged in it

Table 18.2 *Characteristics of real and virtual images of real objects, produced by a lens*

Real images formed by lenses:	Virtual images formed by lenses:
can be formed on a screen	cannot be formed on a screen
are produced by the convergence of rays	are produced due to the divergence of rays
are located on the side of the lens opposite to the object	are located on the same side of the lens as the object
are inverted relative to the object.	are erect (upright) relative to the object.

Constructing diagrams to scale for images formed by a converging lens

1. Draw two perpendicular lines to represent the principal axis and the lens.

2. Place points, F, to scale in position, to represent the principal foci.

3. Draw the object to scale, in size and position, to stand on the principal axis.

4. Draw lines to represent at least **two** of the following rays **from the top of the object:**

   a **parallel to the principal axis**, and then **through F** after passing through the lens

   b **straight through the optical centre**

   c **through F** and then **parallel to the principal axis** after passing through the lens.

5. **The point where the rays cross** represents the **top of the image.** Draw the image perpendicularly from the principal axis to this point (see Figure 18.6).

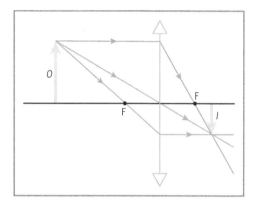

Figure 18.6 *Constructing ray diagrams to scale for images formed by a converging lens*

Important examples are shown in Figure 18.7. The two rays shown from the top of the object in each of these diagrams are those described in **4a** and **4b** above. The points 2F have been included in these diagrams to provide a better understanding of how the position of the object affects the characteristics of the image. These points, however, are unnecessary for the construction of the scale drawings.

Camera\eye – The film or light sensors in a camera, as well as the retina of the eye, are situated between F and 2F so that they can clearly focus the real, inverted image. In the case of the eye, the brain flips the image, so that it does not appear inverted.

Projector – The screen at a cinema is much further than 2F of the projector lens so that it can clearly focus the large, real, inverted image. The object must be inverted so that the image on the screen is upright.

Magnifying glass – This is the only case where the image is **virtual** and **erect**. It occurs when the object is **closer to the lens than its focal length.** That is why when using a magnifying glass to read small text on a page, we must bring the lens close to the words.

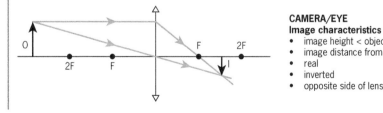

CAMERA/EYE
Image characteristics
* image height < object height
* image distance from lens < object distance from lens
* real
* inverted
* opposite side of lens to object

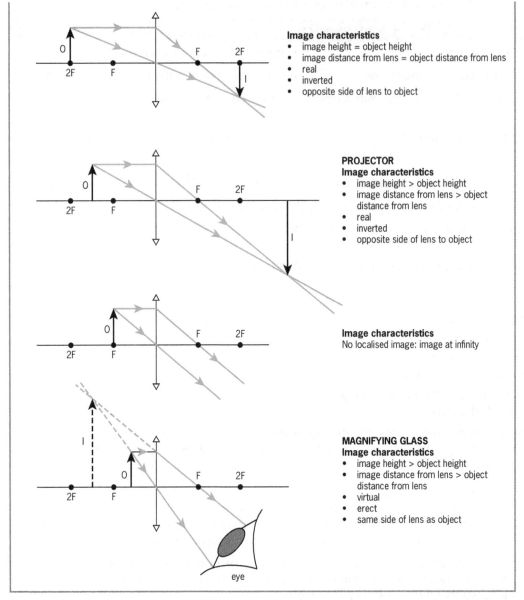

Image characteristics
- image height = object height
- image distance from lens = object distance from lens
- real
- inverted
- opposite side of lens to object

PROJECTOR
Image characteristics
- image height > object height
- image distance from lens > object distance from lens
- real
- inverted
- opposite side of lens to object

Image characteristics
No localised image: image at infinity

MAGNIFYING GLASS
Image characteristics
- image height > object height
- image distance from lens > object distance from lens
- virtual
- erect
- same side of lens as object

Figure 18.7 *Ray diagrams for several object positions relative to a converging lens*

Constructing diagrams to scale involving rays passing through diverging lenses

The lines drawn from the top of the object are illustrated in Figure 18.9 and are as follows:

a **parallel to the principal axis** and then **diverge** as if they **came from F** on the side of the lens where they were parallel

b straight through the optical centre

c **directed to F** on the opposite side of the lens and then parallel to the principal axis after passing through the lens.

Figure 18.8 shows ray diagrams of images produced by diverging lenses. Unlike the case for converging lenses, where the image may be magnified or diminished, real or virtual and erect or inverted, the characteristics of the image formed by a diverging lens are always as follows:

- image height < object height
- image distance from lens < less than object distance from lens

- virtual
- erect
- same side of lens as object.

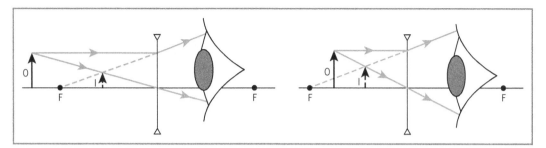

Figure 18.8 *Ray diagrams for object placed further and closer than principal focus of a diverging lens*

For ray diagrams of **diverging lenses** (as well as for a **converging lens** acting as a **magnifying glass**) the emergent rays diverge from each other and it is common practice to place an eye to collect the beam. Since the eye is viewing in the direction along the rays that enter it, it is necessary to represent these rays by **broken lines if they deviate**. Where the two **lines of sight** intersect is the location of the image.

Important rays for diagrams with lenses

Figure 18.9 summarises the rays used to locate images.

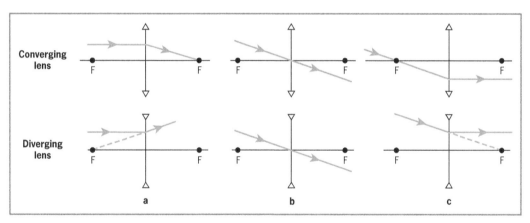

Figure 18.9 *Important rays for diagrams with lenses*

The lens formula

Instead of using ray diagrams drawn to scale to determine the characteristics of images formed by lenses, the lens formula may be applied:

$$\frac{1}{u} + \frac{1}{v} = \frac{1}{f}$$

Table 18.3 *'Real-is-positive' sign convention to be used with the lens formula*

		Positive (+)	Negative (−)
u	distance of object from lens	if the object is **real**	if the object is **virtual***
v	distance of image from lens	if the image is **real**	if the image is **virtual**
f	focal length	if the lens is **a converging lens**	if the lens is **a diverging lens**

* Virtual objects will not be considered at the CSEC examination.

Table 18.4 *Characteristics of the images of real objects are dependent on the sign (+ or −) of v*

image distance, v	nature	orientation	side of lens relative to object
+	real	inverted	opposite side of lens
−	virtual	erect	same side of lens

At this point you should review Table 18.2 for the characteristics of real and virtual images of real objects.

Example 2

A thin rod of length 25 cm stands 75 cm in front of a bi-convex lens of focal length 30 cm. Determine for the image:

a the distance from the lens

b the nature, orientation and side of the lens it is on relative to the object

c the magnification

d the length.

a $\dfrac{1}{u} + \dfrac{1}{v} = \dfrac{1}{f}$

$\dfrac{1}{75} + \dfrac{1}{v} = \dfrac{1}{30}$

$\dfrac{1}{v} = \dfrac{1}{30} - \dfrac{1}{75}$

$\dfrac{1}{v} = \dfrac{1}{50}$

$\therefore v = 50$ cm

b Since v is positive (+), the image is real, inverted and on the opposite side of the lens relative to the object.

c $m = \dfrac{v}{u} = \dfrac{50}{75} = 0.666 \ldots$ (0.67 to 2 sig. fig.)

d $m = \dfrac{I}{O}$

$0.666 = \dfrac{I}{25}$ … (note: use the more precise value and then adjust to 2 sig. fig. for your final answer)

$I = 0.666 \times 25$

$I = 17$ cm

Example 3

The rod of Example 2 is now placed 10 cm in front of the same lens. Determine for the image:

a the distance from the lens

b the nature, orientation and side of the lens it is on relative to object

c the magnification

d the length.

a $\dfrac{1}{u} + \dfrac{1}{v} = \dfrac{1}{f}$

$\dfrac{1}{10} + \dfrac{1}{v} = \dfrac{1}{30}$

$\dfrac{1}{v} = \dfrac{1}{30} - \dfrac{1}{10}$

$\dfrac{1}{v} = -\dfrac{1}{15}$

$\therefore v = -15 \ldots$ (distance of image from lens = 15 cm)

b Since v is negative (−), the image is virtual, erect and on the same side of the lens as the object.

c $m = \dfrac{v}{u} = \dfrac{-15}{10} = -1.5$

magnification = 1.5 (when stating magnification, ignore the negative sign)

d $m = \dfrac{I}{O}$

$1.5 = \dfrac{I}{25}$

$I = 1.5 \times 25$

$I = 38$ cm

Example 4

A nail stands 16 cm in front of a diverging lens of focal length 4.0 cm and perpendicular to its principal axis. The image produced is of length 2.5 cm. Determine:

a the distance of the image from the lens

b the nature, orientation, and side of the lens on which the image is located relative to the object

c the magnification

d the length of the object.

a $\dfrac{1}{u} + \dfrac{1}{v} = \dfrac{1}{f}$

$\dfrac{1}{16} + \dfrac{1}{v} = \dfrac{1}{-4.0}$ (the sign of the focal length of a diverging lens is negative)

$\dfrac{1}{v} = \dfrac{1}{-4.0} - \dfrac{1}{16}$

$\dfrac{1}{v} = -\dfrac{5}{16}$

$v = -\dfrac{16}{5} = -3.2$ (image distance from lens is therefore 3.2 cm)

b Since v is negative (−), the image is virtual, erect and on the same side of the lens as the object.

c $m = \dfrac{v}{u} = \dfrac{-3.2}{16} = -0.20$

magnification = 0.20

---

d $m = \dfrac{I}{O}$

$0.20 = \dfrac{2.5}{O}$

$O = \dfrac{2.5}{0.20}$

$O = 13$ cm

Experiments to determine the focal length of a converging lens

1. Using a distant object

Use a converging lens to focus a distant object such as a building, window or tree onto a white screen by holding the lens so that its **principal axis** is **perpendicular to the screen.**

Since light from the distant object is approximately parallel, and parallel rays always focus on the focal plane, the **distance between the lens and the screen** is the approximate focal length (see Figure 18.10).

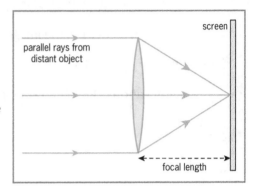

Figure 18.10 *Determining the focal length of a converging lens using a distant object*

2. Using a mirror

A pair of wires crossing over a small hole in a light box is used as a suitable object. The height of the centre of the lens above the desktop should be the same as the height of the crossed wires. The arrangement is set up as shown in Figure 18.11 and the position of the lens is adjusted until the image of the crossed wires is produced next to the object (the mirror must be slightly tilted from the vertical to achieve this). Since the rays have retraced their paths, they must have reflected as a parallel beam. **Parallel rays** always focus on the **focal plane** after passing through a lens, and so the distance between the plane of the lens and the front face of the light box is the focal length.

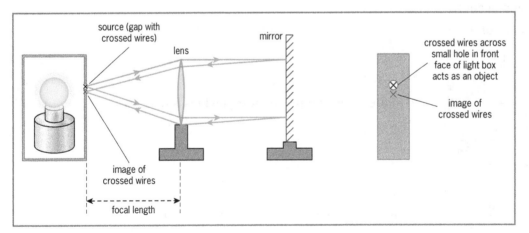

Figure 18.11 *Determining the focal length of a converging lens using a plane mirror*

3. Using the lens formula

The apparatus is set up as shown in Figure 18.12. For several distances between the object and the lens, the position of the screen is adjusted until a clear image of the crossed wires is formed. The object distance, u, and the image distance, v, are measured and recorded for each new focus, and the values of $\dfrac{1}{u}$ and $\dfrac{1}{v}$ calculated and recorded in each case.

A graph is plotted of $\frac{1}{u}$ against $\frac{1}{v}$ and its intercepts are found. From the lens formula, when $\frac{1}{u} = 0$, $\frac{1}{v} = \frac{1}{f}$, and when $\frac{1}{v} = 0$, $\frac{1}{u} = \frac{1}{f}$. The mean value of the x-intercept and y-intercept is therefore the mean value of $\frac{1}{f}$, and from it, f can be calculated. See question 15 at the end of this chapter.

Figure 18.13 shows the experimental arrangement of apparatus for the three methods described above.

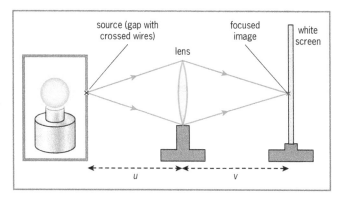

Figure 18.12 *Determining the focal length of a converging lens using a plane mirror*

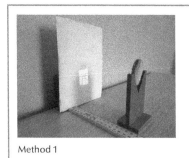

Method 1

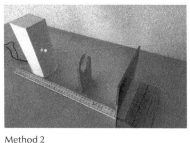

Method 2

Method 3

Figure 18.13 *Finding the focal length of a converging lens*

Recalling facts

1 Distinguish between converging lenses and diverging lenses.

2 Sketch diagrams of the following lenses:
   a convex meniscus
   b plano-concave
   c bi-concave.

3 Name the term that corresponds to each of the following definitions associated with lenses.
   a The point at the centre of a lens through which all rays pass undeviated.
   b The point through which all rays parallel to the principal axis will converge to, or appear to diverge from, after passing through a lens.
   c The line that passes through the centres of curvature of the surfaces of a lens.
   d The surface on which parallel rays will focus after passing through a lens.

4 **a** Distinguish between real and virtual images.
   b Tabulate the following images as either real or virtual. The image formed:
   in a plane mirror/ by a converging lens used as a magnifying glass/ on the retina/ of a straw immersed in a glass of water/ in a lens camera/ by a diverging lens.
   c State **TWO** properties common to all virtual images formed by lenses.

5 Describe the path of the following rays after they pass through a converging lens.

a A ray initially parallel to the principal axis before reaching the lens.

b A ray passing through the principal focus and then reaching the lens.

c A ray directed towards the optical centre.

6 Redraw Table 18.5 and complete column 1 to indicate whether the optical instrument being used may be a camera, a projector or a magnifying glass.

Table 18.5 *Question 6*

Optical instrument	Distance of object from lens
	less than the focal length
	greater than twice the focal length
	between the focal length and twice the focal length

Applying facts

7 Describe how you can make a rough measurement of the focal length of a converging lens if you are given only the lens and a metre rule.

8 An object of height 5 cm is 30 cm from a converging lens and produces a real inverted image 90 cm from the other side of the lens. Determine the height of the image.

9 An object of height 18 mm is placed in front of a converging lens at a distance from the lens of twice its focal length.

a Determine the height of the image.

b What can be said of the distance of the image from the lens relative to the focal length?

The object is now moved further from the lens. How would the following change?

c The height of the image.

d The distance of the image from the lens.

10 A pin of length 2.5 cm is placed to stand vertically in front of a converging lens. The focal length of the lens is 20 cm and its plane is vertical. The pin is placed at the following distances from the lens:

a 60 cm **b** 12 cm.

For each case, **a** and **b**, construct a scale diagram to determine the following characteristics of the image:

i the distance from the lens

ii the height

iii the nature

iv the orientation

v the magnification.

11 An object is placed 8.0 cm in front of a diverging lens of focal length 5.0 cm. Construct a scale diagram to determine the following characteristics of the image:

    a the distance from the lens

    b the nature

    c the orientation

    d the magnification.

12 Justin places a bi-convex lens 4.0 cm from an insect of length 0.80 cm. If the focal length of the lens is 8.0 cm, use the lens formula to determine the following characteristics of the image:

    a the distance from the lens

    b the magnification

    c the length

    d the nature

    e the orientation

    f the side of the lens relative to the object.

13 Sharon has a photo of herself on a small translucent slide. She places a powerful light source behind the slide so that it becomes an illuminated object and brings a converging lens 8.0 cm in front of the photo. A focused image of the photo is formed on a wall 2.0 m from the other side of the lens. Use the lens formula to determine:

    a the focal length of the lens

    b the following characteristics of the image:

        i the orientation

        ii the nature

        iii the magnification.

14 Denzil places an object 8.0 cm in front of a diverging lens of focal length 12 cm. If the height of the object is 5.0 cm, use the lens formula to determine the following characteristics of the image:

    a the distance from the lens

    b the height

    c the nature

    d the orientation

    e the magnification.

15 Ansar carried out an experiment to determine the focal length of a converging lens using the lens formula. He varied the distance u of the object from the lens and in each case measured the distance v at which a focused real, inverted image was formed on a screen. Table 18.6 shows his results for the values of u and v.

a Redraw and complete the table by calculating the values of $\frac{1}{u}$ and $\frac{1}{v}$.

b Plot a graph of $\frac{1}{u}$ against $\frac{1}{v}$. Each axis must begin at zero and must extend at least to 0.040 cm⁻¹.

c Determine the slope of the graph.

d Record the value of the intercept on each axis.

e Calculate the mean intercept.

f Using the lens formula, determine the focal length of the lens.

Table 18.6 *Question 15*

u/cm	v/cm	$\frac{1}{u}$/cm⁻¹	$\frac{1}{v}$/cm⁻¹
210	35		
120	40		
65	56		
55	63		
45	90		
35	210		

19 Wave motion

We live in a world in which everything is constantly being bombarded by waves. All waves are produced by **periodically vibrating (oscillating) bodies** and are capable of **transferring energy** from one point to the next as **periodic disturbances.**

Mechanical waves, such as sound waves or water waves, can only transfer energy through a material medium (matter), but **electromagnetic waves** such as light waves, infrared waves or radio waves can also transfer energy through a vacuum.

Learning objectives

- Differentiate between **pulses** and **wavetrains.**
- Differentiate between **transverse** and **longitudinal** waves, citing important **examples.**
- Be able to produce waves using **ropes**, **springs** and **ripple tanks.**
- Be familiar with the terms **amplitude, speed, phase, wavelength, frequency, period** and **wavefront.**
- Be familiar with drawing **wavefront diagrams.**
- Use important **wave equations.**
- Interpret **displacement vs time, displacement vs position, pressure vs time** and **pressure vs position** graphs relating to waves.
- Understand the phenomena of **diffraction** and **interference.**

Wave pulses and wavetrains

*A wave **pulse** is a single disturbance that propagates from one point to the next.*

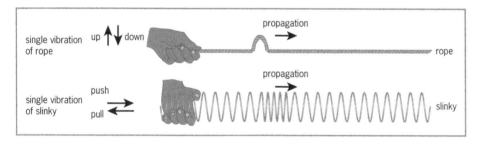

Figure 19.1 *A wave pulse in a rope and in a slinky spring*

*A **wave** (or wavetrain) is a series of periodic disturbances propagating from one point to the next.*

If the end of the rope or slinky in Figure 19.1 is **repeatedly** displaced from its mean position, then back to its mean position, then in the opposite direction from its mean position and then back to its mean position, this would produce a wavetrain (see Figures 19.2, 19.3 and 19.4).

Progressive waves and stationary waves

Waves may be classified as being either **progressive** or **stationary.**

*Progressive waves are those that **transfer energy** from one point to the next.*

Stationary waves can be formed when two identical progressive waves travel along the same line in opposite directions, rendering the net energy transfer to zero. Stationary waves are not on this syllabus.

Transverse and longitudinal waves

Waves may also be classified as being either transverse or longitudinal.

*A **transverse wave** is one that has vibrations perpendicular to its direction of propagation.*

Examples

- The wave produced in a **rope** or **slinky** lying on a horizontal surface and vibrated perpendicularly to its length from one end.
- The wave produced in **water** by an object vibrated perpendicularly upward and downward at its surface.
- An **electromagnetic** wave (radio, infrared, visible light, ultraviolet, X-rays and gamma rays)
- **S waves** that travel through the planet due to an **earthquake.**

Figure 19.2 shows a transverse wave along a rope. **To determine the direction** of motion **of the particles** of the rope at any point along its length, we must know **the direction** of propagation **of the wave.** Since the particles are **only moving up and down**, we can deduce where the rope was a short time earlier and therefore the direction in which the particles must have moved to reach their current positions.

Figure 19.3 shows a transverse wave along a slinky spring. By considering where the spring was a short time earlier, as in Figure 19.2, you can confirm that its particles are directed as indicated.

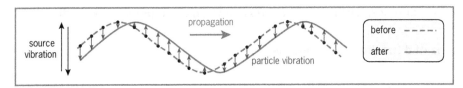

Figure 19.2 *Transverse wave in a rope*

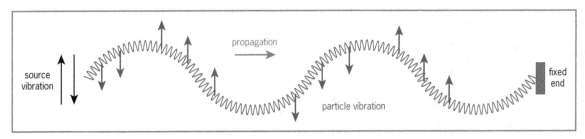

Figure 19.3 *Transverse wave in a slinky spring*

*A **longitudinal wave** is one that has vibrations parallel to its direction of propagation.*

Examples

- The wave produced in a **slinky** spring, lying straight on a horizontal surface and vibrated parallel to its length from one end.
- The **sound** wave produced in a solid, liquid or gas (see Chapter 20).
- **P waves** that travel through the planet due to an **earthquake.**

These waves are characterised by regions of high pressure (**compressions**) and regions of low pressure (**rarefactions**), indicated by C and R respectively in Figure 19.4.

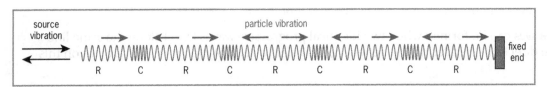

Figure 19.4 *Longitudinal wave in a slinky spring*

Wave parameters

Amplitude

The amplitude of a wave is the maximum displacement of its vibration or oscillation from its mean position (see Figure 19.5).

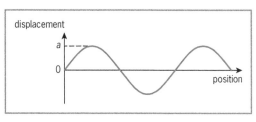

Figure 19.5 *Amplitude,* a

- If the amplitude of a light wave increases, the light becomes brighter.
- If the amplitude of a sound wave increases, the sound becomes louder.

Phase

*Points in a progressive wave are **in phase** if the distance between them along the direction of propagation (path difference) is equal to a **whole number** of wavelengths, λ: 0λ, 1λ, 2λ and so on (see Figure 19.6).*

Points are **out of phase** when this is not the case.

*Points in a progressive wave are **in antiphase** when they are **exactly out of phase**. Their path difference is then equal to $\frac{1}{2}λ$, $1\frac{1}{2}λ$, $2\frac{1}{2}λ$ and so on.*

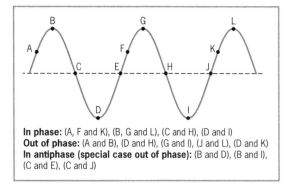

In phase: (A, F and K), (B, G and L), (C and H), (D and I)
Out of phase: (A and B), (D and H), (G and I), (J and L), (D and K)
In antiphase (special case out of phase): (B and D), (B and I), (C and E), (C and J)

Figure 19.6 *Points in phase, out of phase or in antiphase*

Phase difference

Table 19.1 relates to Figure 19.6 and shows that the **phase difference** between two points in a wave is the **fractional part** of the **path difference** between those points when expressed as a number of wavelengths.

Table 19.1 *Phase difference and path difference*

Points	Path difference	Phase difference	Comment
B and L	2λ	0λ (in phase)	Path difference: 0λ, 1λ, 2λ, … implies *in phase*
B and I	1½λ	½λ (exactly out of phase – in antiphase)	Path difference: ½λ, 1½λ, 2½λ, … implies *in antiphase*
B and J	1¾λ	¾λ (out of phase)	

Wavelength, λ

*The **wavelength** is the distance between **successive** points in phase in a wave (see Figure 19.7).*

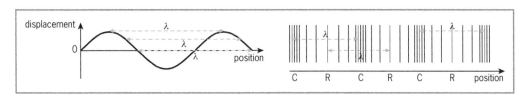

Figure 19.7 *Wavelength of a transverse wave and a longitudinal wave*

Period, T

*The wave **period** is the time for one complete vibration.*

In Figure 19.8 the period is 50 ms = 0.050 s

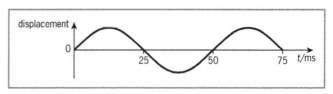

Figure 19.8 *A displacement–time graph for a wave*

Frequency, f

*The wave **frequency** is the number of complete vibrations per second.*

$$T = \frac{1}{f} \quad \text{and} \quad f = \frac{1}{T}$$

***T* must be in seconds (s) for *f* to be in hertz** (Hz) so for the wave displayed in Figure 19.8:

$$f = \frac{1}{T} = \frac{1}{0.050\ \text{s}} = 20\ \text{s}^{-1}\ (\text{or 20 Hz})$$

The frequency of a light wave determines its **colour**.
Red has the **lowest frequency** and **violet** has the **highest frequency** in the visible spectrum.

The frequency of a sound wave determines its **pitch**.
Bass notes have **low frequencies** and **treble** notes have **high frequencies**.

Wavefront

*A **wavefront** is a line perpendicular to the propagation of a wave on which all points are in phase.*

Wavefronts are generally taken through crests of transverse waves and through compressions of longitudinal waves as illustrated in Figure 19.9.

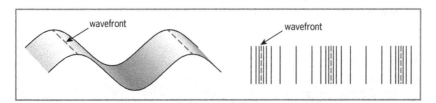

Figure 19.9 *Wavefronts of transverse and longitudinal waves*

Reflection and refraction of rays and wavefronts

A **ray** is the direction of travel of a wave and is therefore always **perpendicular to its wavefronts.** The angles of incidence, reflection and refraction at a surface are the angles between the rays and the **normal** (a normal is an imaginary line perpendicular to the surface at the point of incidence).

Reflection – When a wave reflects, the angle of incidence, *i*, is equal to the angle of reflection, *r*.

Refraction – When a wave refracts into a second medium (say from **medium i** into **medium r**), it changes its direction, θ, relative to the normal (unless the angle of incidence is zero) and it changes its speed, *v*. The following ratios give the **refractive index**, η, of the **second medium relative to the first:**

$$\eta = \frac{\eta_r}{\eta_i} = \frac{v_i}{v_r} = \frac{\sin\theta_i}{\sin\theta_r}$$

Figures 19.10 and 19.11 show the reflection and refraction of rays and of plane wavefronts, respectively. Figure 19.12 shows the reflection of circular wavefronts.

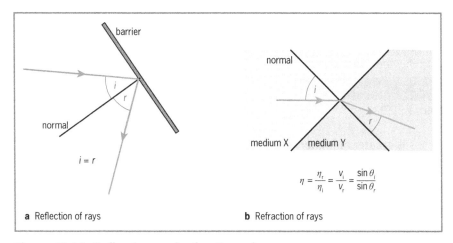

Figure 19.10 *Reflection and refraction of rays*

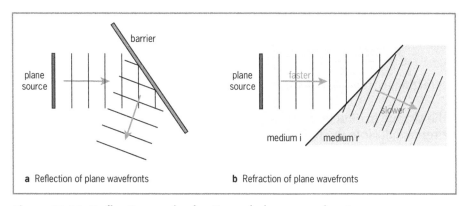

Figure 19.11 *Reflection and refraction of plane wavefronts*

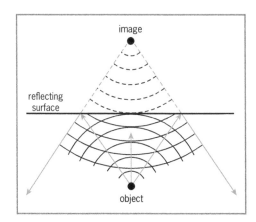

Figure 19.12 *Reflection of circular wavefronts*

Variation of speed of different waves through various media

The speed of light is **higher** in media of **lesser density**. Light therefore travels fastest in a vacuum, only slightly slower in air, and significantly slower through water or glass.

The speed of sound is **higher** through gases of **lesser density**. Molecules of **lesser mass** respond more readily to vibrations than those of greater mass. Sound therefore travels faster through air than through denser gases such as carbon dioxide.

However, the speed of sound is **higher** through **phases of matter** of **greater density** and therefore sound travels fastest in **solids**, less in liquids and least in gases. The **closer packing** of the particles in solids and liquids and the **rigidity of the bonds** between the particles of solids allows vibrations to be transmitted more readily.

Table 19.2 *Speed of sound in steel, water and air*

	Steel	Water	Air at 0 °C
Speed m s⁻¹	5100	1500	330

The speed of sound through gases is **greater** at **higher temperatures** because the increased kinetic energy results in faster vibrations, which can be passed on more readily.

The speed of water waves is **greater** across a **deeper** region.

The speed of the various **frequencies of sound** waves through **air** is the **same**.

The speed of the various **frequencies of electromagnetic** waves through **air** is the **same**.

General wave equations

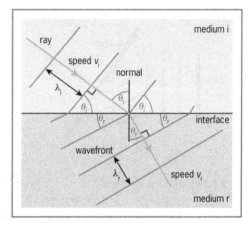

Figure 19.13 *Refraction of waves*

$$v = \lambda f \qquad v = \frac{\lambda}{T}$$

$$\frac{v_i}{v_r} = \frac{\lambda_i}{\lambda_r} = \frac{\sin \theta_i}{\sin \theta_r} = \frac{\eta_r}{\eta_i} = \eta \ldots \text{ refracting to a second medium}$$

Figure 19.13 shows in detail the change in direction of wavefronts when the wave is refracted. Note the following:

- θ_i and θ_r are the angles of incidence and refraction respectively between **rays and the normal** or between **wavefronts and the interface**.

- The speed, v, wavelength, λ, and $\sin \theta$ all **change by the same proportion** when the wave enters a different medium. If v doubles then λ and $\sin \theta$ also double.

- T and f **do not change** when a wave passes from one medium to the next.

- In the third equation above, the ratio of the **refractive indices** has its **subscripts inverted** relative to the other three ratios.

- Fewer errors of substitution are made if subscripts denoting the material of the media are used rather than the subscripts i and r. For example: **a** and **g** to represent **air** and **glass**, respectively.

Example 1

Figure 19.14 shows the profile of a water wave as it flows from deep water to shallow water. The wavelength decreases from 4.0 cm to 2.0 cm and the frequency is 20 Hz. Calculate:

a the speed in the shallow water

b the speed in the deep water.

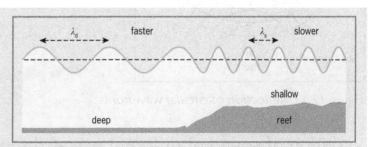

Figure 19.14 *Water wave refracting over shallower depth*

The medium of incidence is 'deep' and the medium of refraction is 'shallow'.

a $v_s = \lambda_s f \qquad v_s = 2.0 \text{ cm} \times 20 \text{ s}^{-1} \qquad v_s = 40 \text{ cm s}^{-1}$

b $v_d = \lambda_d f \qquad v_d = 4.0 \text{ cm} \times 20 \text{ s}^{-1} \qquad v_d = 80 \text{ cm s}^{-1}$

Note: Since $v = \lambda f$, and **frequency is constant**, the wave of **smaller wavelength** is of **lesser speed**.

Figure 19.15 shows a red light wave and a violet light wave travelling in air.

Unlike example 1, which investigated one single wave travelling in two media, these are **two separate waves** travelling in the **same medium.**

They are both electromagnetic waves travelling in air and therefore have the same speed. Since $v = \lambda f$, and **speed is constant**, the wave of **smaller wavelength** is of **higher frequency**.

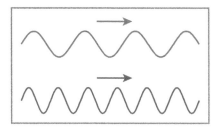

Figure 19.15 *Electromagnetic waves in air have the same speed*

Example 2

A water wave of frequency 0.50 Hz travels at 2.0 m s⁻¹. It reaches a sand bank at an angle of incidence of 60°, and on passing over it, its speed is reduced to 1.25 m s⁻¹. Determine for the wave:

a the wavelength in the deep water

b the period in the deep water

c the frequency as it passes over the sand bank

d the period as it passes over the sand bank

e the wavelength as it passes over the sand bank

f the angle of refraction as it enters the shallow region

g the refractive index on travelling from the deep to the shallow.

Note: The subscripts **d** and **s** have been used to represent deep and shallow, respectively. When a question includes several different quantities, it is best to list the quantities before proceeding with the calculations.

$f = 0.50$ Hz $\qquad v_d = 2.0$ m s⁻¹ $\qquad v_s = 1.25$ m s⁻¹ $\qquad \theta_d = 60°$

a $v_d = \lambda_d f \qquad 2.0 = \lambda_d \times 0.50 \qquad \dfrac{2.0}{0.50} = \lambda_d \qquad 4.0$ m $= \lambda_d$

b $T = \dfrac{1}{f} = \dfrac{1}{0.50} = 2.0$ s

c Frequency does not change when a wave travels to a different medium. The frequency remains at 0.50 Hz.

d Period does not change when a wave travels to a different medium. The period remains at 2.0 s.

e $v_s = \lambda_s f \qquad 1.25 = \lambda_s \times 0.50 \qquad \dfrac{1.25}{0.50} = \lambda_s \qquad 2.5$ m $= \lambda_s$

Alternatively:

$\dfrac{v_s}{v_d} = \dfrac{\lambda_s}{\lambda_d} \qquad \dfrac{1.25}{2.0} = \dfrac{\lambda_s}{4.0} \qquad \dfrac{1.25 \times 4.0}{2.0} = \lambda_s \qquad 2.5$ m $= \lambda_s$

f $\dfrac{\sin \theta_s}{\sin \theta_d} = \dfrac{\lambda_s}{\lambda_d} \qquad \dfrac{\sin \theta_s}{\sin 60°} = \dfrac{2.5}{4.0} \qquad \sin \theta_s = \dfrac{2.5 \sin 60°}{4.0} \qquad \theta_s = \sin^{-1}\left(\dfrac{2.5 \sin 60°}{4.0}\right) = 33°$

g The refractive index on travelling from the deep to the shallow is the refractive index of the shallow **relative to** the refractive index of the deep, i.e. $\dfrac{n_s}{n_d}$

$\dfrac{n_s}{n_d} = \dfrac{\lambda_d}{\lambda_s} \qquad \dfrac{n_s}{n_d} = \dfrac{4.0}{2.5} = 1.6$

Graphs of waves

Displacement–position graph

A displacement–position graph relates the displacement of the vibration of **EACH POINT** in a wave to the distance or position relative to some reference point along the line of propagation at **one instant** in time (**time is held fixed**).

Figure 19.16 shows a wave of speed 2.0 m s⁻¹ at an instant in time. Determine:

a the amplitude   **b** the wavelength   **c** the frequency   **d** the period.

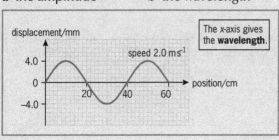

Figure 19.16 *Example 3*

a Amplitude = 4.0 mm

b Wavelength = 40 cm or 0.40 m

c Frequency *f*:

$v = \lambda f$

$f = \dfrac{v}{\lambda} = \dfrac{2.0}{0.40} = 5.0 \text{ Hz}$ (since speed is in m s⁻¹, wavelength must be in metres)

d Period *T*:

$T = \dfrac{1}{f} = \dfrac{1}{5.0} = 0.20 \text{ s}$

Displacement–time graphs

A displacement–time graph indicates the displacement of the vibration of **ONE POINT** in a wave as time continues (**position is held fixed**).

Figure 19.17 shows a wave of speed 20 m s⁻¹. Determine:

a the amplitude of particle P   **b** the displacement of particle P   **c** the period

d the frequency   **e** the wavelength.

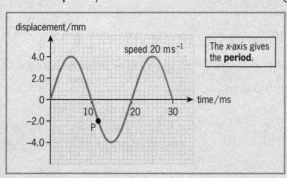

Figure 19.17 *Example 4*

a Amplitude of P = 4.0 mm (the maximum displacement that can be reached by the particle)

b Displacement of P = −2.0 mm

c Period = 20 ms or 0.020 s

d Frequency *f*:

$$f = \frac{1}{T} = \frac{1}{0.020} = 50 \text{ Hz} \dots \text{ to obtain frequency in } \mathbf{Hz} \text{ we } \mathbf{convert \ period \ to \ s} \ (20 \text{ ms} = 0.020 \text{ s})$$

e Wavelength *λ*:

$$v = \lambda f \qquad \therefore \lambda = \frac{v}{f} = \frac{20}{50} = 0.40 \text{ m}$$

Important note

Displacement–position and displacement–time graphs of waves have the shape of transverse waves, but they can represent *both transverse and longitudinal waves* since the vibrations (and therefore displacements) due to both transverse and longitudinal waves vary with time in the same manner. Recall that graphs are a mathematical way of relating two variables – they are not pictures.

Pressure–position and pressure–time graphs

Sometimes you may encounter **pressure–position** graphs and **pressure–time** graphs. These graphs **always represent longitudinal waves**. Recall that these waves have regions of high and low pressures – compressions and rarefactions. Longitudinal waves are sometimes referred to as **pressure waves**.

Note: Figure 19.18 shows that the **mean value** on the vertical axis of these graphs for a sound wave through the air is **atmospheric pressure**; it is **not zero** as is the case for displacement–position or displacement–time graphs.

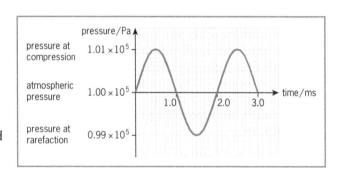

Figure 19.18 *Pressure–time graph of a sound wave*

Investigating waves using a ripple tank

Figure 19.19 shows a ripple tank, a useful apparatus for observing the behaviour of water waves. Important features of the tank and its function are detailed below.

- Water in the tank is between **0.5 cm** to **1.0 cm** in depth.

- The edges of the tank are **gently sloped** like a beach to **reduce reflections** from its walls.

- A **horizontal bar** with a sharp edge is suspended by **elastic bands** so that it just touches the water surface. When made to vibrate, it acts as a **plane source** creating **plane wavefronts**.

- A small **ball-ended dipper** fixed to the beam can be adjusted so that it just touches the water surface. When made to vibrate, it acts as a **point source**, creating **circular wavefronts**.

- An electric motor with a small disc attached to its axle such that the centre of the disc is **not aligned** with the axle, causes the motor, and therefore the bar, to **vibrate** when running. We say that the disc is **off-centered** or **eccentric**.

- A **bright light** is positioned above the tank to cast **shadows** of the wavefronts on a sheet of white paper placed below it. Observing the shadows is easier than observing the ripples of the water surface.

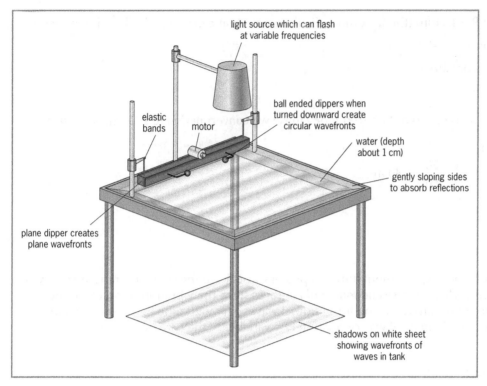

Figure 19.19 *Ripple tank*

- The shadows may be observed as stationary by using a **stroboscope**. One type of a stroboscope is a light which can *flash at variable frequencies selected by the user*. When the flashing frequency is equal to the frequency of the wave, the wavefronts appear stationary. Measurements between the wavefronts can then be made to determine the **wavelength**.

  Another type of stroboscope is a disc with radial slots cut out from it. If the wavefronts are observed by *viewing through the slots*, they appear stationary when the disc rotates at certain frequencies.

- To observe **reflection**, a barrier can be placed in the tank at various angles (see Figure 19.11a).

- To observe **refraction**, a glass plate can be placed in the tank so that the water over the plate is shallower than the water in the rest of the tank (see Figure 19.11b).

- To observe **diffraction** (discussed below), barriers can be set up in the tank as shown in Figure 19.21. As the wavefronts pass through the opening or around the edges they are seen to spread.

- To observe **interference** (discussed later in this chapter), **TWO ball-ended dippers** can be allowed to vibrate in the water at the same time (see Figure 19.25).

  Alternatively, the waves from a **ball-ended dipper** or a **plane dipper** can be passed to a pair of gaps within a barrier (see Figure 19.27). The wavefronts will be seen to interfere after they pass through the gaps.

Diffraction

Diffraction is the spreading of a wave as it passes an edge or goes through a gap.

Speed, wavelength, frequency and period **do not change** as a result of diffraction.

Figure 19.20 shows diffraction of a water wave.

Figure 19.21a indicates that to obtain **strong diffraction** into the complete forward hemisphere, the **gap width** must be

Figure 19.20 *Diffraction of a water wave*

approximately the **same size as the wavelength**. The extent of **diffraction diminishes** as the **gap width** becomes **greater** as shown in Figure 19.21b.

- The wavelengths of light waves are extremely small – approximately 5×10^{-7} m for yellow light. This is much smaller than most gaps commonly encountered. The diffraction of light is therefore not usually observed, and we conclude that 'light travels in a straight line'. Light will diffract, however, if it passes through an extremely small gap.

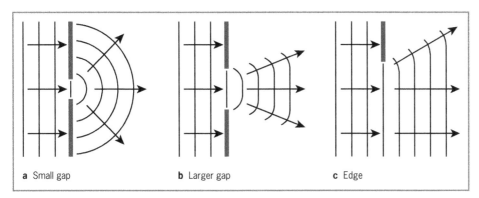

a Small gap **b** Larger gap **c** Edge

Figure 19.21 *Diffraction through gaps and around edges*

- Figure 19.22 shows that **red light diffracts more than violet light** since it has a **longer wavelength**. If sound waves were incident on a gap, **bass notes** will **diffract more than treble notes** for the same reason.
- Radio waves have the largest wavelength of the electromagnetic group (see Chapter 22) and therefore diffract the most. This makes them suitable for radio communication. Longer wavelength radio waves (wavelengths greater than 150 m) can better diffract around hills and buildings on their route to and from receivers and transmitters (see Figure 19.23).

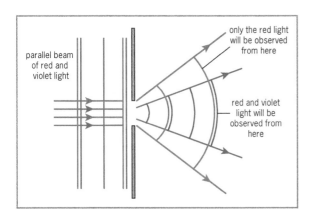

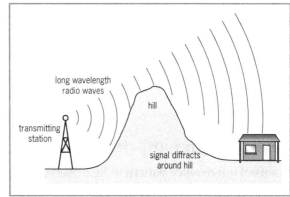

Figure 19.22 *Diffraction of red and violet light*　　Figure 19.23 *Diffraction of radio waves*

Interference

Interference is the phenomenon that occurs at a point where vibrations from two or more waves of the same type superpose on each other (add) to produce a combined vibration of amplitude lesser or greater than any of the individual waves.

The principle of superposition states that when two waves of the same type meet at a point, the resultant displacement at the point is the vector sum of the displacements that the waves would individually produce at that point (see Figure 19.24).

Complete constructive interference occurs at a point where waves superpose in phase to produce a vibration with amplitude equal to the sum of the amplitudes of the individual waves.

For a **transverse wave**, it is where **two crests** or **two troughs meet.**

For a **longitudinal wave**, it is where **two compressions** or two **rarefactions** meet.

*Complete destructive interference occurs at a point where waves superpose in **antiphase** (exactly out of phase) to cancel each other, producing an amplitude of zero.*

For a **transverse wave**, it is where a **crest** meets a **trough**.

For a **longitudinal wave**, it is where a **compression** meets a **rarefaction**.

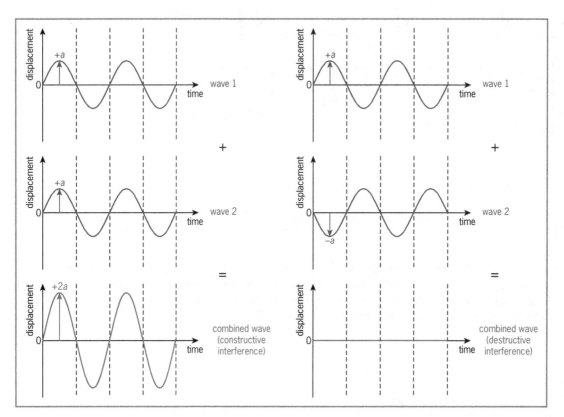

Figure 19.24 *Displacement–time graphs displaying superpositioning during constructive and destructive interference*

Coherence and interference patterns

For a **fixed interference pattern** to be observed, the sources must be **coherent**. This means that they must be of the same frequency and **always** vibrating **in phase or with some constant phase difference**.

Figure 19.25 shows wavefronts produced by ball-ended dippers S_1 and S_2 vibrating due to a **single** motor, S, at the surface of **water in a ripple tank**. The waves are coherent since their vibrations are the same as the frequency of the motor and a fixed interference pattern is obtained.

Constructive interference occurs at all points along the imaginary lines of the **antinodes** (labelled AN). Along the antinodal lines, the water at any point oscillates from a **deep trough to a high crest**.

Destructive interference occurs at all points along the imaginary lines of the **nodes** (labelled N). Along the nodal lines there is **no vibration**. The amplitude there is zero and the **water is calm**.

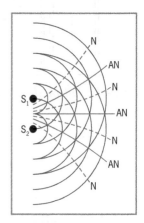

Figure 19.25 *Interference of waves from two coherent sources*

Figure 19.26 shows the interference pattern produced in a pond by two sources vibrating coherently.

Except for lasers, light sources emit wavetrains in random pulses. S_1 and S_2 cannot be separate lamps since their emissions would be random and will not be coherent. A **single monochromatic** light source, S, should be used as shown in Figure 19.27 and the emitted waves should be **diffracted** through slits S_1 and S_2 to produce coherent wavefronts. Even as the source, S, emits randomly, the emissions from S_1 and S_2 will be in phase or have the same constant phase difference (i.e. they will be **coherent**).

Figure 19.26 *Interference of water waves from two coherent sources clearly showing the nodal lines*

In the region where waves interfere, the combined displacement is different than the individual displacements but the **wave characteristics** such as speed, wavelength, frequency and direction **are all maintained.** Consider what occurs when two light beams from torch lights are shone through the night sky so that they cross each other: each beam will continue in the same direction unaffected by the other. The same will occur if the paths of two sound waves or two water waves were to cross each other. Figure 19.28 shows circular wavefronts propagating through each other.

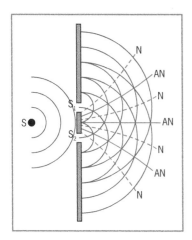

Figure 19.27 *Interference of waves from ONE primary source*

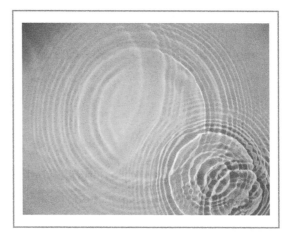

Figure 19.28 *Circular wavefronts propagating through each other*

Example 5

Figure 19.29 shows plane wavefronts at an instant in time represented by the crests of transverse waves produced in water by plane sources of the same frequency. The amplitude of each wave is 4.0 cm. Determine the displacement, s, at each of the points A, B and C.

A: Constructive interference, crest meets crest

$s = 4.0$ cm $+ 4.0$ cm $= 8.0$ cm

B: Constructive interference, trough meets trough

$s = -4.0$ cm $+ -4.0$ cm $= -8.0$ cm

C: Destructive interference, crest meets trough

$s = 4.0$ cm $+ -4.0$ cm $= 0$ cm

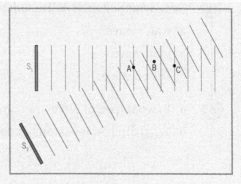

Figure 19.29 *Example 5*

Effect of path difference from two coherent sources in phase

Constructive interference occurs wherever the **path difference** from the **coherent sources** to the point is equal to a **whole number of wavelengths**, 0λ, 1λ, 2λ, ... Consider Figure 19.29.

At A: $S_2A - S_1A = 9\lambda - 7\lambda = 2\lambda$

At B: $S_2A - S_1A = 10\frac{1}{2}\lambda - 8\frac{1}{2}\lambda = 2\lambda$

Destructive interference occurs wherever the path difference is equal to $\frac{1}{2}\lambda$, $1\frac{1}{2}\lambda$, $2\frac{1}{2}\lambda$ ...

At C: $11\frac{1}{2}\lambda - 10\lambda = 1\frac{1}{2}\lambda$

Example 6

Figure 19.30 shows a receiver **R** detecting waves of wavelength 2.0 m and amplitude 5.0 cm transmitted by coherent sources S_1 and S_2 in phase. The wavefronts represent the crests of the wave. Determine the type of interference occurring at R and the amplitude of the vibration at R.

We need to investigate the path difference.

Path difference = (7 m – 2 m) = 5 m.

Since the wavelength is 2 m, this represents $2\frac{1}{2}\lambda$. The phase difference is therefore $\frac{1}{2}\lambda$ (the fractional part of the path difference) and so there is destructive interference at R. The diagram shows that a crest from S_2 is meeting a trough from S_1.

Amplitude of vibration at R = 5.0 cm + –5.0 cm = 0.0 cm

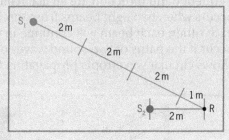

Figure 19.30 *Example 6*

Recalling facts

 1 Name the type of wave in each case described below.

   a Each point along its line of propagation vibrates in a direction parallel to the direction of travel of the wave.

   b Each point along its line of propagation vibrates in a direction perpendicular to the direction of travel of the wave.

 2 List TWO examples of each of the following:

   a mechanical waves

   b electromagnetic waves.

3 State whether each of the following waves is transverse or longitudinal:

   a water wave

   b light wave

   c sound wave.

 4 **a** In what type of wave are compressions and rarefactions found?

   b What are compressions and rarefactions in relation to waves?

5　**a** Define the following terms relating to waves:

　　i amplitude

　　ii wavelength

　　iii period

　　iv frequency.

　b Give equations relating the following:

　　i period, T and frequency, f

　　ii wavelength, λ, speed, v, and frequency, f

6　**a** State with reason in which of the following media is the speed of sound faster:

　　i air or water?

　　ii hydrogen or carbon dioxide?

　　iii hot air or cool air?

　b Does light travel faster in air or in water?

　c How does the speed of a water wave change as it passes from over a reef to a deeper region?

7　A light wave travels from air to glass. What is the effect on each of the following?

　a Its frequency.

　b Its wavelength.

　c Its speed.

8　**a** Define diffraction.

　b Sketch a diagram showing plane wavefronts diffracting through a gap of width:

　　i equal to its wavelength

　　ii equal to THREE times its wavelength.

　c Why is the diffraction of radio waves so strong compared to that of light waves?

9　**a** Define:

　　i interference

　　ii constructive interference.

　b What condition/s must be met for vibrations from a source to be coherent?

　c With the aid of a wavefront diagram, describe how a single monochromatic light source can provide TWO secondary coherent sources.

Applying facts

10　Figure 19.31 shows a water wave at an instant in time as it travels to the right. In what direction is each of the labelled particles moving at the instant shown?

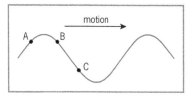

Figure 19.31 *Question 10*

11 Complete Table 19.3 for the following pairs of points in Figure 19.32. Path length between points and phase difference should be given in terms of *number of wavelengths*.

Table 19.3

Pair of points	Path length	Phase difference
B and D		
A and C	1λ	0λ
D and E		
B and E		

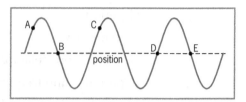

Figure 19.32 *Question 11*

12 A light wave of wavelength 5.0×10^{-7} m and speed 3.0×10^8 m s^{-1} in air enters a block of glass where the speed is reduced to 2.0×10^8 m s^{-1}. The angle of refraction in the glass is 25°. Determine:

a the frequency in air

b the frequency in glass

c the period

d the wavelength in glass

e the angle of incidence

f the refractive index of glass.

13 Figure 19.33 shows a *displacement–time* graph for a wave of speed 500 m s^{-1}.

a Determine:

  i the amplitude

  ii the period

  iii the frequency

  iv the wavelength.

b State with reason whether it can be determined from the graph if the wave is transverse or longitudinal.

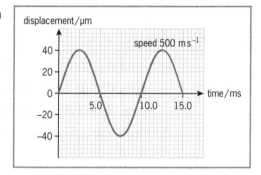

Figure 19.33 *Question 13*

14 Figure 19.34 shows a *displacement vs position* graph for a wave of frequency 1.750 kHz.

a Determine:

  i the amplitude

  ii the wavelength

  iii the speed

  iv the period.

b State for particle X:

  i its instantaneous displacement

  ii the amplitude

  iii its mean displacement during each vibration.

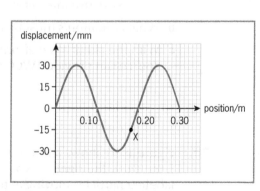

Figure 19.34 *Question 14*

15 Figure 19.35 shows a graph of a wave travelling through warm air at 350 m s^{-1}.

  a Determine:

  i the atmospheric pressure

  ii the pressure at a rarefaction

  iii the wavelength.

  b How can we justify that this wave must be a longitudinal wave?

  c Based on its speed in air, what type of longitudinal wave is it mostly likely to be?

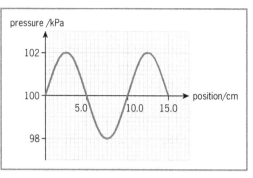

Figure 19.35 *Question 15*

16 Figure 19.36 shows waves emitted by coherent point sources S$_X$ and S$_Y$ which are in phase. They are both of amplitude 12 cm and wavelength 20 cm. Determine:

  a the type of interference occurring

  i at point P

  ii at point Q

  b the amplitude of the vibrations

  i at point P

  ii at point Q.

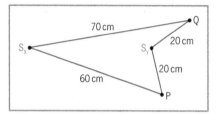

Figure 19.36 *Question 16*

20 Light waves

Near the beginning of the 18th century there were two rival classical theories of the nature of light – the **wave theory** proposed by **Christiaan Huygens** in his publication of 1690 and the **corpuscular (particle) theory** proposed by **Isaac Newton** in 1704.

Learning objectives

- Describe the rival theories of light put forward by **Christiaan Huygens** and **Isaac Newton**.
- Discuss the role of Foucault and Young in **dismissing Newton's corpuscular theory**.
- Provide details of **Young's double-slit experiment**.
- Briefly discuss the role of Planck and Einstein in providing insight into **quantum theory** and **wave–particle duality**.

Rival theories of light

Christiaan Huygens – Wave theory

In 1690 Christiaan Huygens suggested that light was a **wave** capable of propagating through a material called the *aether* which he believed was **invisible**, **weightless** and **filled all space**. It was believed that all waves required a material medium to propagate and the aether justified why light can pass through a vacuum.

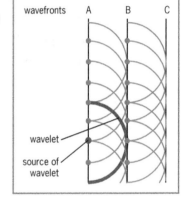

Figure 20.1 *Huygens' wave theory*

Huygens proposed that **each point along a wavefront**, such as A in Figure 20.1, **acts as a source of new 'wavelets'**. After a short time *t*, each of these secondary wavelets advances by the same amount and a new wavefront B is formed on the surface of tangency of the individual wavelets from sources on wavefront A. After a time 2*t*, the wavefront C is the surface of tangency of all the wavelets produced from sources on wavefront B. The advancing wavefront is therefore always perpendicular to the direction of propagation of the wave.

Huygens' wave theory remains useful in explaining phenomena such as reflection, refraction, diffraction and interference.

Wave theory, however, **fails to explain photoelectric emission**, a phenomenon whereby electrons are emitted from materials that have been subjected to electromagnetic radiation. Details of photoelectric emission are not required by this syllabus.

Isaac Newton – Particle (corpuscular) theory

Isaac Newton proposed in 1704 that light was **shot out from a source as particles**. He believed that the 'light particles' travelled in straight lines, since due to their very high speed, they were only very slightly affected by gravity.

The theory **explained** the phenomena of **reflection** and **refraction** but could **not explain diffraction** and **interference**.

There were several problems with the theory.

- The diffraction and interference of light had not as yet been observed and so Newton argued that light cannot be a wave. However, Young later showed experimentally that light does diffract and interfere and therefore must be a wave.
- The theory suggested that light should travel faster in water than in air. This was later disproved experimentally by Foucault.

- 'Particles' emitted by hotter bodies should have greater kinetic energies. However, Newton could not justify why the velocity of light from such bodies was not greater.

Thomas Young – Wave theory

Thomas Young showed experimentally in 1802 that light from two very narrow slits diffracted and produced a pattern of bright and dark fringes on a screen (see later in this chapter). He argued that the **fringes** were the result of the **'principle of superposition'** (interference) of **light acting as waves**. However, Newton's corpuscular theory still lived on, mainly due to his reputation as a dominant physicist and mathematician.

Léon Foucault – Wave theory

Léon Foucault showed experimentally in 1850 that **light travelled faster in air than in water**. This was contrary to what the corpuscular theory suggested and seemed at the time to be the deciding factor which tilted the balance in favour of the wave nature of light.

Max Planck, Albert Einstein – Wave–particle duality

- **Max Planck** in 1900 put forward the **quantum theory**, combining the wave and particle theories. He suggested that electromagnetic wave energy was emitted as **tiny packets ('particles')** known as **quanta** and that this energy was proportional to the **frequency of the wave.**

- **Albert Einstein** in 1905 showed that electromagnetic radiation above certain frequencies can eject electrons from the surface of metals on which it is incident. This phenomenon, known as **photoelectric emission, cannot be explained by wave theory**, but only by particle theory. On the other hand, particle theory could not explain phenomena such as diffraction and interference, which were clearly due to waves.

  Einstein has also shown that there is an equivalence between matter and energy (**particle** and electromagnetic **wave**) in accordance with his famous equation $\Delta E = \Delta mc^2$ (see Chapter 34).

It was later postulated by **Louis deBroglie** that fast-moving electrons (**particles**) behave as **waves** with an **equivalent wavelength** dependent on their momentum!

So today we think of light as being **particle-like** as well as **wave-like** in nature – each wave pulse is a *packet (particle)* or **quantum** of energy, known as a **photon**.

Young's double-slit interference experiment

Figure 20.2 shows the arrangement of the apparatus which can be used to carry out Young's double-slit experiment. The experiment provides **evidence for the wave nature of light since interference is a phenomenon of waves.** Similar experiments can be carried out to demonstrate the interference of other types of waves, for example sound waves (see Chapter 21).

- A **line filament lamp** emitting **white light**, having its filament aligned with a vertical slit, S, acts as the primary source of light. A **colour filter** is placed in front of the lamp so that the light emerging from it is of one colour (monochromatic).

  Alternatively, a **monochromatic** source such as a **sodium discharge lamp** placed behind a narrow vertical slit can be used instead of the line filament lamp with the filter (see Figure 20.3).

- A pair of vertical slits **about 0.5 mm** apart is positioned **about 0.5 m** from the monochromatic arrangement.

- Wavefronts diffract through the slits and propagate to the translucent screen, placed **between 1 m and 2 m** ahead of the slits, to form bright and dark fringes of light on the screen.

- With all external lighting blocked and with the only lighting being from the primary source, S, the fringes may be examined from behind the translucent screen with the aid of a magnifying glass.

The primary source S must be monochromatic (of one frequency)

If the source, S, emitted several frequencies, each colour would produce its own pattern. If S is of white light, the **central fringe** will be **white** and **on either side** of it will be **sets of coloured fringes** (blue to red) with blue being closest to the central fringe.

The secondary sources S_1 and S_2 must be coherent

S_1 and S_2 are **coherent** sources since they depend on a common primary source, S. It is necessary that they emit waves that are **always in phase,** or have some **constant phase difference**, so that they produce the fixed pattern of bright and dark fringes.

If S_1 and S_2 were separate lamps, their **emissions would be incoherent**, causing varying patterns to be formed at an extremely high rate and so preventing the formation of detectable interference fringes.

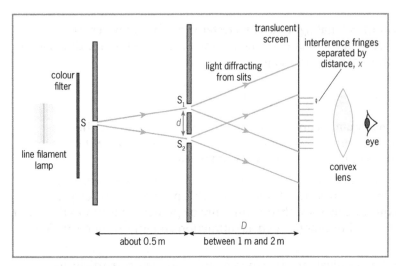

Figure 20.2 *Arrangement of apparatus for Young's double-slit experiment (not to scale)*

Bright and dark fringes

As was shown in Figures 19.25 and 19.27, two coherent sources can produce waves to form a fixed pattern. Figure 20.3 explains how bright and dark fringes may be formed in a Young's double-slit experiment in a similar way.

Bright fringes – When the waves from S_1 and S_2 meet the screen **in phase**, their path difference is a whole number of wavelengths, 0λ, 1λ, 2λ, 3λ etc. They undergo **constructive interference**, and a bright fringe is formed.

Bright fringes are observed at B_0, B_1 and B_2:

$$\text{path difference at central bright fringe} = B_0S_2 - B_0S_1 = 0\lambda$$
$$\text{path difference at 1st bright fringe} = B_1S_2 - B_1S_1 = 1\lambda$$
$$\text{path difference at 2nd bright fringe} = B_2S_2 - B_2S_1 = 2\lambda$$

Dark fringes – When the waves from S_1 and S_2 meet the screen **exactly out of phase**, their path difference is $\frac{1}{2}\lambda$, $1\frac{1}{2}\lambda$, $2\frac{1}{2}\lambda$ etc. They undergo **destructive interference**, and a dark fringe is formed.

Dark fringes are observed at D_1 and D_2:

$$\text{path difference at } D_1 = D_1S_2 - D_1S_1 = \frac{1}{2}\lambda$$
$$\text{path difference at } D_2 = D_2S_2 - D_2S_1 = 1\frac{1}{2}\lambda$$

The graph of Figure 20.3 shows that the brightness of the fringes decreases as the distance along the screen from its centre increases, and that the fringe separation is approximately constant.

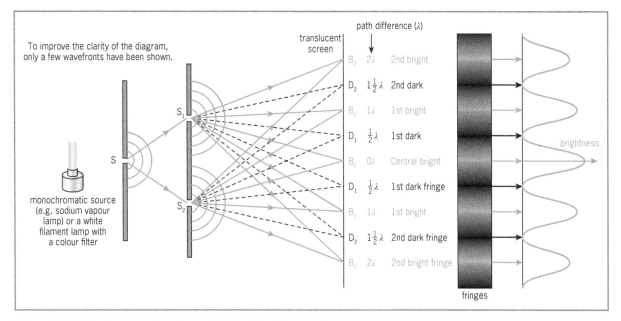

Figure 20.3 *Understanding Young's double-slit experiment*

Factors affecting the fringe separation

The fringe separation, *x*, is increased by:

- increasing the distance, *D*, between the slits and the screen
- increasing the wavelength, λ, of the source
- decreasing the distance, *d*, between the slits S_1 and S_2.

An easy way to remember this is by recalling the equation $\dfrac{x}{D} = \dfrac{\lambda}{d}$ $\quad \therefore x = \dfrac{\lambda D}{d}$

In the first equation, the *small d* is grouped with the extremely *small wavelength* of light and the *large D* is grouped with the *larger fringe separation x*.

Recalling facts

1. Newton's corpuscular theory and Huygens' wave theory were two rival theories of the nature of light existing near the beginning of the 18th century. Briefly describe each of these theories.

2. State TWO phenomena that cannot be explained by Newton's corpuscular theory.

3. **a** What important experiment concerning the nature of light was carried out by Thomas Young?

   b Which of the rival theories did Young's experiment support?

   c Briefly explain how Young's experiment supported the theory mentioned in **b**.

4. **a** State how each of the following contributed to the current theory of light:

   i Max Planck

   ii Albert Einstein.

   b What is meant by the principle of *wave–particle duality*?

5 Sketch the arrangement of the apparatus used in a Young's 'double-slit' experiment on light.

6 The following questions concern a Young's experiment on light.

a How would the pattern on the screen be affected if the primary source was not monochromatic?

b How would the pattern on the screen be affected if two lamps were used instead of a single lamp producing two secondary sources by diffraction through narrow slits?

7 Explain how bright and dark fringes are produced on the screen in a Young's experiment on light.

Applying facts

8 How many wavelengths is the difference in optical path length between each of the secondary sources and the:

a central bright fringe?

b 1st bright fringe?

c 5th bright fringe?

d 2nd dark fringe?

9 An interference pattern is produced in a double-slit experiment using a yellow sodium vapour lamp. Determine the effect on the separation, x, of the bright fringes when:

a the distance, D, between the double slits and the screen is decreased

b the distance, d, between the double slits is increased

c the yellow source is switched to a blue source (a source of shorter wavelength, λ).

21 Sound waves

Sound waves constantly surround us and play an important role in our lives. They provide our most common form of communication and have several other uses including measuring distance, materials testing and imaging. Humans are capable of hearing sounds over a range of frequencies, but there are frequencies below and above this range which our ears cannot detect.

Learning objectives

- Understand how sound waves are **produced** and **propagated**.
- Verify **experimentally** that **sound cannot pass through a vacuum**.
- Relate **pitch and loudness** to **frequency and amplitude**.
- Recall the range of frequencies of **audible** sound, **infrasound** and **ultrasound**.
- Determine **experimentally** the **speed of sound** in air.
- Perform calculations involving the speed of sound including **thundercloud proximity** and **echoes**.
- Cite **evidence** that sound waves **reflect, refract, diffract and interfere**.

Production and propagation of sound

Sound waves are **produced** by mechanical vibrations, for example the vibrations of a loudspeaker cone (see Figure 21.1), if these vibrations occur in an **elastic medium** through which the waves can **propagate**. Sound can therefore travel through solids, liquids and gases, but not through a vacuum.

Figure 21.2 shows sound waves produced by a vibrating speaker and propagated through air, an elastic medium. As the speaker cone moves to the right it forces the air molecules there closer together, creating a region of **higher density** and therefore of **higher pressure**. This force is relayed from particle to particle to the right, away from the speaker. As the cone moves to the left, that compressed region expands, creating a region of **lesser density** and therefore of **lower pressure**. This force is relayed from particle to particle towards the speaker. The vibration therefore creates a series of **compressions, C**, and **rarefactions, R**, as discussed in Chapter 19.

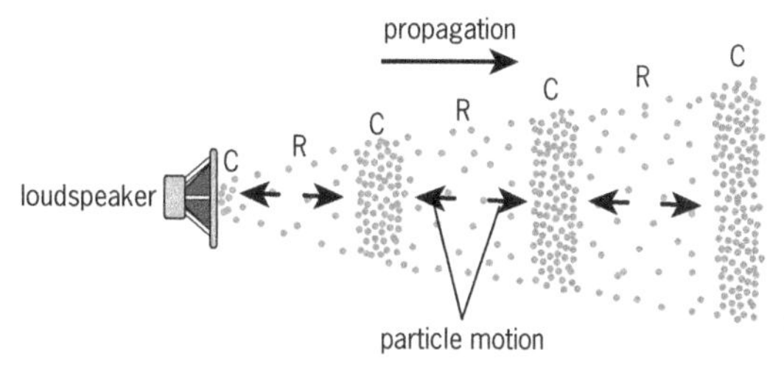

Figure 21.1 *Vibrating cone of loudspeaker*

Figure 21.2 *Sound wave from a loudspeaker*

Tuning fork

A tuning fork is a two-pronged, U-shaped fork made of an elastic metal (see Figure 21.3). When struck it vibrates, creating compressions and rarefactions in the surrounding air. It produces a note of a **fixed frequency** and pitch dependent on its dimensions and on the material from which it is made.

Tuning forks, generally of frequency 512 Hz (C on the musical scale), are used by doctors to test human hearing.

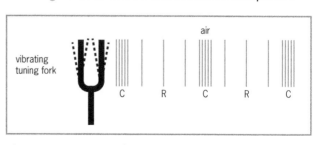

Figure 21.3 *Tuning fork*

To verify that sound cannot propagate through a vacuum

Figure 21.4 shows an electric bell suspended in a **sealed** glass jar. With the bell switched on, observers can see and hear it ringing. As the air is pumped out, the sound level diminishes to almost zero, but it is still seen to be functioning. A slight ringing may be heard since small vibrations may be transmitted by the rubber bands and wires which suspend the bell, but for the most part, there is no longer a material medium to allow the propagation of sound from the bell to the glass jar and then to the surrounding air.

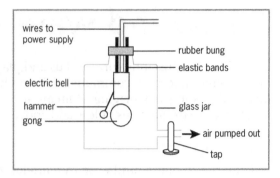

Figure 21.4 *Sound cannot propagate through a vacuum*

Pitch and loudness (volume) from wavefront diagrams

Figure 21.5 shows wavefronts produced by three sound waves A, B and C. As discussed in Chapter 19, they have the **same speed** since they are all propagating through air.

A wave of greater frequency produces a higher pitch sound

The frequency and pitch of A and B are the same

* A and B have the same wavelength and the same speed, so they must also have the same frequency and pitch.

$f = \dfrac{v}{\lambda}$ ... same speed and wavelength in the numerator and denominator implies same frequency

A has a higher frequency and pitch than C

* A has a smaller wavelength than C but the same speed, so A must have the higher frequency and pitch.

$f = \dfrac{v}{\lambda}$ ... **smaller wavelength** in the denominator implies **higher frequency**

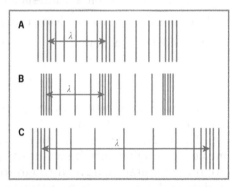

Figure 21.5 *Pitch and volume from wavefront diagrams*

A wave of greater amplitude produces a higher volume (louder) sound

The volume of B is greater than that of A

* The compressions and rarefactions of B have greater and lesser pressures respectively than those of A.
* This **greater change in pressures** between the compressions and rarefactions of B causes the ear drum to be pushed and pulled with **greater amplitude**, and therefore produces a **louder** sound.

Pitch and loudness (volume) from displacement–time graphs

The graph of Figure 21.6 shows a varying sound produced by a signal generator.

* Since A and C have **greater amplitude** than B and D, they are **louder**.
* Since B and C have a **higher frequency** (shorter period) than A and D, they have a **higher pitch**.

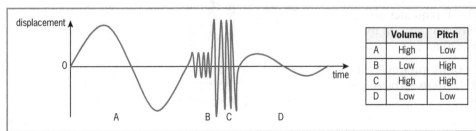

	Volume	Pitch
A	High	Low
B	Low	High
C	High	High
D	Low	Low

Figure 21.6 *Pitch and loudness from displacement–time graphs*

Classification of the frequencies of sound

Table 21.1 shows how sound is classified in terms of its frequency. Not all frequencies of sound are detectable by the human ear.

Table 21.1 *Classification of sound by frequency*

Classification	Frequency range
Infrasound	Below 20 Hz
Audible range of humans	Between 20 Hz and 20 kHz
Ultrasound	Above 20 kHz

Infrasound is produced and detected by elephants, whales, rhinoceroses, giraffes, alligators and other animals for communication. These **low frequency** sounds can **travel long distances** before being absorbed by the medium through which they travel – elephants can stomp their feet to send infrasound messages through the ground across a distance of up to 10 km!

The maximum frequency in the audible range of most humans **decreases with age** to about **15 kHz**. Normal **speech** lies in the range **300 Hz** to about **4000 Hz** since our ears are relatively sensitive at these frequencies.

Ultrasound has several uses, some of which are mentioned below.

- **Communication and hunting capability** – Ultrasound is produced and detected by bats, mice dolphins, seals and other animals for communication. Dogs and cats can also detect ultrasound.

- **Measuring distance** – Bats determine distances by emitting ultrasound and assessing the time in which the echoes return. Depth sounding (discussed later) uses a similar technique.

- **Sonography (diagnostic imaging)** – Ultrasound can be used to scan **soft tissue**. Doctors use a probe to direct ultrasound into a patient. It is partly reflected when it strikes the boundary between materials of different density. The **reflected waves**, received by the same probe, are **analysed by a computer** to produce an image on a screen.

  Since ultrasonic waves are of higher frequency (shorter wavelength) than audible waves, there is relatively **less diffraction** and therefore the **image is better resolved** (better focused).

  Prenatal scanning (Figure 21.7), **measuring blood flow**, as well as the **examination** of the **eye** and internal organs including the heart, can be carried out using ultrasound. Ultrasonic waves are **non-ionising** and are therefore **relatively safe** in **comparison to X-rays**. X-rays are high energy, **ionising** waves which damage and kill body cells and which can cause cancers to develop.

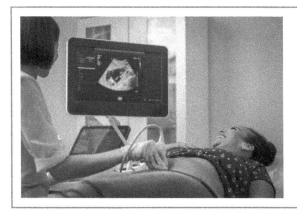 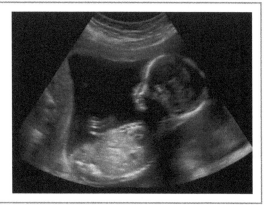

Figure 21.7 *Prenatal scanning*

- **Ultrasonic therapy (treatments)** – Ultrasonic waves of up to 4 MHz may be used to clean teeth and to break up kidney stones, gall stones and cataracts by their **intense vibrations**.

  Focused beams of **high intensity** ultrasound are used to heat and **kill cancer** cells.

- **Materials testing** – Ultrasound imaging is used to determine the **thickness** of materials and to **detect flaws** in solid objects such as **metal castings**.
- **Killing bacteria** – Ultrasonic waves directed onto sewage kills bacteria. It is also used together with antibiotics to kill bacterial cells in the body.
- **Cleaning** – Small objects such as electronic components may be sprayed with a cleanser and then subjected to ultrasonic waves. The high frequency vibration shakes off the dirt and grease. Medical tools and equipment are also cleaned in this manner.

Experiments to determine the speed of sound

Measuring the speed of sound by direct methods

On the games field

Observer A stands at one end of the games field and fires a pistol. Observer B, at a distance s from A, uses a stopwatch to measure the time, t, for the sound of the shot to travel to him/her from the pistol; the stopwatch is **started on seeing the smoke** from the pistol and **stopped on hearing the shot** it fires.

The speed, v, of sound in air is then calculated using the equation $v = \dfrac{s}{t}$.

The accuracy of this experiment is heavily dependent on the reaction time of observer B in starting and stopping the watch. To reduce the inaccuracy, the distance, s, should be increased.

In the laboratory

The apparatus is set up on a long work bench as shown in Figure 21.8. The distance, s, between the microphones is measured. **Microphone A is connected to the start terminal of the millisecond timer** and **microphone B** is connected to its **stop terminal**. The two blocks of wood are struck together and a sound pulse is emitted through the air. As it passes microphone A, the timer starts, and as it passes microphone B, the timer stops, leaving the time, t, on the display screen. The speed, v, of sound in air is then calculated using the equation $v = \dfrac{s}{t}$.

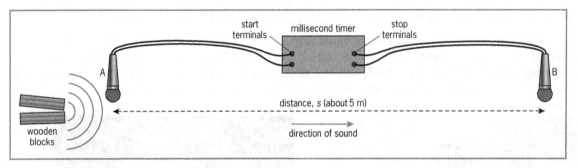

Figure 21.8 *Speed of sound by direct method*

Measuring the speed of sound using echoes

Observer A has two blocks of wood and observer B has a stopwatch. They stand together at a distance, s, of more than 50 m, from a wall that can produce a strong echo: a tall, hard, smooth, vertical wall (Figure 21.9). Observer A claps the blocks at such a rate that the returning echo coincides with each succeeding clap. Observer B then starts the stopwatch on hearing one of the echoes and simultaneously

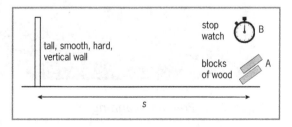

Figure 21.9 *Speed of sound using echoes*

starts counting from zero. He increments his count with the return of each succeeding echo until he reaches a count of N (about 20). The sound would then have travelled a distance $2s$, **to the wall and back**, N times in the measured time, t. The speed of sound, v, is calculated from the equation $v = \dfrac{N(2s)}{t}$.

The distance between the observers and the wall should not be too short since this will make it difficult to distinguish between succeeding echoes. A suitable number for N is about 20 since this will significantly reduce the error due to reaction time on starting and stopping the watch. If the distance from the observers to the wall is 50.0 m and the time recorded is 6.00 s, the speed of sound in air is 333 m s^{-1}, calculated as follows:

$$v = \frac{N(2s)}{t} = \frac{20 \times 2 \times 50.0}{6.00} = 333 \text{ m s}^{-1}$$

Thundercloud proximity

Thunder and lightning occur simultaneously. Due to the extremely high speed of light, it can be assumed that an observer sees lightning as soon as it is produced. Sound travels much more slowly and it generally takes a few seconds for us to hear the thunder. The speed of sound through air is generally between 330 m s^{-1} and 340 m s^{-1}, so **it takes about 3 seconds to travel 1 km**. An observer who **counts the seconds** between seeing the flash and hearing the sound, can approximate the distance from the thunder cloud (in km) by **dividing the count by 3**.

Example 1

During a thunderstorm it took 5.0 s after seeing a lightning flash for the thunder to be heard. Given that the speed of sound in air 340 m s^{-1}, how distant is the thundercloud from the observer?

$$v = \frac{d}{t}$$

$$340 = \frac{d}{5.0}$$

$$340 \times 5.0 = d$$

$$1700 \text{ m} = d$$

Example 2

An officer fires a cannon 510 m from a cliff and 3.0 s later observes the echo it produces. Determine the speed of sound from this data.

$$v = \frac{d}{t}$$

This is an echo, so $v = \dfrac{2x}{t}$ (distance to cliff and back is $2x$)

$$v = \frac{2 \times 510}{3.0} = 340 \text{ m s}^{-1}$$

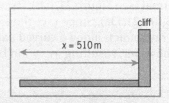

Figure 21.10 *Example 2*

Depth sounding

Ultrasonic waves can be used to detect the depth of water. Owing to its high frequency relative to audible sound, it undergoes relatively **slight diffraction** and produces a more **focused beam**. An ultrasonic pulse is emitted by a **transmitter, T**, as shown in Figure 21.11, and a **timer is started** simultaneously. A **receiver, R**, detects the reflected pulse and records the time. A computer then calculates the depth, x, of the water using:

$$v = \frac{2x}{t}$$

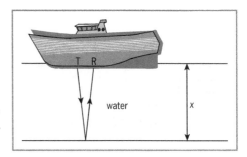

Figure 21.11 *Depth sounding using echoes*

where t is the measured time and v is the velocity of sound in water, which is 1500 m s^{-1}.

$$1500 = \frac{2x}{t}$$

$$\frac{1500t}{2} = x$$

$$750t = x$$

Behaviour of sound waves

Sound waves reflect

The echo produced when there is a loud blast in front of a cliff is evidence that sound waves reflect.

The reflection of sound waves can also be verified by the experiment outlined below using the apparatus shown in Figure 21.12. A hard, vertical surface to act as a reflector is fixed so that its plane is perpendicular to a horizontal desktop. A small clock or stopwatch which produces a loud tick is fixed inside the closed end of a cardboard tube, A. A lies on the table with its open end touching the reflecting surface. A normal is constructed as shown, and a second tube B, open at both ends, is arranged to lie on the table with one end at the base of the normal as shown. An observer listens to the ticking of the clock from the other end of B as the tube is rotated as indicated. It is observed that when the angle of incidence is equal to the angle of reflection, the reflected wave is heard loudest.

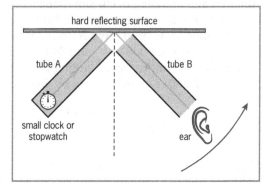

Figure 21.12 *Reflection of sound waves*

Sound waves refract

The air in contact with the ground is **cooler at night**. A sound wave travelling upwards will **increase in speed** as it enters layers of **warmer air**. As it enters the warmer, horizontal layers, it refracts away from the normal (see Figure 21.13b), its wavelength increases and its wavefronts become more separated (see Figure 21.13a). Since rays (lines of propagation), are always perpendicular to wavefronts, the sound ray **refracts along a curved path**, returning to the Earth. **More sound energy** is therefore **directed to the observer**, resulting in sound being **more audible at night**.

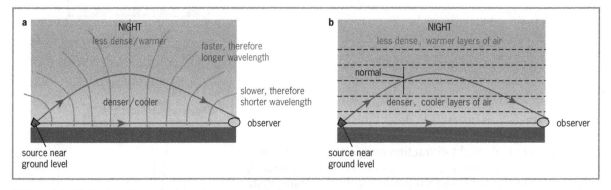

Figure 21.13 *Refraction of sound waves at night*

The refraction of sound waves can also be shown using the apparatus of Figure 21.14. Sound from a speaker is received by a microphone connected to an oscilloscope after it has travelled through a balloon filled with carbon dioxide. The waveform seen on the oscilloscope has greatest amplitude when the microphone is at a particular distance from the balloon, indicating that the sound wave is being focused by refraction as it enters and leaves the balloon of carbon dioxide.

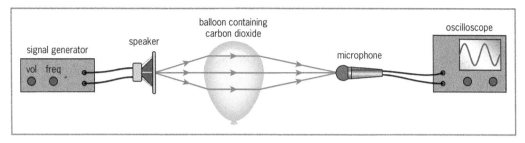

Figure 21.14 *Sound refracting through balloon containing carbon dioxide*

Sound waves diffract

Due to the relatively large wavelengths of sound waves, we can hear around corners as sound waves diffract through windows and doorways and around buildings. Bass notes have larger wavelengths than treble notes and therefore diffract to a greater extent.

Sound waves interfere

A **signal generator** connected to loudspeakers, as shown in Figure 21.15, is adjusted to emit a **constant frequency** in the audible range. Speakers S_1 and S_2 are **coherent sources**.

An experimenter walking along the line AB will observe alternate points where the sound detected is maximum and then where it is minimum. Where the waves from S_1 and S_2 meet the experimenter **in phase**, a **loud** note is heard, and where they meet exactly **out of phase**, **there is silence**.

Note the similarity between this experiment and Young's double-slit experiment described in Chapter 20.

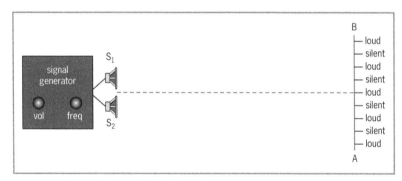

Figure 21.15 *Interference of sound waves*

Recalling facts

1 a How are sound waves produced?

   b Why is it that sound cannot propagate through outer space?

2 A note increases in loudness. State how this affects each of the following characteristics of the sound:

   a frequency

   b amplitude

   c wavelength.

3
  a Create a table showing the classification of the frequencies of sound in terms of the audible range of humans, ultrasound and infrasound.

  b How does the range of frequencies detectable by a person change as he/she becomes older?

4
  a What type of body tissue are ultrasonic scans used for?

  b What advantage do ultrasonic waves have over lower frequency sound waves in providing scans?

  c Briefly describe how an ultrasonic image is produced at a clinic.

  d List TWO medical uses of ultrasonic imaging.

  e Why are ultrasonic scans much safer than X-ray scans?

  f How is ultrasound used to provide therapy for kidney stones and cataracts?

5 Describe how you can demonstrate that sound waves can refract.

Applying facts

6 Figure 21.16 shows a graph of notes A and B. State with reason which note has

  a the higher pitch?

  b the higher volume?

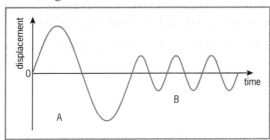

Figure 21.16 *Question 6*

7 Figure 21.17 shows wavefronts in air of sound waves A and B. State with reason which note has

  a the higher volume?

  b the higher pitch?

Figure 21.17 *Question 7*

8 The distance between a thundercloud and an observer is 2.45 km. Given that the speed of sound in the region is 350 m s⁻¹, calculate the time interval between seeing a flash of lightning and hearing the associated crash of thunder from the cloud.

9 A ship uses an echo sounding device to send an ultrasonic signal towards the seabed. The speed of sound in the sea water is 1500 m s^{-1} and the depth of the water is 450 m.

   a Calculate the time for the signal to return to the receiver on the ship.

   b Why does a depth sounding device utilise ultrasound rather than audible sound?

10 Jessica stands 70 m in front of a tall smooth vertical wall and claps her hands at such a rate that the returning echoes coincide with each succeeding clap. Zachary starts a stopwatch at the sound of a clap and counts from 0, incrementing his count by 1 with each returning echo until he reaches a count of 20. He records the time of the process as 8.0 s. Calculate the speed of sound from this data.

11 An official of a 100 m race stands at the finishing tape and starts his watch when he hears the blast from the pistol at the starting position. He stops the watch when the winner reaches the finishing tape and records a time of 11.0 s. Given that the speed of sound on the day is 334 m s^{-1}, calculate the correct time of the winner.

12 Ben stands 900 m in front of a tall vertical cliff and fires a shot from a rifle (Figure 21.18). Alex hears two shots 3.0 s apart. How far is Alex from Ben given that the speed of sound is 340 m s^{-1}?

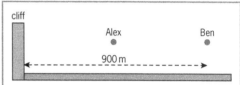

Figure 21.18 *Question 12*

Analysing data

13 Amelia measures the wavelength of sound waves in air corresponding to varying frequencies from a signal generator. Her results are shown in Table 21.2.

   a Redraw the table and add a column giving the values of $1/\lambda$ for each value of λ.

   b Plot a graph of f against $1/\lambda$.

   c Calculate the slope of the graph.

   d Use the graph to determine the speed of sound in air.

Table 21.2 *Question 13*

Frequency, f/Hz	Wavelength, λ/m
500	0.690
1000	0.335
1500	0.227
2000	0.173
2500	0.132

22 Electromagnetic waves

Waves can be classified as **mechanical** or **electromagnetic**.

Sound waves and water waves are common examples of mechanical waves. These waves are **produced by oscillating particles** and require a material medium for their propagation. **Oscillating particles of matter** (atoms or molecules) then vibrate within the medium **transferring the energy**.

Electromagnetic waves are **produced by oscillating electrical charges**. However, unlike mechanical waves, electromagnetic waves do not require a material medium for their propagation, and instead **transfer energy** by **oscillating electric and magnetic fields**.

Learning objectives

- Define an electromagnetic wave.
- State the **general properties** of all electromagnetic waves.
- Be familiar with the frequencies and wavelengths of the **electromagnetic spectrum**.
- Identify the **production** and methods of **detection** of the various types of electromagnetic waves.
- Identify important **properties** and **uses** of the various types of electromagnetic waves.

Electromagnetic waves are a group of transverse waves consisting of an electric field and a magnetic field which vibrate perpendicular to each other and to their direction of propagation (see Figure 22.1).

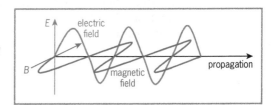

Figure 22.1 *Electromagnetic wave*

General properties of all electromagnetic waves

- They are **transverse** waves.
- They can **propagate through a vacuum**, unlike mechanical waves.
- They travel at the **same speed** of 3.0×10^8 m s^{-1} through a vacuum or through air.
- They consist of varying **electric and magnetic fields**.

As for all waves, electromagnetic waves can:

- transfer energy
- reflect
- refract
- diffract
- exhibit the phenomenon of interference.

The electromagnetic spectrum

Figure 22.2 shows the electromagnetic spectrum. It is a **continuous spectrum**, composed of all frequencies and wavelengths. These waves are emitted as tiny packets known as **photons** whose energy is proportional to their frequency; the **higher the frequency**, the **greater the energy** of the photon.

Each category of electromagnetic energy contains waves of a particular **frequency range** and **wavelength range**. Microwaves are a subset of radio waves and the range of wavelengths of γ-rays overlaps that of X-rays. X-ray machines produce X-rays with wavelengths as small as 10^{-11} m.

Since $v = \lambda f$, we can obtain the frequencies given in the top row of Figure 22.2 by dividing the speed of light in air or a vacuum (3×10^8 m s^{-1}) by the corresponding wavelength in the bottom row.

These different frequencies and wavelengths give **unique properties** to the wave. Within the category of visible light, each smaller frequency range is detected as a different colour. In order of increasing frequency, these colours of the visible spectrum are red, orange, yellow, green, blue, indigo and violet (ROYGBIV). Each colour has a subset of frequencies representing slightly different tones of those colours. The same is true for every section of the spectrum, so for example, there is a range of radio waves of many different frequencies and wavelengths over which different radio stations make their broadcasts. Similarly, there is a range of X-rays of many different frequencies and wavelengths, each having different energies and hence different penetrating powers.

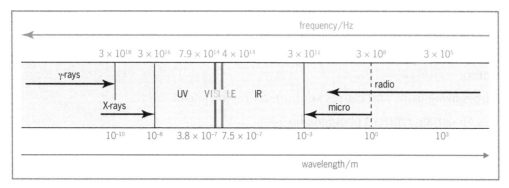

Figure 22.2 *Wavelengths and frequencies of the electromagnetic spectrum*

Table 22.1 shows the various categories of electromagnetic radiation, together with the relative sizes of their wavelengths and how they are produced and detected.

Table 22.2 shows the properties of the various categories of the radiation.

Table 22.1 *Production of electromagnetic waves*

Type	Relative wavelength	Produced by	Detected by
Radio		radio transmitters – aerials (metal rods) emit radio waves due to **electric currents oscillating** within them	radio receivers – aerials connected to electrical circuits convert the wave received to an oscillating electric current
Infrared		all bodies **above 0 K (−273 °C)**	skin of animals, photographic film, thermistor (see Chapter 26, page 272)
Light (RO GBIV)		bodies **above 1100 °C**, LEDs (see Chapter 26, page 273)	photographic film, photocells, eye, LDRs (see Chapter 26, page 272)
Ultraviolet		**very hot bodies**, such as the Sun, welding torches, electric sparks, lightning	photographic film, photocells, fluorescent materials
X-rays and gamma rays		**X-rays: electrons accelerated** through **high voltages bombarding metal targets** **Gamma-rays:** changes in nuclei of **unstable** atoms	photographic film, zinc sulfide screen, Geiger–Müller detector (see Chapter 32)

Table 22.2 *Properties of electromagnetic waves*

Type	Properties
Radio waves	• **Diffract more** than the other electromagnetic waves since they are of longer wavelengths. • Microwaves (radio waves of shorter wavelengths) have a **heating effect** on water and fat molecules.
Infrared radiation	• Detected as **warmth** when incident on our skin. • Prints on certain types of **photographic** paper (film).
Light waves	• Detected by the eye. • Comprised of seven colour groups (red, orange, yellow, green, blue, indigo, violet). • Prints on certain types of **photographic** film.
Ultraviolet radiation	• Produces **fluorescence** of certain materials (see uses of ultraviolet radiation later in this chapter). • **Destroys living tissue** and causes sunburn and skin cancer. • Prints on certain types of **photographic** film.
X-rays and gamma rays	• Can **destroy body cells** since they are **highly ionising** and **penetrating**. • Their extremely short wavelength causes their **diffraction to be negligible** compared to other waves. • Prints on certain types of **photographic** film.

Uses of electromagnetic radiation

Radio waves

Communication networking – Radio waves, including microwaves, are used for radio, television and telephone communications between transmitters and receivers at fixed stations and satellites.

• **Cell phones and FM** radio broadcasts use **shorter wavelength** radio waves including **microwaves**. These **do not diffract much** around hills and across the curvature of the Earth, and so are transmitted from **station to station** en route to their destination.

• **Radar systems** use microwaves to track and locate objects by transmitting signals, which then **reflect** from the object and return to a nearby receiver, giving the information necessary to calculate the location and speed of the object.

• The **longer wavelength radio** waves ($\lambda > 150$ m) **diffract significantly** around hills, mountains and large buildings, providing good communication between receivers and transmitters.

Heating – Microwaves of wavelength 12 cm are used for **heating** foodstuffs. This wavelength corresponds to a frequency of 2.5 GHz, the natural frequency of vibration of water and fat molecules. As the microwave frequency is forced onto the molecules at their natural frequency of vibration, the molecules **vibrate with larger amplitude** and therefore the **temperature rises**.

This process, known as **resonance**, also occurs when someone pushes you on a swing with the same frequency as its natural frequency of vibration; you then swing with increasing amplitude.

Radio astronomy – This is the study of the universe by investigation of radio waves emitted from energetic activities such as those associated with black holes and pulsars.

Infrared radiation

Heating devices – Ovens, toasters and heat radiators emit infrared radiation.

Heat-seeking missiles – These follow aircraft by using sensors to detect the infrared radiation emitted from hot engines.

Treatment of certain muscular disorders – Infrared radiation from special lamps can penetrate skin and act on the muscles below to ease chronic or acute pain.

Thermography – Infrared cameras produce images that reveal regions of different temperatures as shown in Figure 22.3. Bodies at higher temperatures emit waves of higher frequency. The radiation is detected by infrared sensors and passed to a computer program which produces a **colour-coded image** of the **variations in temperature**.

A **thermogram** (see Figure 23.4) is an image produced by infrared radiation showing the variation of temperature of blood flow on or near the surface of a patient's skin. These images are also colour-coded by a computer.

Figure 22.3 *Infrared image showing variation in temperature*

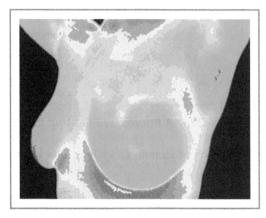

Figure 22.4 *Thermogram investigation of breast cancer*

Some animals detect infrared radiation – Snakes, frogs, some fish and blood-sucking insects such as mosquitoes can hunt in the dark since they can 'see' the infrared radiation emitted from the bodies of their prey.

Transmission of information through fibre-optic cables – Telecommunication systems, including the internet, send digital signals as infrared pulses through fibre-optic cables. Infrared radiation is absorbed less than visible light by the glass fibres and can therefore transmit information over longer distances before the need for amplification.

Remote controls – When a button on a remote control is pressed, an infrared beam pulses with a unique code to carry out a particular function.

Visible light waves

Fibre-optic cables – Light through fibre-optic bundles can illuminate cavities such as the stomach and colon and then return through other fibre-optic bundles to a camera, which produces an image for a doctor to examine.

LASER – **L**ight **A**mplification by **S**timulated **E**mission of **R**adiation. Laser beams are powerful and narrow. They are used to target and dispose of airborne missiles. The **intense heat** they produce make them useful for **welding** and **drilling** metal. In the medical field, lasers are used to break apart kidney stones, to 'weld' detached retinas, to correct eye curvature problems and to remove tumours. Lasers are also used to read bar codes and compact discs (CDs), and to provide the output of some computer printers.

Ultraviolet

Fluorescence is the process in which high-frequency radiation such as ultraviolet radiation is absorbed by certain substances and is then emitted as visible light energy of lower frequency.

- **Detergents** contain fluorescent substances so that clothes washed with them look 'whiter than white' – they absorb ultraviolet radiation from the Sun and then emit the energy as visible waves.

- **Bank notes** have a **marking** made from fluorescent material which becomes visible when ultraviolet radiation is incident on them. This helps to discourage the circulation of **fraudulent** bank notes (Figure 22.5).

- A **fluorescent lighting** tube contains mercury vapour, which emits ultraviolet radiation when a current is passed through it (Figure 22.6). The ultraviolet radiation is absorbed by a fluorescent coating on the inside of the tube, which then emits the energy as visible light.

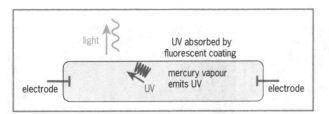

Figure 22.5 Detecting fraudulent notes using ultraviolet radiation

Figure 22.6 *A fluorescent lighting tube*

Production of vitamin D – Ultraviolet light from the Sun is important in the production of vitamin D in the body.

Skin treatment – Ultraviolet radiation of a particular frequency range is used in treating certain skin disorders such as those associated with smallpox, vitiligo and psoriasis.

Killing germs – High frequency ultraviolet radiation is used extensively in **water purification** since it gets rid of germs without adding chemicals. It is also used to **disinfect equipment** and to **increase the shelf life of fruits and vegetables**. Ultraviolet radiation can **kill mould, mildew, bacteria and viruses** which are suspended in the air.

Killing insects – Insects can see ultraviolet radiation and are attracted to the ultraviolet light emitted by a '*bug zapper*'. As they come into contact with the mesh of the 'zapper', they are electrocuted.

X-rays and γ-rays

X-ray and **γ-ray photography** can **detect flaws in metal castings**. X-rays are often used in **scanning passenger luggage** as a security measure at the airport.

γ-ray photography is used in **detecting tumours**. γ-rays, like X-rays, are absorbed by tumours and bone. The patient is given a dose of a γ-emitter which travels through the body to the site of the tumour. Radiation emitted by the source is detected by a sensitive **γ-camera** and produces an image just as an X-ray beam would – the difference is that the rays come from inside the patient. See Chapter 33.

X-ray photography is useful in detecting **tumours** and **broken** and **arthritic bones**, by distinguishing between soft and hard body tissue.

By introducing a small amount of **iodine** into the bloodstream, X-rays can produce **images of blood vessels,** and by giving the patient a dose of **barium sulfate,** images of the **stomach and intestinal tract** can be obtained. The iodine and barium compound are of **higher density** and so **absorb** the radiation to provide the required **contrast** for a good image.

A **computer axial tomography** scan (CAT or CT scan) is a **combination of X-rays** from different angles taken axially through the body to quickly provide a **3D image**, synthesised by a computer. It is particularly useful in detecting problems resulting from accidents where **instant diagnosis** is required.

Figure 22.7 shows a conventional X-ray machine and X-ray images.

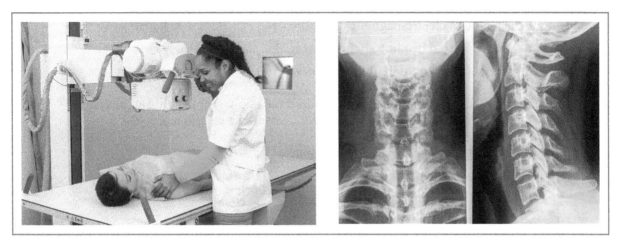

Figure 22.7 *Conventional X-ray photography and X-ray images*

Sterilisation – X-rays and γ-rays are used to sterilise medical equipment and foodstuffs. The gamma **source** cannot contaminate the equipment or food since it is never in contact with it.

Therapy – Fine beam X-rays and gamma rays are useful in destroying cancerous growths. Since they also destroy 'good' cells, this must be a focused and well-planned operation to reduce side effects.

γ-ray tracing – see Chapter 33.

X-ray crystallography – a branch of science where X-rays are used to analyse the structures of crystals.

Recalling facts

1. All waves are produced by vibrations. What vibrations produce electromagnetic radiation?

2. What is the name given to the *tiny packets of energy* emitted as electromagnetic radiation?

3. All waves can reflect, refract, diffract, interfere and transfer energy. List THREE additional properties of electromagnetic waves.

4. What is the relation between the energy and frequency of electromagnetic radiation?

5. Electromagnetic waves can be divided into the following groups:
   visible light waves   X-rays and gamma rays   radio waves
   ultraviolet waves   infrared waves

   a List these groups in order of increasing wavelength.

   b Name the group with the highest frequency.

   c Which group can diffract most strongly?

   d The visible spectrum has wavelengths between 3.8×10^{-7} m and 7.5×10^{-7} m. State the colour pertaining to each of these values.

   e Which colour of the visible spectrum:

      i refracts most?

      ii diffracts most?

   f Name the group that has the lowest frequency.

6 State how each of the groups mentioned in question 5 can be produced.

7 Which type of electromagnetic radiation:

   a is not detected by photographic material?

   b is detected by the skin of animals as warmth?

8 Solar radiation mainly consists of infrared radiation, visible light and ultraviolet radiation. Which of these produces fluorescence of certain materials?

9 State TWO uses of each of the following:

   a radio waves including microwaves

   b infrared radiation

   c visible light waves (other than for seeing objects)

   d ultraviolet radiation

   e X-rays and gamma rays.

Applying facts

The speed of electromagnetic waves in air of 3.0×10^8 m s^{-1} should be used in the following questions.

10 Calculate the frequency of each of the following as they travel through air:

   a radio waves of wavelength 0.400 km

   b X-rays of wavelength 1.5×10^{-10} m.

11 Calculate the wavelength of each of the following as they travel through air:

   a FM radio waves of frequency 104.1 MHz

   b ultraviolet radiation of frequency 5.0×10^{15} Hz.

12 Microwaves of frequency 5.0×10^9 Hz are emitted from a weather station in wavetrains of duration 1.5×10^{-6} s. Determine the following quantities for these waves:

   a the period

   b the wavelength

   c the number of complete waveforms in a single wavetrain

   d the distance travelled by a wavetrain before a succeeding wavetrain is emitted, if they are released every 2.0 s.

13 A cell phone transmits messages by means of microwaves of frequency 900 MHz.

   a Determine the wavelength of its transmissions.

   b Explain why it may be necessary for its transmissions to pass through intermediary stations before reaching a distant destination.

   c State with reason if an AM radio broadcast at 1 MHz would diffract more around hills and buildings than a FM radio broadcast at 100 MHz.

Section D
Electricity and magnetism

In this section

Static electricity

- The study of an imbalance of electric charge residing on a material.

Current electricity

- The study of the motion of electric charges within conductors as alternating and direct currents.

Electrical quantities

- This chapter deals with the relation between electrical quantities that determine the energy and power delivered and consumed within electrical circuits.

Circuits and components

- Here you will learn of various electrical components, how to connect them in electric circuits, and how to be competent in calculating electrical quantities associated with those circuits.

Electricity in the home

- A look at electrical wiring systems within the home, including the use of international colour codes, fuses, switches, circuit breakers, lighting circuits and ring mains (socket) circuits.

Electronics

- Here you will learn of semiconductor diodes – components that can convert AC to DC, and you will study the field of digital electronics using electrical components known as logic gates.

Magnetism

- This is the study of magnetic fields and forces resulting from magnetic dipoles of the fundamental particles comprising a material.

Electromagnetism

- The study of forces resulting from the interaction of electric and magnetic fields.

Electromagnetic induction

- The study of the generation of voltages produced by the relative motion of a conductor within a magnetic field.

23 Static electricity

All materials contain positively and negatively charged particles known as protons and electrons, respectively. A body is neutral if its positive and negative charges are equal in number. If it contains excess positive charge, it is said to be positively charged, and if it contains excess negative charge, it is said to be negatively charged. It is common practice to show only the **net charge** on a body and therefore charges are not shown on a neutral body.

In metals, the protons remain fixed in position but some of the electrons have considerable mobility. In insulating materials such as wood and rubber, not only are the protons fixed, but electron transfer is also poor. Although charge does not flow through insulators, when the surfaces of **different** insulators are **rubbed** together, **electrons** transfer from one onto the other, causing each to obtain equal but opposite **static charges**.

Learning objectives

- Explain and describe experimentally the **charging of insulators by friction**.
- Describe and demonstrate the **forces** that electric charges exert on each other.
- Describe and explain charging by **induction**.
- Define an **electric field** and illustrate the use of electric **field lines**.
- Describe the **hazards** and useful **applications** of electric charge.

Charging insulators by friction

Figure 23.1 shows a glass rod and a polythene rod being rubbed by a dry cloth. Some of the electrons on the surface of the glass transfer onto the cloth and some of the electrons on the cloth transfer onto the surface of the polythene rod.

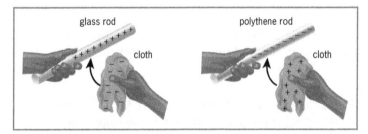

Figure 23.1 *Charging by friction*

The **glass** rod accumulates a net **positive** charge, and the **cloth** it is rubbed with, a net **negative** charge, indicating that cloth has a higher affinity for electrons than does glass. The **polythene** rod accumulates a net **negative** charge, and the **cloth** it is rubbed with, a net **positive** charge, indicating that polythene has a higher affinity for electrons than does cloth.

Other examples of charging by friction

- Human hair becomes positively charged when rubbed with a rubber or plastic comb; the comb becomes negatively charged.
- Human skin becomes positively charged when rubbed with polyester clothes; the clothes become negatively charged.
- Objects made of Perspex (commercial name for acrylic sheets) or of cellulose acetate become positively charged when rubbed with a dry cloth; the cloth becomes negatively charged.
- Vehicles become negatively charged due to air friction; the air becomes positively charged.

Experiment to determine the forces exerted by charges on each other

Two balls, each about 6 mm in diameter and made of aluminium foil, are separately suspended by thin nylon thread. One ball is placed in contact with the positive terminal of a battery so that it becomes positively charged, and the other with the negative terminal of the battery so that it becomes negatively charged, by contact.

A glass rod, charged positively by friction, is brought near to each of the balls as shown in Figure 23.2. It is observed that the positive rod attracts the negative ball but repels the positive one showing that **oppositely** charged objects **attract**, and **similarly** charged objects **repel**.

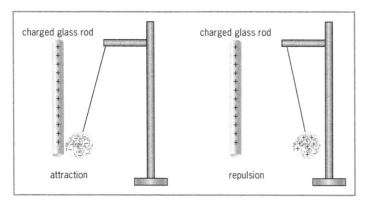

Figure 23.2 *Experiment to determine the forces exerted by charges on each other*

Connecting a charged body to earth

The earth can act as an infinite **source of electrons** or **sink for electrons**. Figures 23.3a and b each show a charged insulated metal dome connected via a switch and ammeter to the earth. When the switches are closed, each ammeter detects a current for a brief period.

The earth acts as a **source** of electrons as it neutralises dome A by transferring electrons to it, and as a **sink** for electrons as it neutralises dome B by receiving electrons from it.

This experiment also shows that the **current** detected by the ammeter **is a flow of charge**.

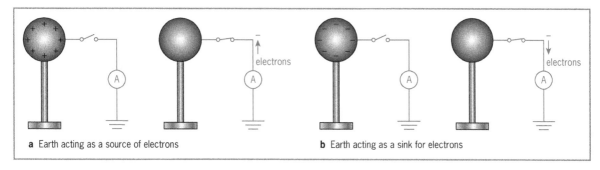

Figure 23.3 *Earth acting as a source of electrons and a sink for electrons*

Attracting an uncharged body by the process of induction

*Electrostatic induction is the process by which electrical **properties** are transferred from one body to a next **without physical contact**.*

Consider a charged polythene rod being brought near to a small piece of dry paper as shown in Figure 23.4. The negative charge on the rod will repel electrons from the side of the paper nearest to it, **inducing** a net positive charge on the side facing the rod and a net negative charge on its far side. Of the charges on the paper, the positive charges are closer to the negative polythene rod, and therefore the paper is attracted to the rod and sticks to it.

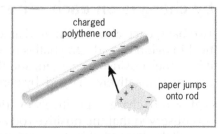

Figure 23.4 *Attracting an uncharged body by induction*

In **humid** regions the paper will stay attached to the rod for only a brief period. Electrons from the rod will conduct to the paper through the layer of **moisture** that coats the materials. This reduces the positive charge on the near side of the paper and the attraction soon becomes insufficient to support its weight.

A charged object can also induce opposite charge in liquids. If a negatively charged balloon is placed near to a slow stream of water from the kitchen tap, the path of the stream will be seen to **curve towards the balloon**. This occurs since the negative charge on the balloon **induces** a positive charge in the water next to it and there is then an attraction between the opposite charges.

Figure 23.5 *Examples of electrostatic induction*

Charging a metal dome by induction

Figure 23.6 shows how a **positively** charged rod can **induce negative** charge on a metal dome.

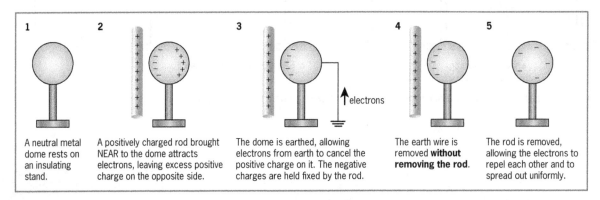

Figure 23.6 *Positive rod inducing negative charge on metal dome*

Figure 23.7 shows how a **negatively** charged rod can induce **positive** charge on a metal dome.

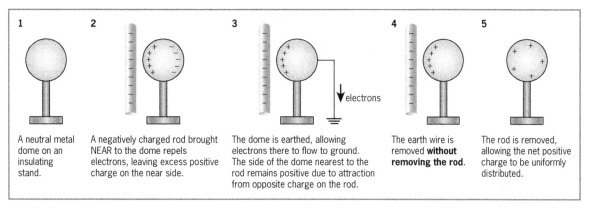

Figure 23.7 *Negative rod inducing positive charge on metal dome*

Charging a metal dome by contact

Unlike charging by induction, charging by **contact** produces a body of **similar charge** to the charge of the body it is in contact with (see Figure 23.8).

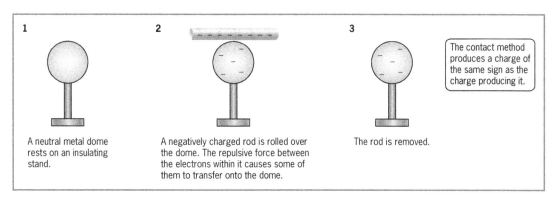

Figure 23.8 *Charging a metal dome by contact*

Electric fields

*An **electric field** is the region in which a body experiences a force due to its charge.*

*The **direction of an electric field** at a point is the direction of the force caused by the field on a positive charge placed at the point.*

Electric field lines

These are **imaginary** lines which are used to represent the strength and direction of an electric field over a region. They have the following properties.

- They **never touch**.
- They behave as though there is a **longitudinal tension** within each line.
- They behave as though there is a **lateral repulsion** between lines close side-by-side.
- The field strength is uniform where the lines are evenly spaced and parallel.
- They are directed from positively charged bodies to negatively charged bodies.
- The field is strongest where the lines are most concentrated.

Creating a uniform electric field

By connecting the terminals of a battery to a pair of **parallel metal plates** as shown in Figure 23.9, equal but opposite charge is transferred to the plates creating an electric field between them. The field is uniform in the **central region** between the plates where the field lines are **parallel and evenly spaced**. In this region each field line is repelled from the ones above and below it and therefore remains straight.

Towards the **edges**, the field is **not uniform**. Field lines at the top are laterally repelled only from lines below, and field lines near the bottom are laterally repelled only from lines above.

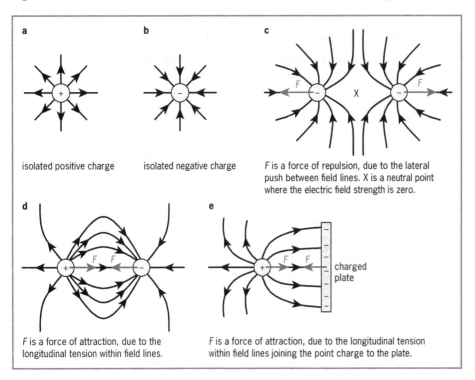

Figure 23.9 *Creating a uniform electric field*

Examples of electric fields

Figure 23.10 shows electric field lines around and between charged bodies.

a

isolated positive charge

b

isolated negative charge

c

F is a force of repulsion, due to the lateral push between field lines. X is a neutral point where the electric field strength is zero.

d

F is a force of attraction, due to the longitudinal tension within field lines.

e

charged plate

F is a force of attraction, due to the longitudinal tension within field lines joining the point charge to the plate.

Figure 23.10 *Electric fields around and between charged bodies*

Point action

The charge on a conductor resides at its **surface** and is most **concentrated on sharply curved** surfaces, especially around **wires** and **points**. Consider a pin attached to the positive terminal of a battery. It will become charged with a high concentration of positive charge, particularly at its point. The concentrated charge results in a strong **electric field** in the surrounding region, which attracts and rips electrons from nearby air molecules and repels the residual positive ions producing an **electric wind** (Figure 23.11).

This ionisation of the air due to the high concentration of charge at a point is known as **point action**.

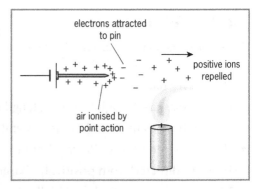

Figure 23.11 *Point action*

Problems and hazards of static charge

Electrostatic shock

- **If you are earthed** when touching a **charged object**, you will receive an **electric shock** as electrons flow through your body between earth and the object.

  It is common for doorknobs to be charged by friction. If you touch a charged doorknob with your bare hand as you stand with bare feet in contact with the floor, you will receive an electric shock as electrons flow between the doorknob and the floor.

  The bodies of vehicles in dry climates can become charged due to friction with the air and with other moving objects. If you touch a charged car, you will receive an electric shock as it discharges through your body. By connecting a **flexible conducting strip** or **chain** between the car and the road, the charge can be **continuously discharged** preventing possible electric shock (Figure 23.12).

- **If you are charged** as you touch an **earthed object** you will receive an **electric shock** as electrons flow through your body between earth and the object.

  As you walk across a nylon carpet, you can become charged due to friction. If you then touch an earthed object, you will receive an electric shock as electrons flow between you and the object.

  Similarly, you may be charged by friction between your clothes and the upholstery as you sit on the seat of a moving vehicle. On disembarking, you may receive an electric shock as you touch the ground.

Figure 23.12 *Anti-static conducting strip*

Attraction of small particles

Charged objects attract **dust** and other small particles including **bacteria** as they induce opposite charge on them as demonstrated in Figure 23.4. For this reason, objects such as computer screens and hospital equipment should be **connected to earth** so that they constantly discharge and so remain **neutral**.

Explosions due to sparking through flammable gases

Dangerous explosions can occur when flammable gases become ignited by sparks produced as electrical charges jump between oppositely charged bodies.

As fuel is added to the storage tanks of oil tankers, aircraft or gasoline stations, there is the possibility of the fuel and the pipes through which it travels becoming oppositely charged due to friction. As these charges build, they may spark across the flammable vapour in the tank, causing it to ignite and explode. To avoid sparking, the tank and pipes of the refuelling trucks are connected by thick conducting materials as shown in Figure 23.13. The discharge then occurs through the thick conducting cable rather than through the flammable vapour.

Similarly, when adding fuel to a car, the **metal nozzle** at the end of the flexible hose is placed into, and **in contact** with, the filler spout on the car. Any static charge in the gasoline caused by friction with the hose can directly conduct to the body of the vehicle instead of sparking through the flammable vapour.

Aircraft tyres experience tremendous friction during landing. To prevent buildup of static charge, they are made with **conducting material** so that any charge on the wheels will be quickly discharged to the ground.

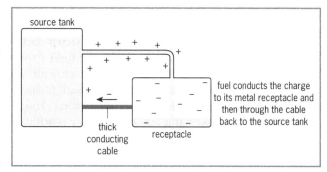

Figure 23.13 *Preventing explosions due to electric discharge*

Lightning

A cloud becomes charged due to friction between the **air, water droplets, ice crystals** and **hail** rising and falling in convection currents within it. The base of the cloud generally becomes negatively charged, and the top, positively charged, as shown in Figure 23.14. Sparks can occur between opposite charges within the cloud.

Negative charge at the base of the cloud repels electrons from the Earth's surface, resulting in a net positive charge accumulating at the surface. When the potential difference (see Chapter 25) between the Earth and the base of the cloud is sufficiently large, electrons and negatively charged ions (heavier charged particles) rush from the cloud to the ground. These high-speed particles crash into neutral air molecules, knocking electrons from them and creating pairs of oppositely charged ions. The result is an avalanche of positively and negatively charged ions, rushing to the cloud and Earth, respectively. The discharge current is typically in the range 5000 to 20 000 A, but can be significantly higher.

Electrical energy transforms into **heat and light energy** (lightning) which transforms further into **sound energy** as the air rapidly expands, increasing the pressure and producing a **sonic shock wave – thunder.**

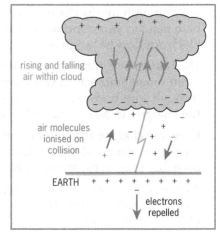

Figure 23.14 *Lightning*

Precautions against the hazards of lightning

Stay away from:

- **tall** objects (examples: tall trees, mountains, and buildings not protected by lightning rods) since lightning tends to strike through the shortest distance
- **metallic** objects since these are readily charged by induction
- **pointed** objects (examples: scissors, knives etc.) since charge concentrates at sharply curved surfaces.

Useful applications of static charge

Lightning conductors

A lightning conductor consists of one or more **metal spikes** connected by a **thick metal strip** to a **thick metal plate** buried below the surface of the Earth in damp soil. The metals used include **copper** and **aluminium**, which are both good conductors of electricity (Figure 23.15).

Tall buildings are vulnerable to lightning strikes since they can be close to clouds, so lightning conductors are fixed to high points on these buildings. The **negative** charge on the base of a nearby cloud **induces positive** charge at the **spikes**. This positive charge is highly concentrated due to the **sharp** curvature of the rod and spikes and creates an **electric field** strong enough to produce **point action** that **ionises** the nearby **air**. The positive ions produced rush to the base of the cloud, reducing its negative charge and making it less dangerous. The negative ions rush to the spikes and rod where they readily conduct to the Earth via the **thick conducting** strip and plate.

Should a bolt of lightning strike directly at the rod, the electrons will easily conduct through the thick conducting strip and plate and so avoid passing through the building. If

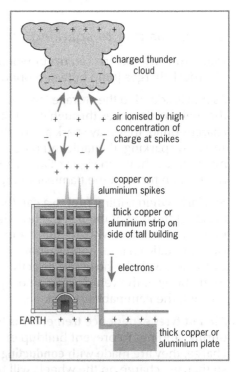

Figure 23.15 *A lightning conductor*

lightning strikes a building which is not protected by a lightning rod, the **charge flows through the moisture** in its structure, **instantly vaporising** it to produce **strong pressures** which may break it apart.

Electrostatic dust precipitators

Figure 23.16 shows one design of an electrostatic precipitator which can **reduce pollution** of the atmosphere caused by industrial processes. Smoke containing dust, carbon and other toxic particles rises through a **metal grid** kept at a high **negative** potential (voltage). **Electrons** from the grid **adhere to the particles** as they rise through it, making them negatively charged. These particles are then attracted to the **positive** cylindrical metal **anode** on the wall of the chimney, where they become attached. The anode plates are **periodically vibrated** so that the pollutant particles detach, fall into a trough, and are collected for disposal.

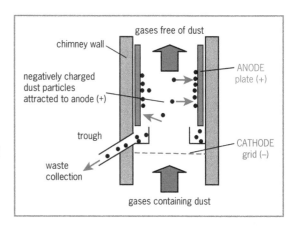

Figure 23.16 *Electrostatic dust precipitator*

Electrostatic spraying

Electrostatic spraying is widely used in the auto manufacturing industry. Paint in the nozzle of the spray gun passes over a needle point raised to a high positive potential (voltage). The paint droplets lose electrons to the needle and become positively charged. **Charging of the paint drops** is also caused by **friction with the nozzle** of the spray gun. As the positive droplets approach the metal body of the car, they **induce opposite charge** and are attracted to it. The metal to be painted can be connected to a **negative** potential to increase the attraction of the droplets (see Figure 23.17).

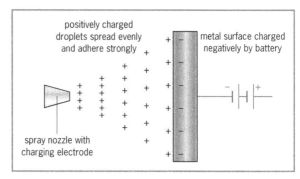

Figure 23.17 *Electrostatic spray painting*

Advantages

- Similarly charged droplets repel each other, **spreading uniformly** and giving good coverage.
- The droplets are opposite in charge to the metal and therefore **adhere** well to it.
- **Shadow regions** (those not in the direct line of the spray) also receive the paint since the electric field can redirect the droplets.

Insecticides, **herbicides**, and **fertilisers** can also be applied by electrostatic spraying. A solution of the chemical to be used is sprayed as a fine mist onto the plants, where it sticks firmly to the leaves and stems due to the opposite charge it induces. Figure 23.18 shows that the droplets are negatively charged by friction as they leave the spray nozzle. As they approach the leaves of the plants, they induce positive charge there and are attracted.

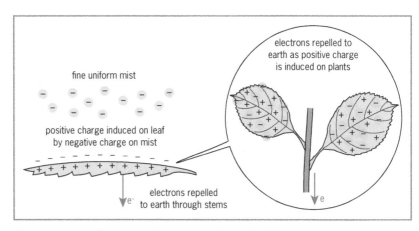

Figure 23.18 *Electrostatic spraying in agriculture*

Electrostatic photocopiers

Figure 23.19 illustrates the role of electrostatic charge in the action of a photocopier. The item to be copied is turned face down on the glass of the photocopier.

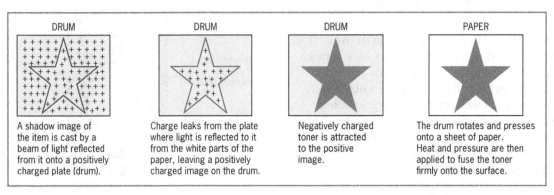

DRUM	DRUM	DRUM	PAPER
A shadow image of the item is cast by a beam of light reflected from it onto a positively charged plate (drum).	Charge leaks from the plate where light is reflected to it from the white parts of the paper, leaving a positively charged image on the drum.	Negatively charged toner is attracted to the positive image.	The drum rotates and presses onto a sheet of paper. Heat and pressure are then applied to fuse the toner firmly onto the surface.

Figure 23.19 *Electrostatic photocopiers*

Recalling facts

1. **a** Describe in terms of the motion of positive or negative charges, what happens in each of the following cases:

   i a glass rod is rubbed with a silk handkerchief

   ii a polythene rod is rubbed with a fur sash.

   b Explain why a copper spoon held with bare hands is not charged when rubbed with the silk or fur.

2. **a** What is meant by the term electrostatic induction?

   b Explain how a negatively charged rubber balloon can attract a stream of water flowing from a tap.

3. Describe an experiment for each of the following to show that:

   a current is a flow of charge

   b like charges repel.

4. **a** Explain why someone touching the metal body of a car after it has made a long journey may experience an electric shock.

   b Why is the shock mentioned in **a** more likely on a dry day?

5. Explain why some vehicles have a flexible conducting strip or a chain hanging from their metal body to the road below.

6. **a** Define:

   i an electric field

   ii the direction of an electric field.

b Draw electric field lines to show the electric fields produced in the following situations:

  i an isolated negative particle

  ii a positive and a negatively charged particle in close proximity

  iii two negative particles in close proximity

  iv two facing metal plates of opposite charge.

7 Explain each of the following:

  a how a cloud can accumulate a charge

  b how a charged cloud can lose its charge by a lightning strike.

8 **a** Explain what happens to the atmosphere close to the copper spikes of a lightning conductor fixed to a building when a charged cloud passes overhead.

  b If the cloud discharges, how does the lightning conductor protect the building?

Applying facts

9 Explain why an uncharged, small, metallised pith ball placed between two charged, vertical and parallel metal plates (A and B in Figure 23.20), will repeatedly bounce to and from the plates.

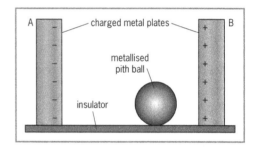

Figure 23.20 *Question 9*

10 **a** Draw a diagram of the electric field existing between the ground and a cloud having a negatively charged base.

  b In which direction will a positively charged particle move if released in the field?

11 Justin walks up a nearby hill holding a garden fork during a thunderstorm. Give THREE reasons why this increases his chances of being struck by lightning.

12 Jeremy buys a type of paint containing an additive that causes it to be easily charged by friction. He uses the paint to spray the door of his car.

  a Explain why the paint:

   i adheres well to the door

   ii does not clog together, but instead gives an evenly distributed coat

   iii can readily reach the underside of the door handle.

  b State ONE other use of electrostatic spraying.

24 Current electricity

Some materials allow electrically charged particles to flow easily within them. When connected together with a battery to form a **complete circuit**, the charged particles flow as an **electric current**.

Learning objectives

- Distinguish among **conductors**, **semiconductors** and **insulators**, giving examples of each.
- State that an electric current in a metal consists of a flow of **electrons**, but in other conducting media it can consist of flows of **positive and negative charges**.
- Define electric **current**, together with its **SI unit**.
- Differentiate between **electron flow** and **conventional current**.
- State the SI **unit of charge** and apply the relation $Q = It$ to solve problems.
- Differentiate between **direct** and **alternating** currents and voltages.
- Analyse **current–time** and **voltage–time** graphs to deduce the **period** and **frequency** of alternating currents.

Conductors and insulators

Conductors are materials through which electrical charges can flow freely.

The following are examples of conductors.

- **Metals** – These contain free electrons that have broken away from their atomic orbits and are not bound to any particular atom.
- **Graphite** – Although a non-metal, graphite also has free electrons, and is widely used for 'make and break' contacts in electrical circuits.
- **Molten ionic substances and solutions of ionic substances** – These have positive and negative ions which can move in opposite directions when connected in an electrical circuit.

Insulators are materials in which electrical charges do not flow freely.

Non-metallic solids are usually good insulators because most of them do not contain charged particles that can move freely.

Semiconductors are a class of materials with an electrical conductivity in a range between that of conductors and insulators.

An understanding of charge carriers in semiconductors is beyond the scope of the CSEC® Physics course.

Table 24.1 *Examples of good conductors, semiconductors and insulators*

Good conductors	Semiconductors	Insulators
silver, gold, copper, aluminium, graphite	silicon, germanium	glass, mica, quartz, rubber, plastic, PVC, nylon, polythene

Charge and electric current

Electric charge is the property of matter that can cause it to experience a force due to an electric or magnetic field.

An **electric current** is the rate of flow of electric charge.

$$\text{current} = \frac{\text{charge}}{\text{time}} \qquad \text{i.e.} \quad I = \frac{Q}{t} \quad \text{or} \quad Q = It$$

*The **coulomb, C**, is the SI unit of electric charge.*

1 coulomb is the electric charge that passes through a cross-section of a conductor when 1 ampere flows through it for a period of 1 second.

*The **ampere, A**, is the SI unit of current.*

1 ampere is the current that passes through a cross-section of a conductor when the rate of flow of electric charge through it is 1 coulomb per second.

Since $Q = It$: $1\ C = 1\ A\ s$ Since $I = \dfrac{Q}{t}$: $1\ A = 1\ C\ s^{-1}$

Note: If the number of mobile charged particles (N) is known, together with the charge (q) on each particle, then the total charge (Q) is given by $Q = Nq$.

Electron flow and conventional current

Consider a stream of positively charged particles **directed towards** a sheet of metal; the sheet will accumulate a **positive** charge. If instead, electrons (negatively charged particles) were **emitted** by the sheet of metal, the sheet will also become **positively** charged. Positively charged particles flowing to the object therefore have the same effect as negatively charged particles moving from the object.

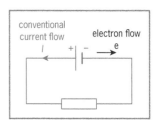

In metals, the only free charges are electrons and they have only one direction of motion in an electrical circuit. However, in molten ionic substances and in ionic solutions, there are positive and negative ions which can flow in opposite directions. Since, with respect to charge, positive flow in one direction is equivalent to negative flow in the opposite direction, the conventional direction of current was chosen as the direction of positive flow to describe the direction of current for all situations (see Figure 24.1).

Figure 24.1 *Electron flow and conventional current*

Electron flow is the direction of flow of NEGATIVE charge. It is opposite in direction to the flow of conventional current.

Conventional current flows in the direction in which a POSITIVE charge would flow if free to do so. It flows in the direction of the electric field.

Example 1

A current of 50 μA flows past a point in a circuit in a time of 8.0 ms. Determine:

a the charge flowing past the point

b the number of electrons flowing past the point, given that the charge on an electron is -1.6×10^{-19} C.

a $Q = It$

  $Q = 50 \times 10^{-6} \times 8.0 \times 10^{-3}$

  $Q = 4.0 \times 10^{-7}$ C

b $Q = Nq$

  $4.0 \times 10^{-7} = N \times 1.6 \times 10^{-19}$

  $\dfrac{4.0 \times 10^{-7}}{1.6 \times 10^{-19}} = N$

  $2.5 \times 10^{12} = N$

Note: The negative sign of the charge is ignored.

Alternating and direct current

An **alternating current (AC)** is a current that repeatedly reverses direction with time.

A **direct current (DC)** is a current that does not change direction with time.

Currents are the result of voltages (see Chapter 25). For each of the voltage–time graphs of Figure 24.2 the corresponding current–time graph would have the **same pattern**.

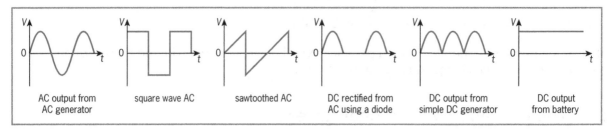

| AC output from AC generator | square wave AC | sawtoothed AC | DC rectified from AC using a diode | DC output from simple DC generator | DC output from battery |

Figure 24.2 *Examples of different types of AC and DC*

Figure 24.3 shows the voltage–time graph for the AC electrical mains supply in Barbados. This wave pattern has a **sinusoidal waveform**.

For an alternating current or voltage:

- Each **cycle** is represented by one crest and one trough on the graph.

- The **period, T**, is the time taken for one complete oscillation (cycle). Its unit is the second, s.

- The **frequency, f**, is the number of complete oscillations (cycles) per second. Its unit is the hertz, Hz (1 Hz = 1 s^{-1}).

- Frequency and period are related by the equation $f = \dfrac{1}{T}$

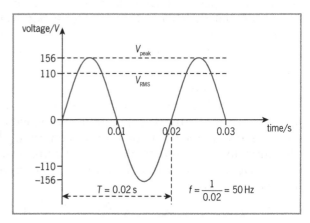

Figure 24.3 *AC mains supply in Barbados*

- The **peak value** is the maximum value that the current or voltage acquires during a cycle.

- The **RMS value** of an AC current or voltage is the value of the **direct current or voltage** that delivers the **same power** to a device. It is about 70% of the peak value.

  RMS is an abbreviation for "*square Root of the Mean of the Square* of all values of current or voltage during each cycle". Calculations involving conversions between peak values and RMS values are not required by the CSEC syllabus. However, students should be aware that AC is generally discussed in terms of its RMS value (which is approximately 70% of its peak value).

  When we speak of an AC voltage in general, we are referring to the RMS voltage, not the peak voltage, since it is the RMS voltage which determines the power consumed by the device.

Example 2

The period of an AC supply is 20 ms. What is its frequency?

$$f = \frac{1}{T} = \frac{1}{0.020} = 50 \text{ Hz} \quad \text{(to 2 sig.fig.)}$$

Note: We must convert milliseconds to seconds to obtain frequency in Hertz.

Recalling facts

1 Differentiate among conductors, insulators and semiconductors, giving TWO examples of each.

2 What conducting material is generally used to provide make and break contacts in electrical circuits?

3 Define the following electrical terms:

a current

b the ampere.

4 Differentiate between the directions of conventional current and electron flow.

5 a Distinguish between direct current and alternating current.

b Give an equation to show the relationship between the frequency and period of an alternating current.

Applying facts

Data to be used in the following questions:

$$charge\ on\ an\ electron = -1.6 \times 10^{-19}\ C$$

6 Calculate the charge flowing during a lightning strike when a current of 75 kA flows for a period of 2.0×10^{-4} s.

7 Determine the time required for a charge of 2.4 kC to be transferred through a component when a current of 2.0 A flows through it.

8 Electrons of total charge 12 μC travel across a vacuum tube in a time of 3.0 ms. Calculate:

a the current in the tube

b the number of electrons flowing across the tube.

9 5.0×10^{14} electrons flow through an electrical component in a time of 25 ms. Calculate:

a the total charge flowing through the component

b the current during the 25 ms.

10 a Sketch a graph of current *I* against time *t* for an alternating current of peak value 10 A and period 16.7 ms. Your graph should show two cycles of the AC.

b Determine the frequency of the AC to 2 significant figures.

c If the resistance of the circuit is 20 Ω, determine the peak voltage of each cycle.

d Using the same axes of the graph drawn in **a**, indicate the approximate position of the RMS current of the AC.

11 Table 24.2 shows the relationship between the total charge (*Q*) which flows past a point in a conductor as time (*t*) progresses.

Table 24.2 *Question 11*

Time, *t*/s	15	27	42	60	70	80
Charge, *Q*/C	3.8	7.0	10.0	15.0	18.0	20.0

a Plot a graph of *Q* against *t*.

b Calculate the gradient of the graph.

c What physical quantity does the gradient represent?

d How many electrons flow past the point in 50 s?

25 Electrical quantities

Electrical energy is an important form of energy and is present in several energy transformations of our daily lives.

The cells of batteries transform chemical energy to electrical energy, rotating coils in magnetic fields of generators transform kinetic energy into electrical energy, and photovoltaic cells of solar panels transform light energy into electrical energy.

The electrical energy can then be transformed to other forms of energy, such as kinetic energy in a motor, thermal energy in a heater, thermal and light energy in a lamp, and even back to chemical energy if is redirected into a rechargeable battery.

Learning objectives

- Be familiar with common **electrical quantities**.
- Define **electromotive force** and **potential difference**, and their SI unit, the **volt**.
- Use the relation, energy = voltage × charge: $E = VQ$
- Define **electrical power** and its SI unit.
- Use the relation, power = voltage × current: $P = VI$
- Compare the consistency of units for $P = VI$ and $P = \dfrac{E}{t}$
- Discuss the importance of **conserving electrical energy** and the means of doing so.

Table 25.1 shows common electrical quantities and their corresponding SI units.

Table 25.1 *Common electrical quantities*

Quantity	Common symbol	SI unit	
power	P	watt	W
energy	E	joule	J
voltage	V	volt	V
current	I	ampere	A
charge	Q	coulomb	C
time	t	second	s

Electromotive force and potential difference

Electrical components generally **resist the transfer of charge** and therefore **energy is needed** to drive charged particles through them.

*The **electromotive force (emf)** of a cell is the energy used (or work done) in transferring unit charge around a complete circuit, including through the cell itself.*

$$\text{electromotive force} = \frac{\text{energy}}{\text{charge}}$$

*The **potential difference (pd)** between two points in a circuit is the energy used (or work done) in transferring unit charge between those points.*

$$\text{potential difference} = \frac{\text{energy}}{\text{charge}}$$

Note the similarity between electromotive force and potential difference.

They are both voltages and therefore we can write in symbols (see Table 25.1) for each:

$$V = \frac{E}{Q} \quad \text{or} \quad E = VQ$$

where E is the **electrical potential energy**.

*The **volt, V**, is the SI unit of voltage.*

1 V is the potential difference existing between two points in a circuit when 1 J of energy is required to transfer 1 C of electric charge between those points.

$$1\,V = \frac{1\,J}{1\,C} \qquad \therefore \quad V = J\,C^{-1}$$

Relation between electromotive force and potential difference

The chemical reactions occurring within a cell result in a transformation of **chemical potential energy** to **electrical potential energy**, which is necessary to drive an electrical current through the resistance of components such as filament lamps. Figure 25.1 shows an electrical circuit where these transformations are occurring. The circuit symbols for a cell and for a filament lamp have been introduced here to demonstrate the relationship between electromotive force and potential difference (a more detailed list of circuit symbols can be found in Chapter 26).

Chemical reactions within the cell create a **potential difference** across its terminals. A cell of emf 1.5 V raises the potential at one end of the string of lamps so that it is 1.5 V higher than the potential at the other end of the string. Current then flows from the terminal at the higher potential, through the filament lamps, to the terminal at the lower potential. On returning to the cell, the potential is once again raised due to chemical reactions within it.

If the filament lamps are of different resistance, they will require a different share of the pd. For the lamps, X and Y, shown in Figure 25.1, the pd across Y is 0.9 V and that across X is 0.6 V, indicating that lamp Y has a greater resistance. The sum of the potential differences is equal to the emf.

The **greater the emf** of the cell, the **greater is the potential difference** it creates, and the greater will be the rate of flow of charge through the filament lamps.

The circuit is analogous to a pump **increasing the potential energy** (and therefore potential) of water as it drives it to the top of a hill. This potential energy will then allow the **current** of water to flow down the hill across surfaces such as rocks and grass that can offer different **resistance** to its flow.

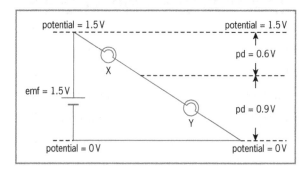

Figure 25.1 *Relation between emf and pd in a circuit*

Electrical power

Electrical power is the rate of transfer of electrical energy.

$$\text{power} = \frac{\text{energy}}{\text{time}}$$

$$P = \frac{E}{t}$$

$$P = \frac{VQ}{t} \quad \text{(since } E = VQ\text{)}$$

$$P = VI \quad \left(\text{since } \frac{Q}{t} = I\right)$$

The **watt** is the SI unit of power.

1 watt is the rate of transfer of 1 joule per 1 second: $1 \text{ W} = 1 \text{ J s}^{-1}$

Since $P = VI$, we can also say: $1 \text{ W} = 1 \text{ V A}$

Example 1

A pd of 5.0 kV exists across an electrical component through which a charge of 2.0 μC flows in a time of 500 ms. Determine:

a the current that flows through the component

b the energy used by the component

c the power consumed by the component.

a $Q = It$

$$I = \frac{Q}{t}$$

$$I = \frac{2.0 \times 10^{-6}}{0.500}$$

$$I = 4.0 \times 10^{-6} \text{ A}$$

b $E = VQ$

$$E = 5.0 \times 10^3 \times 2.0 \times 10^{-6}$$

$$E = 1.0 \times 10^{-2} \text{ J}$$

c $P = VI$

$$P = 5.0 \times 10^3 \times 4.0 \times 10^{-6}$$

$$P = 2.0 \times 10^{-2} \text{ W}$$

Alternatively:

$$P = \frac{E}{t}$$

$$P = \frac{1.0 \times 10^{-2}}{0.500}$$

$$P = 2.0 \times 10^{-2} \text{ W}$$

Example 2

Figure 25.1 shows one type of a vacuum tube. As the cathode (negative electrode) is heated by the heater, electrons are ejected from it by a process known as **thermionic emission**. The circuit is switched on for 20 ms, causing the ejected electrons around the cathode to accelerate across the vacuum to the anode (positive electrode) as a current of 5.0 mA.

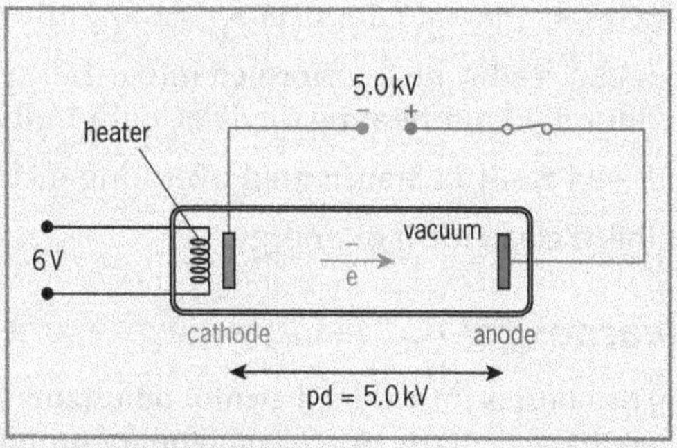

a Describe the energy transformation occurring as electrons leave the cathode, accelerate across the vacuum, and strike the anode.

Figure 25.2 *Example 2*

Given that the mass and charge of an electron are 9.1×10^{-31} kg and -1.6×10^{-19} C respectively, determine:

b the charge flowing between the electrodes

c the number of electrons that flowed between the electrodes

d the electrical potential energy consumed by the process

e the electrical potential energy of an electron on the cathode

f the maximum kinetic energy of a single electron on striking the anode

g the maximum speed of an electron on striking the anode.

a electrical potential energy $\longrightarrow$ kinetic energy $\longrightarrow$ thermal energy

b $Q = It = 5.0 \times 10^{-3} \times 20 \times 10^{-3} = 1.0 \times 10^{-4}\,C$

c $Q = Nq$

$$\therefore N = \frac{Q}{q} = \frac{1.0 \times 10^{-4}}{1.6 \times 10^{-19}} = 6.25 \times 10^{14}$$

Note: The negative sign of the electron's charge is ignored.

d $E = VQ = 5 \times 10^3 \times 1.0 \times 10^{-4} = 0.50\,J$

e $E_1 = Vq = 5 \times 10^3 \times 1.6 \times 10^{-19} = 8.0 \times 10^{-16}\,J$

alternatively: $E_1 = \dfrac{\text{total energy}}{\text{number of electrons}} = \dfrac{E}{N} = \dfrac{0.50}{6.25 \times 10^{14}} = 8.0 \times 10^{-16}\,J$

f Since the electrical potential energy of each electron is transformed to kinetic energy of that electron, the maximum kinetic energy acquired by an electron is also $8.0 \times 10^{-16}\,J$.

g $E_k = \dfrac{1}{2}mv^2$

$$\frac{2E_k}{m} = v^2$$

$$v = \sqrt{\frac{2E_k}{m}}$$

$$v = \sqrt{\frac{2 \times 8.0 \times 10^{-16}}{9.1 \times 10^{-31}}} = 4.2 \times 10^7\,m\,s^{-1}$$

Importance of electrical energy

Reasons for using electrical energy

Electricity plays an important role in our everyday lifestyle.

- It can **readily be transformed** into other types of energy such as thermal, light, sound and kinetic energy in our heating devices, light bulbs, speakers and electric motors.
- It can **easily be transmitted** over long distances.
- It is a **clean** form of energy.

Reasons for not using energy from fossil fuels

- Fossil fuels introduce **harmful pollutants** into the environment, which lead to increasing health care cost, particularly due to **pulmonary problems** and diseases.
- Some of the pollutants are **greenhouse gases** that increase the threat of **global warming**.
- **Limited reserves** of fossil fuels together with geopolitical factors are causing **rising and fluctuating fuel costs**.

Ways to reduce reliance on fossil fuels

- Utilise **alternative sources** of energy (see Chapter 5).
- **Conserve** energy.

Ways to conserve electrical energy consumption in the home

- Install high efficiency, certified appliances.
- Use energy efficient lighting such as LED or fluorescent bulbs.
- Install photovoltaic (PV) systems to produce electricity from sunlight.
- Switch off lights and other electrical equipment when not in use.
- Install solar water heaters to produce thermal energy from solar radiation.
- Cook with covered saucepans to prevent wastage of thermal energy to the surrounding air.
- When boiling, adjust the heat source to the minimum required to maintain boiling.
- Wash only full loads in the washing machine and use energy saving cycles.
- Use clothes lines or racks instead of electric dryers for drying clothes.
- Avoid frequent opening of the refrigerator door and ensure that its rubber gasket is not worn.
- **Ensure that your home is efficiently designed** to reduce air conditioning costs by having:
  - double-glazed windows, especially in rooms that are air conditioned
  - proper insulation of roofs and ceilings
  - walls built with hollow blocks (air is a poor conductor)
  - hoods, awnings and curtains at windows to block solar radiation
  - outer walls painted white to reflect radiation
  - inner walls painted white to reduce emission of radiation.

Recalling facts

  1   **a**   Differentiate between electromotive force and potential difference.
       b   Name and define the SI unit of potential difference.

  2   Describe the energy transformation occurring:
       a   in a cell producing a current
       b   in a lamp through which current flows.

  3   State THREE reasons why electricity is an important form of energy.

  4   State:
       a   THREE ways that the use of fossil fuels has a negative impact
       b   TWO major steps which can be taken to reduce the use of fossil fuels.

  5   Give FIVE ways by which we may conserve the use of electrical energy in our homes.

Applying facts

6 Determine the charge transferred through a circuit when a 1.5 V cell delivers 3.0 kJ of energy to its components.

7 Work of 165 kJ is done in supplying a current of 5.0 A through a heater for a period of 5.0 minutes. Determine:

  a the energy supplied

  b the charge passing through the heater

  c the pd across the heater

  d the power consumed by the heater.

8 20 mC of charge flows through a component in a circuit in a time of 5.0 s causing an energy transfer of 0.24 J. Calculate:

  a the power consumed by the device

  b the current which flows

  c the potential difference across the component.

9 The rate of energy transfer as electrons flow across a vacuum tube between an anode and a cathode is 10 J s^{-1}. The absolute value of the total charge flowing across the tube is 8.0 mC and the current is 2.0 mA. (Given: mass of electron = 9.1×10^{-31} kg and charge of electron = -1.6×10^{-19} C)

  a Determine:

    i the number of electrons flowing across the tube

    ii the time taken

    iii the potential difference between the electrodes

    iv the energy transformed.

  b Describe the energy transformation occurring as the electrons are ripped from the cathode, accelerate across the vacuum and then crash into the anode.

  c Determine for a single electron:

    i its electrical potential energy

    ii the maximum kinetic energy it can attain

    iii its maximum possible speed.

10 A 12 V electric motor of efficiency 100% lifts an object of mass 4.0 kg to a height of 15 m in a time of 20 s.

  a Determine:

    i the power supplied by the motor

    ii the current in its coils.

  b Calculate the current in the coils if the efficiency of the motor was 80%.

11 Students investigated the relationship between the energy dissipated in a resistor and the charge flowing through it. They placed a fixed voltage across the resistor. The energy dissipated in a given time was measured by a joulemeter, and the charge flowing in that time was calculated using the readings from an ammeter and stopwatch. Table 25.2 was prepared.

Table 25.2 *Question 11*

Energy E/J	15	30	45	60	75	90
Charge Q/C	2.8	4.6	7.3	10.0	13.0	16.0

a Plot a graph of *charge* against *energy* using the values in Table 25.2.

b Calculate the gradient of the graph.

c What does the gradient of the graph represent?

d Determine the voltage across the resistor.

e Determine the energy dissipated when a current of 2.4 A flows for 5.0 s.

26 Circuits and components

Electric current from a cell or generator can only flow in a **complete circuit**. Circuit diagrams are a useful way to show the arrangement of components in an electrical circuit. Each component of the diagram is represented by a circuit symbol.

Components of a circuit may be connected in series or in parallel. Series circuits only provide **one path** for the current to flow, but **parallel circuits** have **junctions** where the **current divides** and then **rejoins** after passing through **branches**.

Cells, batteries (groups of cells) and electrical generators are used to provide the power necessary to drive current around a circuit against the resistance offered by its components.

Learning objectives

- Draw **circuit diagrams** using **circuit symbols** representing various components and devices.
- Understand why **ammeters have negligible resistance** and **voltmeters have infinite resistance**.
- Understand how to **connect ammeters** and **voltmeters** with correct **polarity**.
- Differentiate between **series** and **parallel connections** in an electrical circuit and the relation between the currents and voltages associated with them.
- Combine **cells** in **series** and in **parallel** and state the **charge rating** of a cell in **amp-hours**.
- Distinguish between **primary** and **secondary cells** and compare the **zinc-carbon** dry cell to the **lead-acid** cell stating the advantages of each.
- Draw a circuit showing how a secondary cell can be **recharged from the AC mains supply**.
- Explain the concept of **resistance** and relate it to **length** and **cross-sectional area**.
- Calculate the **total resistance** of resistors in **series** and of resistors in **parallel**.
- Use a variable resistor as a **rheostat** and as a **potential divider**.
- Define **Ohm's law** and relate it to the *I–V* **characteristic** of various types of conductors.
- Understand the relation between **resistance and pd ratios** for resistors in series.
- Perform **calculations** involving voltage, current, resistance, energy, and power for **complete circuits**.

Figure 26.1 shows the circuit symbols for some common circuit components.

Some important circuit symbols

- **Voltmeter** – A device that measures the potential difference between two points.
- **Ammeter** – A device that measures the current flowing through a point.
- **Galvanometer** – A device that detects, measures and determines the direction of very small electric currents.
- **Resistor** – A component that opposes the flow of current.
- **Rheostat** – A resistor with a resistance that can be altered, usually by a slider control.
- **Light-dependent resistor (LDR)** – A resistor having a resistance that decreases with increased light intensity.
- **Thermistor** – A resistor having a resistance that changes with temperature (the resistance generally decreases with increased temperature).
- **Diode** – A device that normally allows current to flow through it in one direction only.

- **Light-emitting diode (LED)** – A diode (electrical valve) that emits light when current flows through it in a particular direction (see Chapter 28).
- **Fuse** – A metallic resistance wire placed in series with a device, which protects a circuit from excessive current by becoming hot, melting and disconnecting it.
- **Circuit breaker** – An electromagnetic switch which protects a circuit from excessive current by disconnecting it.
- **Transformer** – A device that increases (steps up) or decreases (steps down) an alternating voltage.

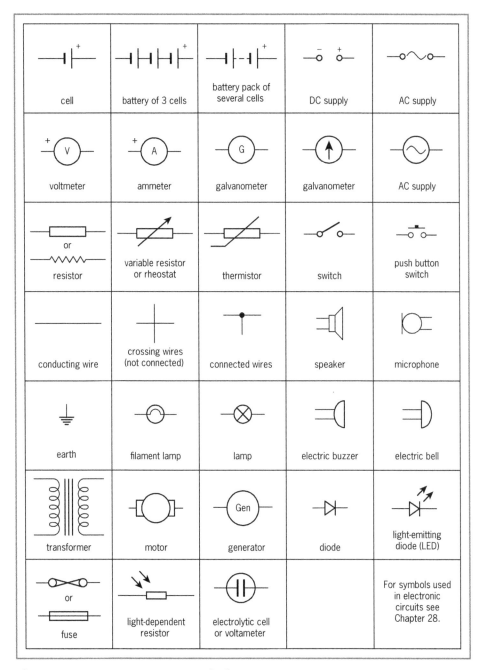

Figure 26.1 *Common circuit symbols*

Series and parallel connections

Connection of ammeters and voltmeters

Ammeters are connected in **series** with a component to measure the current **through it**. They must therefore have **negligible resistance** so as not to affect the current they are measuring. An **ideal ammeter** has **zero resistance**.

Voltmeters are connected in **parallel** with a component to measure the potential difference **across it**. Current will divert through the voltmeter and this can alter the current through (and therefore the pd across) the component being investigated. The voltmeter should therefore have a **very high resistance** to keep this diversion to a minimum. An **ideal voltmeter** has **infinite resistance**.

For some devices or components such as ammeters and voltmeters, the polarity of their terminals must be considered when connecting them in a circuit. Figure 26.2 shows that current **leaving the + terminal of a battery** must always **enter through the + markings** on an ammeter or voltmeter.

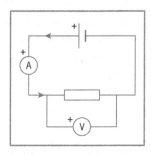

Figure 26.2 *Connection of ammeters and voltmeters with correct polarity*

Current flow in series and parallel circuits

Figure 26.3 shows how current flows in series and parallel connections in circuits. Current flows from the **positive to the negative** terminal of the battery through the **external circuit** (circuit outside the battery).

In the **series** circuit of Figure 26.3a, current has only **one path** around the circuit. The current is therefore the same at all points in the circuit and each of the ammeters will register the same reading.

In the **parallel** circuit of Figure 26.3b, current leaving the positive terminal of the battery divides into two **branches** on meeting point X. Since charge must be conserved, the charge flowing into the branch point must be equal to the charge leaving it. The current registered by ammeter A_1 is therefore equal to the sum of the currents registered by ammeters A_2 and A_3. These branch currents will then rejoin at the next branch point Y and flow through ammeter A_4 before returning to the cell through its negative terminal.

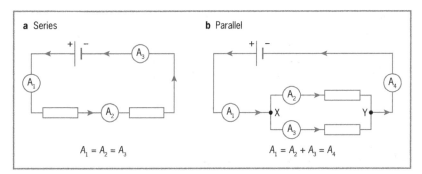

Figure 26.3 *Current in series and parallel circuits*

Potential difference (voltage) across components in series and parallel circuits

Figure 26.4 shows the pds existing in series and parallel circuits. Since voltmeters are always connected in parallel with the component they are measuring, the wires connecting them in the figures below are printed in black to distinguish them from the circuits they are investigating. In both circuits, the **cell** sets up a **potential difference** (potential drop) **equal to its emf** across the **group** of resistors.

In the series circuit of Figure 26.4a, part of the fall in potential occurs through the first resistor and the remainder through the second. Each resistor only obtains a **share of the total pd** across the combination. The emf of the cell is therefore equal to the reading, V_E, and this is equal to the sum of the readings V_1 and V_2.

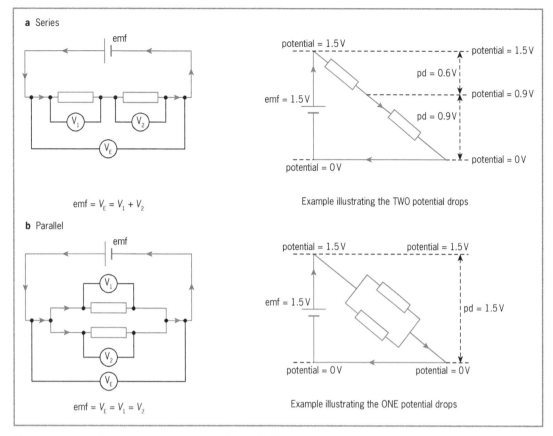

Figure 26.4 *Voltage across series and parallel connections*

In the parallel circuit of Figure 26.4b, the **pd across the group** of resistors measured as V_E is **the same** as the **pd across the individual** resistors measured as V_1 and V_2. There is only **ONE potential drop** across the **parallel section**.

Figure 26.5 shows a circuit where a combination of two resistors connected in parallel is joined in series to a third resistor. Note that in this case, there are **TWO potential drops** across the resistors. The first drop is across the parallel combination and the second drop is across the resistor in series with it.

$$V_E = V_\| + V_3$$

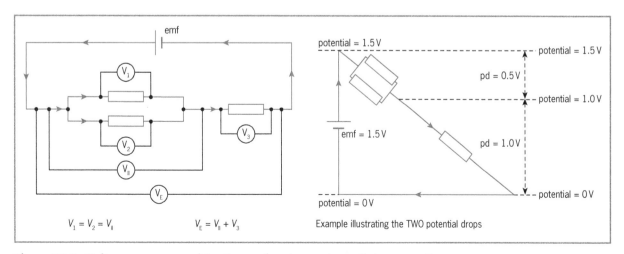

Figure 26.5 *Voltage across combinations of series and parallel connections*

Joining cells to make batteries

Figure 26.6 shows total emfs produced by cells in series and by cells in parallel.

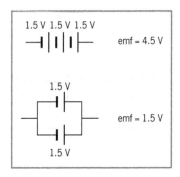

- **Series connection** – When cells are joined in series, the total emf is the sum of their emfs.

- **Parallel connection** – Sometimes it is useful to connect cells of equal emf in parallel. The total emf is then the same as the emf of one of the individual cells.

Although parallel connection of cells does not increase the total emf, it has the following advantages:

Figure 26.6 *Combined emf of cells in series and in parallel*

- To produce a particular current from a **two-cell** parallel combination, the **rate of reaction** in each cell would only be **half as high** since each cell would have to produce just **half the current**. Similarly, for a three-cell parallel combination, it would be one third as high.

  Combining cells in parallel is therefore useful if the current demand of the circuit is greater than that which can be produced by an individual cell. Several AAA 1.5 V cells can be used in parallel instead of one larger D 1.5 V cell (see later in this chapter for types of cells).

- Since the current demand on each cell is less, parallel connection of cells will last longer before the battery runs flat. This is particularly convenient when hunting at night or when using under water lights for diving. If, for example, each cell used individually will run flat in ½ hour, then two of the same cells in parallel will last 1 hour. The two cells in total deliver the **same energy** for the hour, but the user avoids having to replace the flat cell after 30 minutes.

Charge stated in amp-hours (A h)

The charge stored in a battery is often stated in amp-hours (A h) instead of in coulombs (C). It is then easier for users to estimate how long they will last. For example, a battery rated at 80 A h can deliver 1 A for 80 hours or 4 A for 20 hours.

Since $Q = It$

$\quad$ 1 A h = 1 A (60×60) s = 3600 A s = 3600 C

Primary and secondary cells

A *primary cell* is a source that converts chemical energy into electrical energy by a *non-reversible* chemical reaction.

$$A + B \rightarrow C + D + \text{electrical energy}$$

A *secondary cell* (also known as an *accumulator* or storage cell) is a source that converts chemical energy to electrical energy by a *reversible* chemical reaction.

$$A + B \rightarrow C + D + \text{electrical energy}$$

$$\text{electrical energy} + C + D \rightarrow A + B$$

To **charge** a secondary cell, **current must flow** through it in the **opposite direction** to that in which it delivers current during discharge. This means that current enters through the positive terminal of the cell and leaves through its negative terminal when connected to a charging device.

$\qquad\qquad$ **Discharging:** $\quad$ chemical energy $\rightarrow$ electrical energy

$\qquad\qquad$ **Charging:** $\quad\quad$ electrical energy $\rightarrow$ chemical energy

The zinc-carbon dry cell

Figure 26.7 shows a zinc-carbon dry cell.

- Chemical reaction between the **ammonium chloride** and the **zinc** produces the electricity.
- The **manganese oxide and carbon** mixture is used to prevent **polarisation** (the collection of hydrogen bubbles forming around the carbon rod), which would add to the unwanted **internal resistance** of the cell.

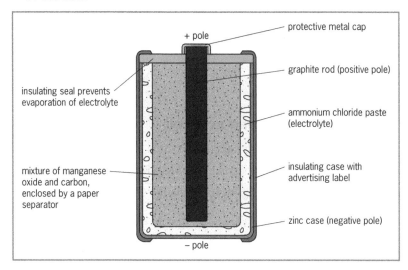

Figure 26.7 *The zinc-carbon dry cell*

The lead-acid accumulator – a secondary cell

The lead-acid accumulator is the battery generally used in vehicles. Unlike the zinc-carbon dry cell, it is **rechargeable** and can produce the **high currents** required by the starting motor of a vehicle. Although the structure of the lead-acid cell is not a requirement of the CSEC syllabus, you must be able to compare its characteristics to those of the zinc-carbon dry cell (see Table 26.1). The following Figures will give you a better understanding of some of the accumulator's characteristics.

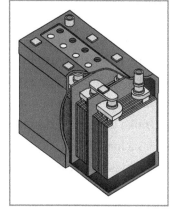

Figure 26.8 *Cross-section through a lead-acid accumulator*

Figure 26.8 shows a lead-acid accumulator. Figure 26.9 a shows one lead-acid cell, and Figure 26.9 b shows how a typical accumulator of terminal voltage 12 V is comprised of SIX 2 V cells joined in series.

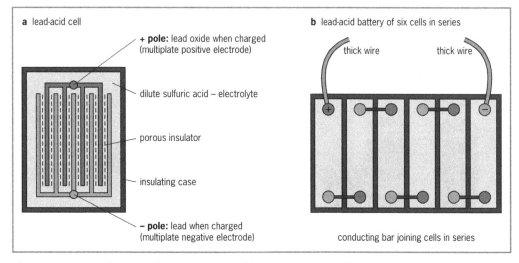

Figure 26.9 *Lead-acid cell and lead-acid battery of six cells*

Table 26.1 compares the features of the zinc-carbon cell and the lead-acid cell.

Table **26.1** *Comparison of zinc-carbon cell and lead-acid cell*

Characteristic	Zinc-carbon dry cell	Lead-acid cell
Rechargeability	not rechargeable	rechargeable
Terminal voltage	1.5 V	2.0 V
Internal resistance	high (0.5 Ω)	low (0.01 Ω)
Maximum current	maximum current drawn is typically less than 1 A	> 400 A
Electrolyte	ammonium chloride paste	dilute sulfuric acid
Portability	small and light	large and heavy

Advantages of the zinc-carbon dry cell over the lead-acid cell

- **More portable** since it is **small and light**, whereas the lead-acid accumulator is large and heavy.
- **Can be inverted** (turned upside down) without spillage because its electrolyte is in the form of a paste.
- Batteries of various voltages can **easily be compiled** by arranging the cells in series or in parallel.
- **Less costly.**

Advantages of the lead-acid cell over the zinc-carbon dry cell

- Unlike the zinc-carbon cell, these can produce much **larger currents** due to a **higher rate of reaction:**
  - the **ions have greater mobility** because the **electrolyte is a liquid** and not a paste
  - the electrodes consist of **several plates**, allowing **greater surface area** for reaction.
- Has a much **lower internal resistance.**
- Unlike the zinc-carbon cell, it is **rechargeable.**
- Has a **longer life.**
- Provides **more electrical energy** than the dry cell.

Different types of dry cells

Alkaline cells have potassium hydroxide as their electrolyte.

Lithium-ion cells are used in UPS (uninterrupted power supply) systems to provide energy to offices during power outages and power fluctuations. They are also used in laptop computers, cell phones and digital cameras. Electric vehicles utilise lithium-ion batteries since they can store more charge than a lead-acid accumulator and they have a longer life.

Figure 26.10 shows examples of different types of dry cells. The cylindrical ones shown here are all 1.5 V.

Figure **26.10** *Examples of different types of dry cells*

Charging an accumulator from the AC mains supply

Figure 26.11 shows how an accumulator can be charged from the AC mains supply.

- A **step-down transformer** (see Chapter 31) is used to **reduce the voltage** from the mains supply.
- A **diode** is used to **rectify the AC to DC** (see Chapter 28).
- The variable resistor is adjusted to bring the pd across the battery to just above 12 V.
- The charging current is directed **into the positive terminal** of the battery.

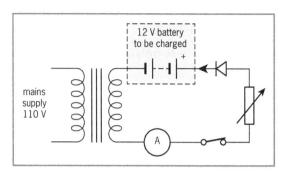

Figure 26.11 *Charging an accumulator*

A device uses a power of 48 W and operates on a current of 4.0 A from a supply storing 120 A h of charge. Determine:

a the voltage of the supply

b the charge stored by the supply in coulombs

c the time for which the average current of 4.0 A flows.

a $P = VI$ $\therefore V = \dfrac{P}{I} = \dfrac{48}{4.0} = 12\ \text{V}$

b $Q = 120\ \text{A h} = 120\ \text{A} \times (60 \times 60)\text{s} = 432\,000\ \text{A s} = 432\,000\ \text{C}$

c $Q = It$ $\therefore t = \dfrac{Q}{I} = \dfrac{432\,000}{4.0} = 108\,000\ \text{s}$ or $\dfrac{108\,000}{(60 \times 60)} = 30\ \text{hours}$

Alternatively, using charge in A h: $t = \dfrac{Q}{I} = \dfrac{120}{4.0} = 30\ \text{hours}$

Resistance

Resistance is a measure of the opposition provided to an electric current; it is the ratio of the pd across a conductor to the current through it.

A **resistor** is a device that has resistance.

Explanation of electrical resistance in metals

When electrons flow through a conductor, they are obstructed by vibrating cations (positively charged atoms) in their paths. These vibrating cations therefore obstruct the rate of flow of the electrons and hence the current.

Effect of length and cross-sectional area on resistance

The resistance, R, of a conducting wire is **proportional** to its **length**, l, and **inversely proportional** to its **cross-sectional area**, A:

$$R \propto \dfrac{l}{A}$$

The resistors shown in Figure 26.12 are made of the same material. R_3 has the greatest resistance and R_1 has the least resistance.

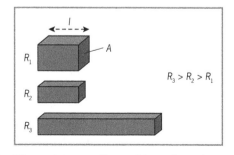

Figure 26.12 *Effect of length and cross-sectional area on resistance*

Adding resistors

Connected in series $\qquad R = R_1 + R_2 + R_3 \ldots$

Connected in parallel $\qquad \dfrac{1}{R} = \dfrac{1}{R_1} + \dfrac{1}{R_2} + \dfrac{1}{R_3} \ldots$

For the case where only TWO resistors are in parallel:

$$R = \frac{\text{product}}{\text{sum}} \qquad\qquad R = \frac{R_1 \times R_2}{R_1 + R_2}$$

- When resistors are connected in parallel, their total resistance is less than that of any of the individual resistors of the combination.
- When two identical resistors are in parallel, their combined resistance is half the value of any of the individual resistors of the combination. If each is of resistance 10 Ω, then the total resistance is 5 Ω.

Example 2

Determine the total resistance of each of the following combinations.

a 4 Ω and 3 Ω in series

b 6 Ω and 4 Ω in parallel

c 4 Ω, 5 Ω and 20 Ω in parallel.

a $R = 4\,\Omega + 3\,\Omega = 7\,\Omega$

b $R = \dfrac{\text{product}}{\text{sum}} = \dfrac{6 \times 4}{6 + 4} = \dfrac{24}{10} = 2.4\,\Omega$

c $\dfrac{1}{R} = \dfrac{1}{4} + \dfrac{1}{5} + \dfrac{1}{20}$

$\dfrac{1}{R} = \dfrac{1}{0.5}$

$R = 2\,\Omega$ ... Remember to flip $\dfrac{1}{R}$ to obtain R

Example 3

Determine the total resistance between X and Y of each of the combinations of filament lamps shown in Figure 26.13, if the resistance of each lamp is 2.0 Ω.

HINT: To fast-track these calculations, the resistance of any set of resistors in series within a branch can be interpreted as the sum of the individual resistances in that branch.

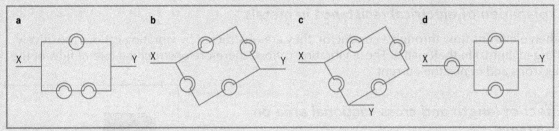

Figure 26.13 *Example 3*

a Interpret as 2.0 Ω in parallel with 4.0 Ω: $R = \dfrac{2.0 \times 4.0}{2.0 + 4.0} = \dfrac{8.0}{6.0} = 1.3\,\Omega$

b Interpret as 4.0 Ω in parallel with 4.0 Ω: $R = \dfrac{4.0 \times 4.0}{4.0 + 4.0} = \dfrac{16.0}{8.0} = 2.0\,\Omega$

Section D: Electricity and magnetism

c Interpret as 6.0 Ω in parallel with 2.0 Ω: $R = \dfrac{6.0 \times 2.0}{6.0 + 2.0} = \dfrac{12.0}{8.0} = 1.5\ \Omega$

d Interpret as 2.0 Ω in series with (2.0 Ω in parallel with 2.0 Ω) $R = 2.0 + \dfrac{2.0 \times 2.0}{2.0 + 2.0} = 3.0\ \Omega$

The variable resistor

Figure 26.14 shows a variable resistor being used as a **rheostat** and as a **potential divider**.

Rheostat – This **controls the current** in a circuit. When used as a rheostat, only terminals A and C are connected in the circuit. By adjusting the position of the slider, the total resistance and therefore the current in the circuit can be varied; the voltage (**pd**) across a component connected in series with the rheostat can be **varied** but **can never be zero**.

Potential divider – This puts a **fraction of a voltage** across a component. When used as a potential divider, all three terminals are connected in the circuit. A voltage is applied to the variable resistor between the terminals A and B, and a part of this voltage is applied to a component connected in parallel with the resistor coil between terminal A and the slider at terminal C. Used in this way, the pd across a component **can be reduced to zero**.

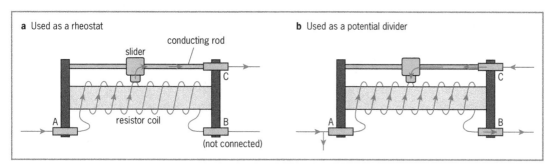

a Used as a rheostat

b Used as a potential divider

Figure 26.14 *Variable resistor used as a rheostat and as a potential divider*

Figures 26.15a and 26.15b show arrangements of a variable resistor being used as a rheostat and as a potential divider, respectively. Figures 26.16a and 26.16b show practical arrangements of the same circuits.

Ohm's law and I–V characteristic curves

*Ohm's law: Provided the **temperature remains constant**, the current, I, through an ohmic conductor is proportional to the potential difference, V, between its ends.*

$V \propto I$

$\therefore V = IR$ … where the constant of proportionality, R, is the resistance of the conductor.

Experiment to determine I–V characteristics of various components

The relationship between the current, I, through a component and the potential difference, V, across it can be investigated using the circuits shown in Figure 26.15. Lamps and diodes behave differently at low voltages than they do at higher voltages and so circuit 26.15b, which can also measure the low voltages, is preferred when investigating these components.

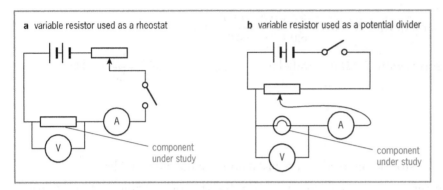

a variable resistor used as a rheostat   **b** variable resistor used as a potential divider

component under study   component under study

Figure 26.15 *Investigating I–V relationships*

An initial pair of readings of *I* and *V* is taken.

- By adjusting the variable resistor, five other pairs of readings of *I* and *V* are taken.
- The connection between the terminals of the battery and the circuit is reversed and the process is repeated.
- A graph of *I* against *V* is plotted.

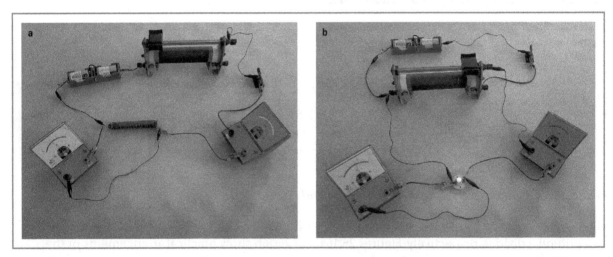

Figure 26.16 *Practical circuits for investigating I–V relationships utilising a variable resistor.*
a as a rheostat. b as a potential divider

Graphs of *I* against *V* for a component are called the **I–V characteristics** of the component.

Figure 26.17 shows the *I–V* characteristic curves for different components.

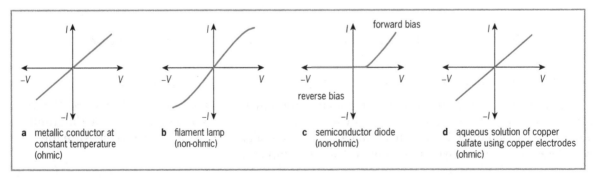

a metallic conductor at constant temperature (ohmic)

b filament lamp (non-ohmic)

c semiconductor diode (non-ohmic)

d aqueous solution of copper sulfate using copper electrodes (ohmic)

Figure 26.17 *I–V characteristic for different components*

- **Metallic conductors at constant temperature (Figure 26.17a) and electrolytes such as copper sulfate having copper electrodes (Figure 26.17d)**

  These are ohmic conductors since a graph of I against V, or V against I, is a **straight line through the origin**.

  For these conductors, since $R = \dfrac{V}{I}$, the **gradient** of an I–V graph is the **inverse of resistance** and the

  gradient of a V–I graph is the **resistance** (see question 8 at the end of this chapter).

- **Filament lamp (Figure 26.17b)**

  This is a non-ohmic conductor. As the voltage is increased across the filament, the electrons in it are pulled with greater force and collide more vigorously with the cations of the metal filament. Their kinetic energy transforms into thermal energy of the cations, causing the vibration of the cations to increase. The increased vibration blocks the flow of the electrons to a greater extent than previously, causing the resistance to increase.

- **Semiconductor diode (Figure 26.17c)**

  When the diode is forward biased (its polarity such that it conducts), a small initial voltage of about 0.5 V is needed before current flows. When it is reverse biased, conduction is almost zero (taken as zero for applications pertaining to the CSEC syllabus).

Relation of pd to resistance for resistors in series

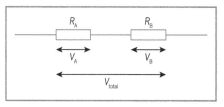

For resistors of resistances R_A and R_B connected in series, having pds V_A and V_B across their ends (see Figure 26.18):

$$\frac{V_A}{V_B} = \frac{R_A}{R_B} \qquad \text{and} \qquad \frac{V_A}{V_{total}} = \frac{R_A}{R_{total}} \qquad \text{and} \qquad \frac{V_B}{V_{total}} = \frac{R_B}{R_{total}}$$

Figure 26.18 *Relation of pd to resistance for resistors in series*

Example 4

With reference to Figure 26.19, calculate:

a the resistance R_A

b the pds V_C and V_D

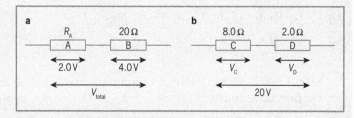

Figure 26.19 *Example 4*

a $\dfrac{R_A}{R_B} = \dfrac{V_A}{V_B}$ $\qquad \dfrac{R_A}{20} = \dfrac{2.0}{4.0}$ $\qquad R_A = \dfrac{2.0 \times 20}{4.0}$ $\qquad R_A = 10\ \Omega$

b V_C: $\dfrac{R_C}{R_{total}} = \dfrac{V_C}{V_{total}}$ $\qquad \dfrac{8.0}{8.0 + 2.0} = \dfrac{V_C}{20}$ $\qquad \dfrac{8.0}{10.0} = \dfrac{V_C}{20}$ $\qquad V_C = \dfrac{8.0 \times 20}{10.0}$ $\qquad V_C = 16\ \text{V}$

$\quad\ V_D$: $V_{total} = V_C + V_D$ $\qquad 20 = 16 + V_D$ $\qquad \therefore 20 - 16 = V_D$ $\qquad V_D = 4\ \text{V}$

Calculations involving complete circuits

Pointers to note when performing calculations involving complete circuits.

- The emf is the sum of the pds in a circuit.
- The values of V, I and R used in calculations must pertain to the same component or group of components. It is advisable to place a double-ended arrow (see the examples that follow) across ALL resistors in the circuit as a reminder of the components that can be used in formulae involving emf or total pd.
- Subscripts can be used to identify components. In the examples that follow, if no subscripts are used, V, I and R pertain to totals; V would then be the emf, I the total current (the current leaving the battery) and R the total resistance.

Example 5

Figure 26.20 shows a metallic resistor of resistance 200 Ω connected to a *thermistor* which has a resistance of 800 Ω at a temperature of 25 °C. Determine the current through the thermistor at this temperature.

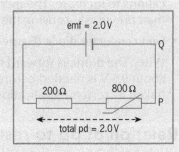

Figure 26.20 *Example 5*

There is only one current in a series circuit; the current in the thermistor is the same as the current in the 200 Ω resistor, and this is the same as the total current from the cell. Since the total pd is known, the current can be found by dividing the total pd by the total resistance.

$V = IR$

$2.0 = I \times 1000$

$I = \dfrac{2.0}{1000}$

$I = 2.0 \times 10^{-3}\ \text{A}$

Example 6

Figure 26.21 shows a circuit containing two lamps, A and B, in series. Determine:

a the current delivered by the cell

b the pd (voltage) across lamp A

c the reading on a voltmeter connected across lamp B

d the reading on a voltmeter connected between points P and Q.

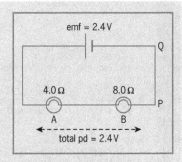

Figure 26.21 *Example 6*

a $V = IR$ $2.4 = I \times 12.0$ $I = \dfrac{2.4}{12.0}$ $I = 0.20\ \text{A}$

b $V_A = IR_A$ $V_A = 0.20 \times 4.0$ $V_A = 0.80\ \text{V}$

c $V_B = IR_B$ $V_B = 0.20 \times 8.0$ $V_B = 1.6\ \text{V}$

d $V_{PQ} = IR_{PQ}$ $V_{PQ} = 0.20 \times 0$ $V_{PQ} = 0\ \text{V}$

Note: The **pd** across a section of **perfectly conducting wire** is **zero**.

Example 7

Figure 26.22 shows a circuit containing two lamps, M and N, in parallel. Determine:

a the resistance of the parallel combination of lamps

b the current output of the battery

c the pd across lamp M

d the pd across lamp N

e the reading on a voltmeter if connected between points, P and Q

f the reading on the ammeter connected in series with lamp M.

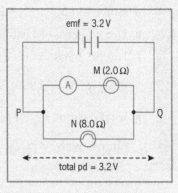

Figure 26.22 *Example 7*

a $R_\parallel = \dfrac{2.0 \times 8.0}{2.0 + 8.0}$

$R_\parallel = \dfrac{16.0}{10.0}$

$R_\parallel = 1.6\ \Omega$

b $V = IR$ $\qquad$ $3.2 = I \times 1.6$ $\qquad$ $\dfrac{3.2}{1.6} = I$

$I = 2.0\ \text{A}$

c $V_M = 3.2\ \text{V}$ ... since the lamp is in parallel with the source

d $V_N = 3.2\ \text{V}$ ... since the lamp is in parallel with the source

e $V_{PQ} = 3.2\ \text{V}$ ... since this section is in parallel with the source

f $V_M = I_M R_M$ $\qquad$ $3.2 = I_M \times 2.0$ $\qquad$ $\dfrac{3.2}{2.0} = I_M$ $\qquad$ $I_M = 1.6\ \text{A}$

Example 8

Figure 26.23 shows a circuit containing a resistor N connected in series to a pair of lamps L and M, which are connected in parallel. Determine:

a the total resistance

b the current from the battery

c the pd registered by the voltmeter

d the pd across the parallel section of lamps L and M

e the pd across lamp M

f the current registered by the ammeter A_1 (i.e. the current through L)

g the current through M.

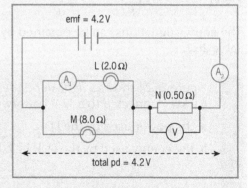

Figure 26.23 *Example 8*

a $R = \dfrac{2.0 \times 8.0}{2.0 + 8.0} + 0.50$ $\qquad$ $R = \dfrac{16.0}{10.0} + 0.50$ $\qquad$ $R = 1.6 + 0.5$ $\qquad$ $R = 2.1\ \Omega$

b $V = IR$ $\qquad$ $4.2 = I \times 2.1$ $\qquad$ $\dfrac{4.2}{2.1} = I$ $\qquad$ $I = 2.0\ \text{A}$

c The total current from the battery (2.0 A) flows through N.

$V_N = I_N R_N$ $\qquad$ $V_N = 2.0 \times 0.50$ $\qquad$ $V_N = 1.0\ \text{V}$

d emf $= V_{||} + V_{N}$ $4.2 = V_{||} + 1.0$ $V_{||} = 4.2 - 1.0$ $V_{||} = 3.2$ V

e The pd across M is the same as the pd across the group of parallel resistors (3.2 V).

f The pd across M is equal to the pd across L (3.2 V) since they are in parallel.

$V_{L} = I_{L}R_{L}$ $3.2 = I_{L} \times 2.0$ $\dfrac{3.2}{2.0} = I_{L}$ $I_{L} = 1.6$ A

g Total current $= I_{L} + I_{M}$ $2.0 = 1.6 + I_{M}$ $2.0 - 1.6 = I_{M}$ $I_{M} = 0.4$ A

   Alternatively: $V_{M} = I_{M}R_{M}$ $3.2 = I_{M} \times 8.0$ $\dfrac{3.2}{8.0} = I_{M}$ $I_{M} = 0.4$ A

Energy and power in complete circuits

If you have a good understanding of examples 5 to 8 you should be competent in the following calculations on energy and power. The rules are the same for the use of V, I and R in the various equations. Table 26.2 shows how we can calculate power and energy from V, I and R.

Table 26.2 *Energy and power*

Power $= \dfrac{\text{energy}}{\text{time}}$	Energy = power × time
$P = VI$	$E = VIt$
$P = I^2R$	$E = I^2Rt$
$P = \dfrac{V^2}{R}$	$E = \dfrac{V^2t}{R}$

Using $P = VI$ with a knowledge of $V = IR$, the other two equations of power can be derived.

$P = VI$ $\therefore P = (IR)I$ $\therefore P = I^2R$

$P = VI$ $\therefore P = V\left(\dfrac{V}{R}\right)$ $\therefore P = \dfrac{V^2}{R}$

The energy equations can be obtained by adding t to each of the power equations, as shown in Table 26.2.

Example 9

Figure 26.24 shows a *light-dependent resistor* of resistance 6.0 kΩ, connected to a 12 V battery. Determine:

a the power used by the LDR

b the energy used by the LDR in 4.0 minutes.

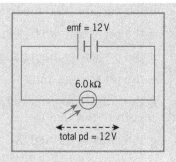

Figure 26.24 *Example 9*

a $P = \dfrac{V^2}{R} = \dfrac{12^2}{6.0 \times 10^3} = 0.024$ W

b $E = Pt$ $E = 0.024 \times (4.0 \times 60)$ $E = 5.76$ J   (5.8 J to 2 sig.fig.)

Example 10

From the circuit shown in Figure 26.25, determine:

a the current flowing through A

b the power used by A

c the power used by B

d the energy used by A in 3 minutes

e the energy used by B in 3 minutes

f the total energy used in 3 minutes.

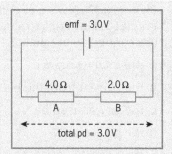

Figure 26.25 *Example 10*

a There is only one current in a series circuit. This can be found using the **total voltage** and the **total resistance**. We cannot use the total voltage with the resistance of A in the equation.

$V = IR$ $3.0 = I(4.0 + 2.0)$ $3.0 = I \times 6.0$ $\dfrac{3.0}{6.0} = I$ $I = 0.50$ A

b $P_A = I^2 R_A$ $P_A = 0.50^2 \times 4.0$ $P_A = 1.0$ W

c $P_B = I^2 R_B$ $P_B = 0.50^2 \times 2.0$ $P_B = 0.50$ W

d $E_A = P_A \times t$ $E_A = 1.0 \times (3 \times 60)$ $E_A = 180$ J

e $E_B = P_B \times t$ $E_B = 0.50 \times (3 \times 60)$ $E_B = 90$ J

f Summing the results of **d** and **e**: $E = 180 + 90 = 270$ J

The total **energy used by the resistors** is **equal** to the **energy delivered by the battery**. Therefore it can also be calculated using the emf of the battery, the current from the battery and the time.

$E = VIt$ $E = 3.0 \times 0.50 \times (3 \times 60)$ $E = 270$ J

Example 11

Figure 26.26 shows resistors in parallel connected to a 3.0 V supply. Determine:

a the pd across lamp M

b the power used by lamp M

c the power used by lamp N

d which lamp is brighter

e the energy used by lamp M in 5 minutes

f the power delivered by the battery

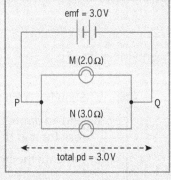

Figure 26.26 *Example 11*

a 3.0 V   (Both lamps are in parallel with the source and so the pd across each is equal to that of the source).

b $P_M = \dfrac{V^2}{R_M}$ $P_M = \dfrac{3.0^2}{2.0}$ $P_M = 4.5$ W

c $P_N = \dfrac{V^2}{R_N}$ $P_N = \dfrac{3.0^2}{3.0}$ $P_N = 3.0$ W

d M is brighter since it uses the greater *power*

e $P_M = \dfrac{E_M}{t}$ $\therefore E_M = P_M \times t$ $E_M = 4.5 \times (5 \times 60)$ $E_M = 1350$ J

f $P = P_M + P_N$ $P = 4.5 + 3.0$ $P = 7.5$ W

Example 12

Figure 26.27 shows a circuit containing a resistor N connected in series to a pair of lamps L and M, which are connected in parallel. Determine:

a the total resistance

b the current from the battery

c the power used by N

d the pd across N

e the pd across M

f the power used by L

g the total power used by the circuit.

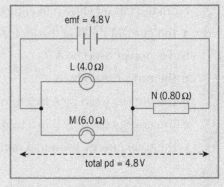

emf = 4.8 V

L (4.0 Ω)

N (0.80 Ω)

M (6.0 Ω)

total pd = 4.8 V

Figure 26.27 *Example 12*

a $R = \dfrac{4.0 \times 6.0}{4.0 + 6.0} + 0.80 \qquad R = 2.4 + 0.80 \qquad R = 3.2\ \Omega$

b $V = IR \qquad\qquad I = \dfrac{V}{R} \qquad\qquad I = \dfrac{4.8}{3.2} = 1.5\ \text{A}$

c The current leaving the battery is the current flowing through N.

$P_N = I_N^2 R \qquad P_N = 1.5^2 \times 0.80 \qquad P_N = 1.8\ \text{W}$

d $V_N = I_N R \qquad V_N = 1.5 \times 0.80 \qquad V_N = 1.2\ \text{V}$

e emf = pd across parallel section + pd across N

$4.8 = V_\| + 1.2 \qquad V_\| = 4.8 - 1.2 \qquad V_\| = 3.6\ \text{V}$ \quad (This is the pd across L and across M)

f $P_L = \dfrac{V_L^2}{R_L} \qquad P_L = \dfrac{3.6^2}{4.0} \qquad P_L = 3.24$ \quad (3.2 W to 2 sig.fig.)

g $P = VI \qquad\qquad P = 4.8 \times 1.5 \qquad P = 7.2\ \text{W}$ \quad (this is the same as the power delivered by the battery)

Recalling facts

 1
  a Sketch a circuit diagram containing a switch, a battery of 3 cells in series and a resistor.

  b Add an ammeter and voltmeter to the circuit to measure the current through the resistor and the pd across it, respectively.

  c Indicate the polarity of the ammeter and voltmeter connections with a + symbol.

2 Explain why ammeters should have a very low resistance and voltmeters should have a very high resistance.

3 TWO cells of the same emf, *E*, are connected to form a battery. What is their total emf if connected:

  a in series?

  b in parallel?

4 The resistance of A is twice the resistance of B. What can be said of the relative magnitudes of the currents flowing in the resistors when connected to a cell:

  a in series?

  b in parallel?

5 The chemical reactions that occur in cell X are reversible but those occurring in cell Y are not. Which of the cells:

    a is an accumulator?

    b is not rechargeable?

    c is a secondary cell?

6 Figure 26.28 shows a zinc-carbon dry cell.

    a What do the labels A to E represent?

    b What is meant by *polarisation* within the cell?

    c How is polarisation of the cell prevented?

    d Which pole is the positive pole?

    e List THREE advantages and THREE disadvantages of a lead-acid accumulator compared to a zinc-carbon dry cell.

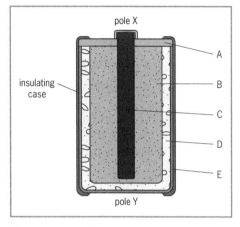

Figure 26.28 *Question 6*

7 **a** State Ohm's law.

    b Sketch graphs of *voltage against current* for each of the following conductors (no values are required but the axes of the graphs should indicate positive and negative voltages and currents):

      i filament lamp

      ii semiconductor diode

      iii metallic conductor at constant temperature.

    c State which, if any, of the conductors listed in **b** are ohmic conductors, giving the reason/s for your choice.

    d What does an increase in the gradient of a *V–I* graph reveal about the resistance of a component?

    e Explain, in terms of the kinetic theory, what happens to the resistance of the component in **b i** as the pd across it is increased.

8 Describe an experiment to determine the resistance of a metallic resistor. Your answer should include:

    a a suitable method

    b a circuit diagram

    c an expected graph (sketch with no values)

    d a statement of how the resistance is calculated.

Applying facts

9 A battery charger is designed to charge a 3 V battery from a domestic 110 V socket in the home. Sketch a circuit diagram of the charging circuit.

10 Sketch a circuit diagram which contains a battery of 3 cells in series and a resistor connected in series to two light bulbs that are connected in parallel. The circuit is to be controlled by a main switch and there should also be individual switches to control each of the two light bulbs.

11 A fully charged 12 V battery with a charge capacity of 200 A h is used to deliver a current of 2.0 A. Determine:

  a the charge stored in coulombs

  b the energy that the battery can deliver

  c the power output of the battery

  d the charge used in 4.0 hours

  e the time taken for the battery to completely discharge.

12 Three resistors A, B and C are made of the same material. A is twice the length of B but has the same cross-sectional area. C is the same length as B but has twice its cross-sectional area. List the resistors in order of decreasing resistance.

13 Determine the total resistance between the terminals *marked by dots* in each of the following circuits of Figure 26.29, given that each resistor is of resistance 3.0 Ω.

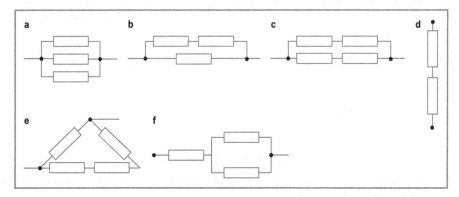

Figure 26.29 *Question 13*

14 Resistors A and B of resistance 3.5 Ω and 14.0 Ω are in series. Determine the pd across each when the total voltage across them is 20 V.

15 Resistors C and D are in series. The pd across C is 4.0 V and that across D is 12.0 V. If the resistance of D is 6.0 Ω, what is the resistance of C?

16 From the information given in Figure 26.30 determine:

  a the pd across the 2.0 Ω resistor

  b the current through the 4.0 Ω resistor

  c the current entering point X

  d the pd between X and Y.

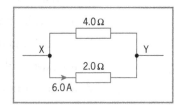

Figure 26.30 *Question 16*

17 Figure 26.31 shows a circuit powered by a 6.0 V battery. The pd across A is 1.0 V. Determine:

  a the current delivered by the battery

  b the pd across C

  c the pd across B

  d the resistance of B

  e the power used by C

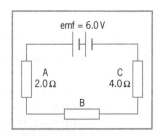

Figure 26.31 *Question 17*

f the energy used by A in 5.0 minutes

g the energy taken from the battery in 5.0 minutes

h the power delivered by the cells.

18 Each cell in Figure 26.32 has an emf of 1.5 V. Determine:

a the pd across XY

b the total resistance

c the current in L

d the current in M

e the power consumed by L

f the energy used by M in 2.0 minutes

g the current delivered by the battery

h the power delivered by the battery.

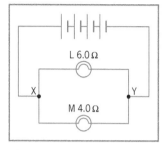

Figure 26.32 *Question 18*

19 The voltmeter across A in Figure 26.33 reads 4.8 V. Determine:

a the resistance of A

b the total resistance of the circuit

c the pd between points X and Y

d the current in the 4.0 Ω resistor

e the current in the 6.0 Ω resistor

f the charge flowing through A in 20 seconds

g the charge flowing from the battery in 20 seconds

h the power delivered by the battery.

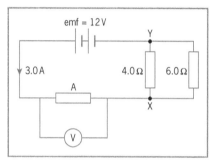

Figure 26.33 *Question 19*

20 Each of the lamps shown in Figure 26.34 is of resistance 2.0 Ω.

a With S_1 and S_2 initially open, calculate:

  i the pd across each of the strings, P and Q

  ii the pd across each lamp

  iii the current through each string

  vi the current through each lamp

  v the power used by each lamp

  vi the total power delivered.

b Describe and explain what happens with the lamps of strings P and Q when S_1 is closed and S_2 remains open.

c S_2 is closed and S_1 is opened.

  i Describe and explain the effect on the brightness of the lamps of string P.

  ii Describe and explain TWO possible occurrences with the lamps of string Q.

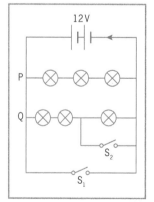

Figure 26.34 *Question 20*

21 Table 26.2 shows readings obtained in an investigation to determine the relation between the potential difference across a component to the current flowing through it.

Table 26.2 *Question 23*

Potential difference (*V*/V)	0.48	1.0	1.4	1.6	2.2	3.4
Current (*I*/A)	0.20	0.40	0.50	0.60	0.70	0.85

a Plot a graph of the potential difference *V* across the component against the current *I* through it.

b Comment on the relationship between *V* and *I*.

c Is the component ohmic? Give a reason for your answer.

d What is the resistance of the component at the following voltages?

 i 0.6 V

 ii 3.0 V.

e What can you deduce about the resistance of the component as the voltage is increased?

27 Electricity in the home

Electricity is delivered to homes as an **alternating current and voltage** for reasons discussed in Chapter 31. In Barbados this voltage is 110 V and alternates at a frequency of 50 Hz. A **live** wire and a **neutral** wire bring the electricity from the power company to a **distribution box** in the home, where it is distributed to several circuits in **parallel**. A third wire, known as an **earth** wire, is also used in the circuitry.

Learning objectives

- Understand the importance of **fuses**, **circuit breakers** and **thickness of wires** in controlling the **safety** of electrical wiring schemes.
- Be familiar with **international colour codes** and the use of the live, neutral and earth wires for electrical wiring.
- Understand the role of the **distribution panel**, **ring mains** circuits and **lighting circuits**.
- Be aware of the **advantages** of the use of **parallel connection** of domestic appliances.
- Discuss the use of the **earth wire** as protection against short circuits.
- Understand why **switches**, **fuses** and **circuit breakers** should be placed in the **live** wire.

Fuses and circuit breakers

A **fuse** (see Figure 27.1) is a short piece of thin wire placed in a circuit in series with one or more devices. When the **current exceeds a certain value**, the fuse becomes hot and **melts** (blows), **breaking the circuit** and thereby **preventing damage** to the device. A suitable fuse should melt at a current slightly higher than the normal operating current of the device. This value is known as the **current rating** of the fuse. See also *Fuses in plugs and appliances* later in this chapter.

A **circuit breaker** is a device which creates an **electromagnetic force** that is strong enough to **open a trip-switch** that breaks the circuit when the **current exceeds a certain value**. It serves the same purpose as a fuse. Figure 27.2 shows circuit breakers in the panel of a distribution box. Their function is explained in Chapter 30.

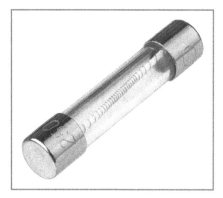

Figure 27.1 *A fuse*

Figure 27.2 *Circuit breakers in panel of distribution box*

Thickness of wires

Wires carrying **high currents** should be **thick** to **prevent overheating** and melting of the insulation, which can result in **short circuits** and can cause **electrical fires**. The following circuits require thick wires.

1. Circuits containing **high-powered electrical appliances** such as water heaters, refrigerators, cookers, washing machines and heavy-duty power tools. Since $P \propto I$, high-powered devices draw high currents.
2. Circuits from the distribution box connected to **power sockets** (see Figure 27.5). **Several appliances**, each drawing their own current, may be using a particular circuit at the **same time**.

Colour codes for wires

Many appliances in the Caribbean are sourced from the UK, US or Canada. Table 27.1 shows the colour codes used in those regions and Figure 27.3 shows wires having the UK colour code. Other colour codes are used in different countries.

Table 27.1 *International colour codes*

	LIVE	NEUTRAL	EARTH
UK	brown	blue	green + yellow
US and Canada	black	white	green + yellow

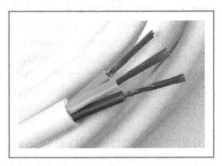

Figure 27.3 *UK colour code*

Distribution box

- In Barbados, the live and neutral wires have a voltage of 110 V between them. In the distribution box, they are each connected to a conducting strip.

- Several circuits, including **ring-mains circuits** (socket circuits) and **lighting circuits**, are taken from the **LIVE** and **NEUTRAL** strips (see Figure 27.4). Each circuit is fitted with a **fuse** or **circuit breaker** in its **live** wire and provides a pd of 110 V to devices it is connected to. An actual distribution box generally has many more circuits exiting. Each room in the home may have its own power socket circuit and lighting circuit.

- The **EARTH strip** is connected to a cable that is **buried in the ground**.

- The **current rating** of the **fuse or circuit breaker** depends on the **total possible current** that should flow in the circuit when several appliances and lights are being used at the same time.

- **Heating devices** such as heaters and electric cookers usually take **large currents**, and so generally have a fuse or circuit breaker **specially assigned** to them in the distribution box.

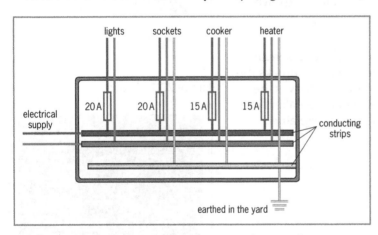

Figure 27.4 *Distribution box using UK colour code*

Advantages of using parallel connection of domestic appliances

Appliances can:

- **be controlled individually**; if connected in series, switching on/off one, will switch on/off all

- **be designed to operate on a single voltage** (for example, 110 V); if appliances were connected in series, they would share the total pd across the group of appliances, and therefore each obtain a smaller pd across their terminals

- **obtain different currents** dependent on their power rating; since $P = VI$, $P \propto I$ for a constant 110 V
- **use individual fuses** that are applicable to their normal operational current; in a series circuit, only one fuse can be used.

Ring-mains circuit (socket circuit)

A ring-mains circuit is illustrated in Figure 27.5.

- **Thick wires** are used in these circuits so that they can carry the total current drawn by **several devices**.
- Each wire forms a **loop (ring)** that is connected to the distribution box.
- The **sockets** are connected to these rings such that, when viewed with the EARTH appearing at the **top and centred**, the **LIVE is on the right**. Verify that this is so for each socket of Figure 27.5.
- Under **normal operating conditions** current flows only in the **LIVE** and NEUTRAL wires.
- If the casing of an appliance becomes electrified due to a **short circuit**, current then flows in the **LIVE** and EARTH wires (see Figure 27.6).

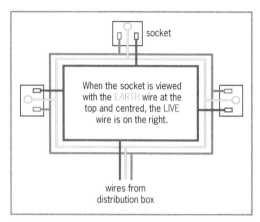

Figure 27.5 *Ring mains with power sockets*

Function of the earth wire

A **short circuit** can occur if a conductor bridges the gap between a circuit and the **casing of an appliance**. If the **casing** is made of **metal**, this can be dangerous unless an **EARTH** wire is attached in the circuit. Should the **casing be electrified** with no earth wire attached, the circuit will continue to work as long as the case is insulated (for example, it rests on rubber supports). However, if the casing is now touched, current may then flow between it and the ground through the person.

To prevent electrical shock, an earth wire is connected **between the casing and the ground** via the distribution box. As soon as the short circuit occurs, the **high current** will flow through the **live** and **earth** wires instead of through the **live** and **neutral** wires. The fuse will blow (or the breaker will trip), making it safe to handle the appliance.

Figure 27.6 illustrates the use of the earth wire in an electric toaster having a metal casing and resting on insulating rubber supports.

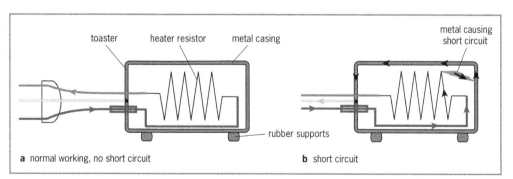

Figure 27.6 *Function of the earth wire*

- No current flows in the earth wire unless a malfunction occurs.
- When current flows in the earth wire, the neutral wire is bypassed.
- If a short circuit is discovered, the appliance should be repaired before the fuse is replaced or the breaker is reset.

Lighting circuit

Lamps generally take small currents. Figure 27.7 shows that lamps are normally connected without the use of an earth wire unless they have a metal casing. Figure 27.8 shows a working model of a lighting circuit.

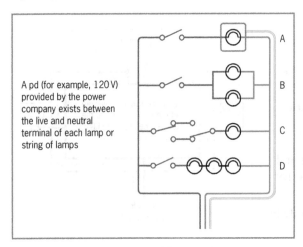

A pd (for example, 120 V) provided by the power company exists between the live and neutral terminal of each lamp or string of lamps

Figure 27.7 *Lighting circuit*

Notes: For the different light strings of Figure 27.7:

A – A single switch controlling a lamp with an earthed metal casing.

B – A single switch controlling two lamps in parallel. A pd of 120 V is across each lamp.

C – A two-way switch to operate a light from the top and bottom of a staircase.

D – A single switch controlling three lamps in series. Strings of lamps in series are designed to have the total pd supplied by the power company across their length. For this string, a pd of 40 V (120 V / 3) is across each lamp if the lamps are similar.

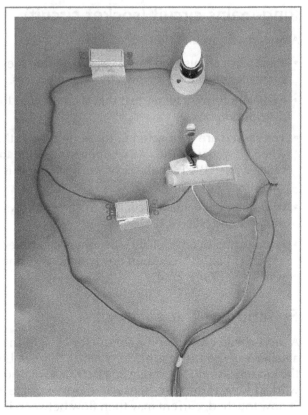

Figure 27.8 *Working model of a 'lighting circuit' showing the use of an earth wire for an appliance with a metal casing*

Placement of fuses, breakers and switches in the live wire

Fuses blow and **breakers trip** when there is either:

- a **short circuit** in the appliance causing the case to be electrified; a current, higher than the normal operating current of the appliance, then flows in the **live and earth** wires, or
- a **power surge** for **reasons external to the appliance**; a current, higher than the normal operating current of the appliance, then flows in the **live and neutral** wires.

By placing the fuse or circuit breaker in the **live wire** (see Figure 27.6), the appliance will be protected in both situations.

Switches should always be placed in the **live wire**. Consider a lamp connected in a circuit in which a switch is placed in the neutral wire. The switch will still be able to switch the circuit on and off. If, however, the live terminal is touched when attempting to change the lamp, the person can receive an electric shock. By placing the **switch in the live wire**, once it is in the 'off' position, current cannot flow through any of the terminals of the appliance. Both of these situations are illustrated in Figure 27.9.

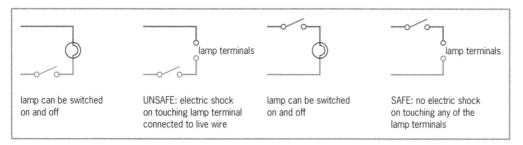

lamp terminals

lamp can be switched on and off

UNSAFE: electric shock on touching lamp terminal connected to live wire

lamp can be switched on and off

lamp terminals

SAFE: no electric shock on touching any of the lamp terminals

Figure 27.9 *Placement of switches in live wire*

Fuses in plugs or appliances

A fuse in the plug of an appliance (Figure 27.10) or in the appliance itself can offer better protection than the fuse in the distribution box, because its current rating is selected specifically for the appliance and not for a group of appliances. Its current rating should only be slightly higher than the current normally taken by the appliance.

If a device normally operates on 2.5 A, a fuse with a current rating of 3.0 A would be more suitable than the larger fuse in the distribution box of about 20 A. Should a current of 5 A flow, the fuse in the distribution box will remain intact, but the one connected in the plug or appliance will blow and save the appliance from damage.

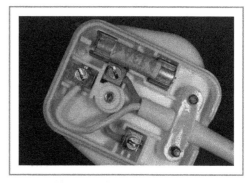

Figure 27.10 *Three-pin plug with fuse in live wire*

Adverse effects of connecting electrical appliances to an incorrect or fluctuating supply

* **Incorrect frequency:** Appliances are normally designed to operate on AC frequencies of either **50 Hz** or **60 Hz**. If operated on an incorrect frequency, they are likely to malfunction or even be damaged.

* **Voltage too high:** If the **voltage** supplied to an appliance is **greater** than its recommended operational voltage, **overheating** may occur and the appliance may be destroyed.

* **Voltage too low:** If the **voltage** supplied to an appliance is **less** than its recommended operational voltage, the appliance can also be damaged. An electric motor supplied with a voltage less than it is designed for will have an increase in current in its coils (due to reasons beyond the scope of this syllabus) that can cause it to overheat and be damaged.

* **Voltage fluctuating:** If the voltage supply is fluctuating, the response of the device will also fluctuate and may be damaged. Fluctuating voltages can be very detrimental to information stored in digital form.

Recalling facts

1
  a Explain the function of a fuse.

  b Name the commonly used electromagnetic device that has the same purpose as a fuse.

2
  a Why do high-powered devices require thick wires?

  b Why are the wires of a ring-mains circuit connected to power sockets generally thick?

c A ring-mains circuit is formed using THREE wires from the distribution box. In which of the wires:

  i is current flowing under normal working conditions?

  ii is current flowing during a short circuit?

  iii is current flowing during a power surge from the electric company?

d In which wire of a domestically wired system are each of the following connected?

  i Fuses.

  ii Switches.

  iii Circuit breakers.

3 **a** A high-powered device has an earth wire connected to its metal casing and includes a fuse in its circuit. Draw a diagram showing the flow of current in this device as its metal case becomes electrified.

b Explain why the short-circuit causes the fuse to blow.

Applying facts

4 Jonelle receives electricity from the mains at a voltage of 120 V. She has a small light display which draws a total of 4.0 A and a vacuum cleaner rated at 1200 W.

a What current is drawn by the vacuum cleaner?

b Fuses available to Jonelle have the following current ratings: 3 A, 5 A, 13 A and 15 A. What fuses should she select to install in the plug:

  i of the light display?

  ii of the vacuum cleaner?

c Why is it that the light display will not be adequately protected by the 20 A fuse connected in the circuit from the distribution box?

5 Becky wants to install several 120 V lamps in her home where the electrical mains supply is 120 V. Each lamp is enclosed in an insulated casing. She requires:

i TWO lamps to be switched on and off together

ii ONE lamp to be operated by a switch from TWO different locations

iii ONE lamp to be individually controlled by a separate switch.

Draw a possible circuit for this arrangement, showing the wires after they exit the distribution box to their connection with the switches and lamps.

6 Chad is thinking of installing a string of FIVE 24 V, 0.8 A lamps in series, each enclosed in an attractive metal casing. The electrical mains supply is 120 V and the lamps are to be switched on and off together. Chad has available to him fuses with the following current ratings: 0.5 A, 1 A, 5 A and 10 A.

a Calculate which fuse has the most suitable current rating to protect the string.

b Calculate the resistance of each lamp.

c Draw a possible circuit, showing the wires after they exit the distribution box. Include in your circuit suitable positions for the fuse, switches and lamps.

28 Electronics

Electronics plays an important part of our daily lives. Our cell phones, computers, musical instruments, automobiles and machinery all use electronic technologies. This field is developing at an increasing rate, providing complex data analysis, processing and synthesis, and producing significant advancements in finance, medicine and science, which could not be achieved otherwise.

Learning objectives

- Describe how a **semiconductor diode** can be used in **half-wave rectification**.
- Differentiate between DC from **batteries** and DC from **alternating current which has been rectified**.
- Recall the circuit **symbols** used for the **logic gates** AND, OR, NAND, NOR and NOT.
- State the function of each of the above logic gates using **truth tables**.
- **Analyse circuits** involving combinations of up to three logic gates.
- Discuss the impact of **electronic and technological** advances in society.

The semiconductor diode and half-wave rectification

A **semiconductor diode** is an **electronic valve** allowing current to flow through it in one direction only. It is made of semiconducting materials, for example **silicon** or **germanium**. Figure 28.1 shows the circuit symbol for such a diode directly above an actual diode. The **bar** in the circuit symbol for a diode and on the actual diode indicates that current cannot enter through that end.

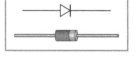

Figure 28.1 *Forward and reverse biased connection of diodes*

A diode that is **forward biased** can **conduct** when the voltage across it is greater than a certain critical value (see Chapter 26, Figure 26.16c). This value is different for diodes made of different materials; for silicon diodes it is 0.6 V. When the diode is **reverse biased**, **conduction is practically zero** (see page 283). Figure 28.2 shows diode B being forward biased and diodes A and C reverse biased. The lamp connected in series with B is therefore the only lamp which is lit.

Rectification

Rectification is the process of converting alternating current to direct current.

Half-wave rectification is the process of converting alternating current to direct current by preventing one half of each cycle from being applied to the load.

An **oscilloscope** is an instrument that produces a graph of **voltage against time**, allowing the **observation** and **measurement** of **varying electrical signals**. Figure 28.3 shows an oscilloscope illustrating the waveform produced due to half-wave rectification of an alternating voltage. An oscilloscope can be used as a **voltmeter**.

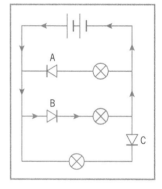

Figure 28.2 *Forward and reverse biased connection of diodes*

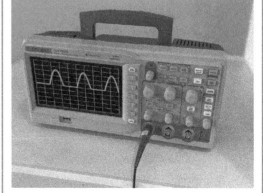

Figure 28.3 *Oscilloscope showing half-wave rectification*

- When an alternating voltage is connected across a load, the pd across the load varies with time as shown in Figure 28.4a.
- By connecting a diode in the circuit as shown in Figure 28.4b, the current can only flow in one direction and so becomes DC. Since $V = IR$, a pd will only exist across the load in the half of the cycle when the current flows.
- Figure 28.4c shows how DC from a battery is different from AC rectified to DC using a diode.

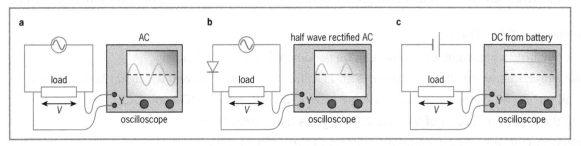

Figure 28.4 *Displaying voltages using an oscilloscope*

Logic gates and digital electronics

Digital electronics is based on circuits that can exist in one of **TWO possible states.** The two states are generally represented by the numbers (digits) 1 and 0. Electronic systems typically represent these two logic states as a **'high'** (5 V) and a **'low'** (0 V) by using electronic pulses.

*A **logic gate** is an idealised or physical electronic device that performs a **logical operation** on its **one or more inputs** to produce a **single output.***

*A **truth table** is a means of disclosing all outcomes of a logic gate or combination of logic gates.*

Logic circuits function on **binary decision making** since there are only two states. An example of a binary decision could be an **AND** decision. Say that you will be chosen for a position in a debate team if you are female and you are a teenager. Assume the specifications of Table 28.1 below.

Table 28.1

	0	1
Gender	male	female
Teenager	no	yes
Chosen	no	yes

Figure 28.5 shows the **truth table** for this **AND** decision and how the logic symbol may be used.

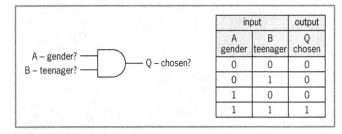

Figure 28.5 *AND gate*

Table 28.2 shows five logic gates and their corresponding truth tables.

Table 28.2 *Some logic gates and their truth tables*

Instead of memorising truth tables for AND, NAND, OR and NOR, we can memorise rules based on just **TWO** questions. This makes completing truth tables much easier.

GATE	AND		NAND		OR		NOR		NOT	
Circuit symbol										
Truth table	input	output	input	output	input	output	input	output	input	output
	0 0	0	0 0	1	0 0	0	0 0	1	0	1
	0 1	0	0 1	1	0 1	1	0 1	0	1	0
	1 0	0	1 0	1	1 0	1	1 0	0	This gate simply switches the logic	
	1 1	1	1 1	0	1 1	1	1 1	0		

RULES

AND:	Are **both** inputs 1?	Yes implies that output is 1	(otherwise it is 0)
NAND:	Are **both** inputs 1?	Yes implies that output is 0	(otherwise it is 1)
OR:	Are **any** inputs 1?	Yes implies that output is 1	(otherwise it is 0)
NOR:	Are **any** inputs 1?	Yes implies that output is 0	(otherwise it is 1)

- The **same question** is used for AND and NAND; it is just that the outputs are switched.
- The **same question** is used for OR and NOR; it is just that the outputs are switched.

Constructing truth tables for a combination of logic gates

The example that follows shows the construction of a truth table for a **combination** of logic gates.

- **Each line**, input to or output from a logic gate should be **uniquely labelled**.
- The **number of inputs** determines the **number of possible combinations** of input states to the circuit and therefore the **number of rows** of 0s and 1s in the table.

The number of rows of different combinations is equal to 2^x, where x is the number of inputs.

For example:

TWO inputs $\rightarrow 2^2 = 4$ different combinations (4 rows)

THREE inputs $\rightarrow 2^3 = 8$ different combinations (8 rows)

- Unless you apply an easy and systematic way in setting up the input part of the table, it is likely that you will have difficulty in including all the combinations. The following approach is recommended and is illustrated in Example 2.
  - The least significant input column (C in this case) has its values alternating: 0, 1, 0, 1, 0, 1, 0, 1.
  - The column immediately preceding it (B) has its values changing: 0, 0, 1, 1, 0, 0, 1, 1.
  - The third column (A) has its values changing: 0, 0, 0, 0, 1, 1, 1, 1.
- The output from each gate is determined by application of its inputs to the rule of that gate.

Example 1

An alarm (A) of a car is to sound (logic 1) if the headlights (H) are on (logic 1) when the ignition (I) is off (logic 0). Draw a truth table and a suitable circuit to perform this operation.

The combination of gates will have inputs, H and I, and final output A.

Interpret the logic as: If headlights are on **AND** ignition is off, then alarm sounds

If headlights = 1 **AND** ignition = 0, then alarm = 1

Since the AND gate can only yield a 1 when both of its inputs are 1, the *ignition* value must be **inverted**.

If H = 1 **AND** (**NOT** I = 1) then A = 1

This can also be written as: H **AND** (**NOT** I) = A ... see circuit below

- Having determined that there are 2 inputs, we draw our table having 4 rows of values ($2^2 = 4$). The input values do not depend on the logic of the circuit and can be inserted in the table as described immediately prior to this example.

- Apply the rule of the NOT gate to each value of I. Its output is an intermediary output and has been labelled $\bar{I}$ since it is the inverse of I. This makes it easier to trace when synthesising the table.

- Apply the rule of the AND gate to each pair of inputs H and $\bar{I}$ to obtain the corresponding outputs A.

Figure 28.6 shows the truth table and the corresponding logic circuit.

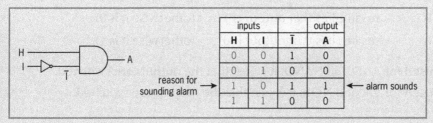

inputs			output
H	**I**	**$\bar{I}$**	**A**
0	0	1	0
0	1	0	0
1	0	1	1
1	1	0	0

reason for sounding alarm →   ← alarm sounds

Figure 28.6 *Example 1*

Example 2

The door of a vault (V) is to unlock (logic 1) when the manager (A) and any of two assistant managers (B or C) press a switch (logic 1) on their desks. Draw a truth table and a suitable circuit to perform this operation.

The combination of gates will have 3 inputs, A, B and C, and final output V.

Interpret the logic as: If A presses the switch **AND** either B **OR** C presses the switch, then V opens

If A = 1 **AND** (B **OR** C) = 1, then V = 1

This can also be written as: A **AND** (B **OR** C) = V ... see circuit on next page

- Having determined that there are 3 inputs, we draw our table having 8 rows of values ($2^3 = 8$). The input values can be inserted in the table as previously described.

- We start by examining the inputs to the OR gate, since its output must be input to the AND gate. Apply the rule of the OR gate to inputs B and C to determine the intermediary output X in each case.

- Apply the rule of the AND gate to each pair of inputs A and X to determine the corresponding outputs V.

The truth table and logic circuit are shown in Figure 28.7.

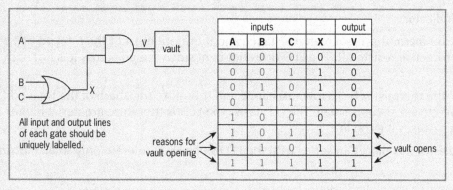

	inputs				output
A	**B**	**C**	**X**	**V**	
0	0	0	0	0	
0	0	1	1	0	
0	1	0	1	0	
0	1	1	1	0	
1	0	0	0	0	
1	0	1	1	1	
1	1	0	1	1	
1	1	1	1	1	

All input and output lines of each gate should be uniquely labelled.

reasons for vault opening → ← vault opens

Figure 28.7 *Example 2*

Using NOR gates or NAND gates to implement NOT gates

By sending the value on a single line to both input terminals of a NOR gate or a NAND gate, we obtain a NOT gate (Figure 28.8). The inputs in such cases can only be TWO (0)s or TWO (1)s.

- **Using a NOR gate** – Recall that for a NOR gate, if any inputs are 1, then the output is 0, otherwise the output is 1.

- **Using a NAND gate** – Recall that for a NAND gate, if both inputs are 1, then the output is 0, otherwise the output is 1.

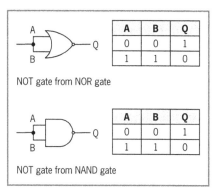

NOT gate from NOR gate

A	B	Q
0	0	1
1	1	0

NOT gate from NAND gate

A	B	Q
0	0	1
1	1	0

Figure 28.8 *Not gate from NOR gate or NAND gate*

Using sensors in digital circuits

LEDs and **thermistors** are **sensors** that are often used in electronic circuitry. Below shows the use of these in **potential divider circuits**. Example 4 of Chapter 26 explained that the share of the total pd for resistors in series is proportional to their share of the total resistance. The values of the pds shown in Figure 28.9 are only possible examples of how the pds may be shared by the components.

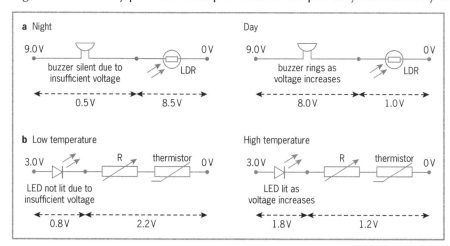

Figure 28.9 *Using sensors in digital circuits*

a Sunrise alarm

In the dark, the LDR has a high resistance and therefore most of the total pd of 9.0 V is across it. The buzzer does not ring since the pd across it is not high enough.

In the day (bright light), the resistance of the LDR is very small and therefore the pd across it is very small. Most of the 9.0 V is now across the buzzer and this higher voltage will cause it to ring.

b High temperature indicator

At low temperatures, the **thermistor** has a high resistance and therefore most of the total pd of 3.0 V is across it and the protective resistor, R. The LED does not light since the pd across it is not high enough.

At high temperatures, the thermistor has a low resistance and therefore the share of the total pd across it and the protective resistor is small. The pd across the LED is therefore greater at higher temperatures and this causes it to be lit.

Note: A protective resistor R is placed in series with the LED since LEDs tolerate only small currents.

Impact of electronic and technological advances on society

Pros

- **Learning** – the internet provides several advantages for learning:
  - access to a wide range of information
  - available any time of the day
  - audio-video is much better than just reading text from a book or from listening to lectures
  - learning programs can be interactive
  - learning programs can be self-paced and specially designed to meet the needs of different capabilities
  - there is no racial, cultural or class discrimination as to who gets access to the information.
- **Communication** – several types of communication have improved social and financial networking:
  - e-mail, messaging and social media apps have greatly enhanced relationships between families, friends and businesses on a global scale, having almost no geographical boundaries and few language boundaries
  - sales markets for goods and services have been expanded, allowing rapid improvement of some rural regions and a reduction in consumer costs across the globe
  - electronic banking has greatly facilitated financial transactions.
- **Health** – physical and mental health has improved for the following reasons:
  - the emergence of advanced technologies in equipment design for hospitals and medical research, including new generation CAT, MRI and gamma scanners, as well as lasers and fibre-optic devices such as endoscopes
  - advanced computer programming design that analyses and synthesises complex data at very high speeds, allowing for rapid developments in medical research
  - a large variety of apps that can be used with cell phones and computers to enhance physical fitness and to monitor health vitals such as body temperature, blood pressure, heart rate, breathing rate and oxygen levels.
- **Business and industry** – increased efficiency and productivity has resulted due to:
  - the emergence of advanced business architecture and systems design
  - improved technologies in equipment relevant to manufacturing and other industries.
- **New job opportunities** – jobs are now available for:
  - hardware and software programmers
  - web designers
  - system designers
  - computer operators including data entry personnel.

Cons

- **Social and societal** – interaction between individuals has been negatively affected due to:
  - reduced physical and social activities
  - the fact that the effects of facial expression, tone of voice and body language cannot be obtained from text messaging
  - loss of privacy, given that many years' worth of personal data may be accessed readily by individuals, businesses and governments
  - exposure to pornographic material, which affects all age groups and may have negative effects on otherwise wholesome relationships.
- **Health** – physical and mental health have been negatively affected due to:
  - increased dependence on devices for work and leisure, which can lead to reduced fitness since there is less tendency to exercise
  - increased incidence of addiction to device and app usage, leaving less time to carry out productive tasks including homework and study
  - headaches and eyestrain due to excessive time spent focusing on a monitor or device screen.
- **Criminal activity** – ready access to data and increased connectivity has caused more people to fall victim to:
  - fraud and monetary theft from bank accounts
  - cyber bullying
  - sexual harassment
  - child trafficking.
- **Loss of jobs** – many types of jobs have ceased to be relevant, due to:
  - automation of some manual and office roles
  - shifting trends such as the move to online retail activity.

Recalling facts

1 Why is a semiconductor diode described as an electronic valve?

2 Draw a circuit diagram for each of the following: a circuit containing a diode, filament lamp and cell connected in series, with the diode in

   a forward bias

   b reversed bias.

3 **a** Define each of the following:

     i rectification

     ii half-wave rectification.

   b Sketch a diagram showing the circuit you would use to observe the varying voltage across a resistor produced by half-wave rectification of an AC domestic supply.

4 **a** Define the term *logic gate*.

   b What TWO integers are used to represent the possible binary states of a digital component?

   c How are the states mentioned in **a** above realised in an electronic circuit?

5 Redraw and complete Table 28.3.

Table 28.3 *Question 5*

Description	Logic gate	Symbol
Output is high only if both inputs are high		
Output is low only when one or more of the inputs is/are high		
Output is high except when both inputs are low		

6 Draw circuit diagrams to show how a NOT gate can be formed from a single:

a NOR gate

b NAND gate.

7 Construct a truth table and draw the circuit symbol for each of the following logic gates:

a NOR

b AND

c NAND

d NOT.

8 Electronic and technological advancements in the 21st century have affected society significantly. For each of the following list THREE ways by which these advances have been:

a beneficial to society

b detrimental to society.

Applying facts

9 Sketch graphs to show the variation of voltage with time across a resistor that is connected in series with each of the following:

a a 12 V DC supply

b a domestic AC supply of frequency 50 Hz and peak voltage 155 V

c an AC supply of period 0.04 s and peak voltage 155 V which has been half-wave rectified.

10 Redraw the circuit shown in Figure 28.10 and use the rules laid out below Table 28.2 on page 301 to write the value, 1 or 0, for each line connected to the gates.

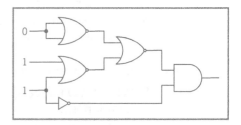

Figure 28.10 *Question 10*

11 An alarm is to sound in a TWO-door car if any of its doors are open when the ignition is engaged. Construct a truth table and draw the corresponding logic circuit to satisfy the requirement using the following assigned logic values.

Doors (D_1, D_2):   open = 1   closed = 0

Ignition (I):   on = 1   off = 0

Alarm (A):   sound = 1   no sound = 0

12   **a** Construct a truth table for each of the following circuits of Figure 28.11.

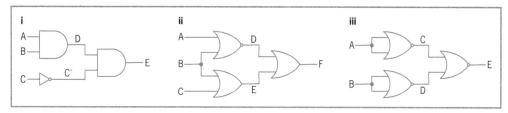

Figure 28.11 *Question 12*

  b By examining the final output of **a iii** above, determine the single gate which the combination represents.

13 Which bulb/bulbs shown in Figure 28.12 will be lit when:

  a S_1 is open and S_2 is closed?

  b S_1 is closed and S_2 is open?

  c S_1 and S_2 are closed?

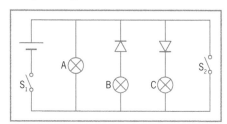

Figure 28.12 *Question 13*

14 Figure 28.13 shows a thermistor in series with an electric bell. At low temperatures, the resistance of the thermistor is much higher than that of the bell, but at high temperatures, its resistance falls to almost zero ohms. Explain what happens as the temperature changes from a low value to a much higher value.

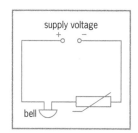

Figure 28.13
Question 14

29 Magnetism

Magnetism is a physical phenomenon produced due to the motion of electric charge. When magnetised, magnetic materials can attract and repel each other and can attract magnetic substances that are initially in an unmagnetised state. Magnetism and electromagnetism (see later chapters) play an important role in many devices used in our daily lives.

Learning objectives

- Differentiate between **magnetic** and **non-magnetic** materials in terms of **atomic magnetic dipoles.**
- Define a **magnetic field** and draw and interpret magnetic fields expressed as **field lines.**
- Understand **magnetic forces** as interactions between magnetic fields.
- Describe how to produce a **uniform magnetic field** using two flat magnetic poles.
- Explain **magnetic induction** and describe the chain of nails experiment.
- Test for **magnetic polarity.**
- Contrast the magnetic properties of **soft-iron** and **steel.**
- Distinguish between **permanent** and **temporary** magnets.
- Map magnetic fields using **iron filings** and using a **plotting compass.**
- Understand why the **Earth has a magnetic field** and how it can be used.

Magnetic and non-magnetic materials

Magnetic materials are those that can be in a magnetised or demagnetised state, but **non-magnetic materials** cannot be magnetised. **Iron, nickel and cobalt,** together with a few alloys and rare earth metals, are ferromagnetic materials – the type of magnetism we commonly encounter.

A bar of magnetic material is comprised of tiny **atomic magnetic dipoles** (atoms that behave as magnets, each having a *north pole* and a *south pole*). If the bar is in an **unmagnetised** state, these dipoles are arranged such that they point in **different directions** and their associated magnetic fields do not exist outside the material. In the presence of a strong magnetic field, however, the **dipoles align with the field** and the bar becomes **magnetised** with a north pole (N) and a south pole (S) at its opposite ends.

Figure 29.1 shows the arrangement of atomic magnetic dipoles (represented as arrows with an arrowhead for each north pole) in an unmagnetised and in a magnetised bar of iron. The magnetised bar behaves as a magnetic dipole with a N pole at one end and a S pole at the other. The N pole and S pole are the ends that tend to attract to the north pole and south pole of the Earth, respectively, and so are referred to as **north-seeking** and **south-seeking** poles.

Except for the atomic dipoles at the ends of the bar, the dipoles form straight lines since each N pole is held in position by the attraction from an adjacent S pole ahead of it. At the ends of the bar, however, they 'fan-out' slightly due to the repulsion between similar atomic magnetic poles at their sides and the fact that there are no adjacent atomic poles of opposite polarity, either ahead or behind them to maintain the straight-line arrangement. (see Figure 29.1).

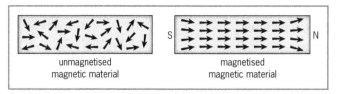

Figure 29.1 *Magnetic material in an unmagnetised state and in a magnetised state*

Understanding magnetic fields

*A **magnetic field** is a region in which a body experiences a force due to its magnetic polarity.*

*The **direction of a magnetic field** is the direction of motion of a freely moving N pole placed in the field.*

Magnetic field lines

Magnetic fields can be expressed in terms of **imaginary magnetic field lines** which have the following characteristics.

- They never cross or touch.
- They are directed from a N pole to a S pole.
- There is a **longitudinal tension** within each line, causing it to behave as a stretched elastic band.
- There is a **lateral repulsion** between lines that are close 'side-by-side'.
- The field is uniform where the lines are evenly spaced and parallel.
- The field is stronger where the lines are more concentrated.

Magnetic field of a bar magnet

Figure 29.2 shows the magnetic field of a bar magnet. The field is not uniform and is most concentrated at its magnetic poles.

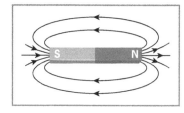

Figure 29.2 *Magnetic field of a bar magnet*

Forces between bar magnets explained in terms of magnetic field lines

Forces of attraction exist between dissimilar poles due to the longitudinal tension within the field lines joining them. Figure 29.3 shows these field lines.

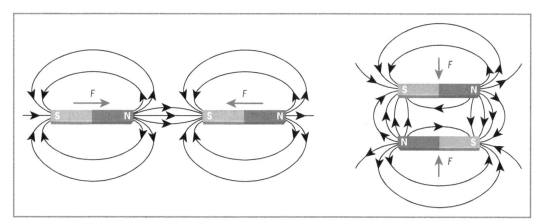

Figure 29.3 *Longitudinal tension within field lines*

Forces of repulsion exist between similar poles due to the lateral push between field lines lying close to each other, side-by-side. Figure 29.4 shows these field lines.

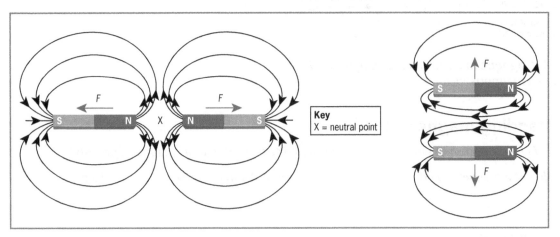

Figure 29.4 *Lateral repulsion between field lines*

A uniform magnetic field

Figure 29.5 shows how a uniform magnetic field can be produced. Unlike the lines at the upper and lower edges of the field, each line in the central region between the poles is being **repelled downward** by the lines above it and **repelled upward** by the lines below it. These lines therefore remain straight. Since they are **parallel and evenly spaced**,

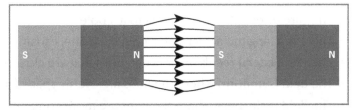

Figure 29.5 *Uniform magnetic field*

the field there is **uniform**. Towards the **edges**, the field is **not uniform**. Field lines at the top are laterally repelled only from lines below, and field lines near the bottom are laterally repelled only from lines above.

Magnetic induction

*Magnetic induction is the process by which magnetic properties are transferred from one body to another **without physical contact**.*

When a magnetic material is in an unmagnetised state, its atomic magnetic dipoles are randomly oriented. On placing it near to a **magnet**, or near to a **conductor carrying a current** (see Chapter 30), it becomes magnetised by induction as its atomic dipoles align with the field in which it is immersed.

When a magnet induces magnetism into a material:

- a **pair of opposite poles** is created
- the inducing pole is always **opposite in polarity to the induced pole** facing it
- induction occurs **at a distance**, that is, before the materials are in contact with each other.

Chain of nails induction experiment

Figure 29.6 shows how several iron nails can be magnetised by induction from a permanent magnet in a demonstration known as the 'chain of nails' experiment. On bringing an unmagnetised nail near to one pole of the magnet, an opposite

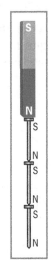

Figure 29.6 *Chain of nails experiment*

pole is induced onto it and the nail is then attracted. The magnetised nail can now aid in inducing magnetism into another nail placed below the arrangement. A 'chain of nails' can be hung from the permanent magnet in this way.

Making a magnet by induction

Figure 29.7 shows how a steel bar can be magnetised by a bar magnet. The bar should be stroked by **one pole** of a permanent magnet several times in the **same direction**, each time **raising** the magnet **high** before performing the next stroke. As the N pole in Figure 29.7 is moved to the right, it pulls on the atomic magnetic dipoles of the steel bar so that their S poles face the right end.

Chapter 30 shows how a magnet can be made by placing it in a current-carrying solenoid (long hollow coil).

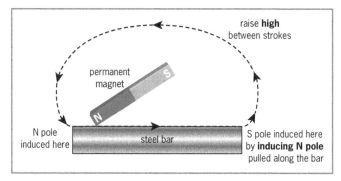

Figure 29.7 *Making a magnet*

Testing for the polarity of a magnet

A magnet is freely suspended by means of a string as shown in Figure 29.8. One end of the body X under test is brought close to one of the known poles of the magnet.

If there is attraction, X may be either of the following:

- a **magnet** whose nearby pole is of opposite polarity to that of the suspended magnet pole, attracting it
- a previously **unmagnetised** piece of **magnetic material** whose end, adjacent to the magnet, has been magnetised by induction with opposite polarity to the inducing pole.

If there is repulsion, X is a **magnet** whose nearby pole is of similar polarity to that of the suspended magnetic pole repelling it.

Since attraction does not produce a unique conclusion but repulsion does, the **true test** for the polarity of a magnet is **REPULSION**.

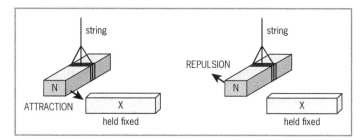

Figure 29.8 *Testing for magnetic polarity*

Experiment to contrast the magnetic properties of iron and steel

- Attach unmagnetised bars of iron and steel, identical in size and shape, to one pole of a strong magnet (see Figure 29.9).

- Dip the free ends of the bars into iron filings and note that much more filings stick to the iron than to the steel. It is therefore **easier to magnetise iron** than it is to magnetise steel.

- Remove the inducing pole of the strong magnet and note that most of the filings fall from the iron but very few fall from the steel. It is therefore **easier to demagnetise iron** than it is to demagnetise steel.

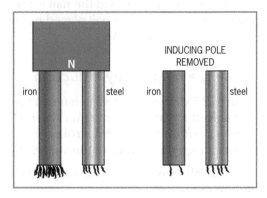

Figure 29.9 *Contrasting the magnetic properties of iron and steel*

Temporary magnetic materials

Iron is magnetically 'soft' and forms **temporary magnets**. Other temporary magnetic materials are low-carbon steel, silicon steel, nickel-iron alloys and iron-cobalt alloys. **Mumetal** is a good nickel-iron alloy.

Temporary magnetic materials form the **cores of electromagnets** (see Chapter 30), which are used for **lifting** iron and steel objects, and are important components in electric **bells**, magnetic **relays**, moving coil **ammeters** and **voltmeters**, **motors** and **generators**, and electromagnetic **circuit breakers**.

Permanent magnetic materials

Steel is magnetically **'hard'** and forms **permanent magnets**. The **strongest** permanent magnets are made of **neodymium**. Other permanent magnetic materials are **magnadur**, **samarium cobalt** and **alnico**.

Permanent magnetic materials are used in **directional compasses**, **moving coil loudspeakers**, **moving coil microphones**, **MRI scanners**, **ammeters and voltmeters**, **motors** and **generators**.

Soft-iron is very permeable to magnetic fields

If a bar of unmagnetised soft-iron is placed between a pair of opposite magnetic poles (see Figure 29.10), the magnetic field will **induce magnetism** within the iron and will be directed through it rather than through the air. Note that the region surrounding the length of the bar has been **shielded** from the magnetic field.

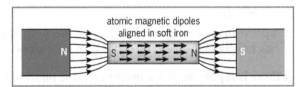

Figure 29.10 *Soft-iron is very permeable to magnetic fields*

Magnetic shielding

Magnetic fields can cause some electrical devices to malfunction. Figure 29.11 shows that by surrounding these devices in a **highly permeable magnetic material** such as **soft-iron** or **mumetal**, the field will be guided around the devices but will not pass through them. Magnetic shielding is particularly important to protect electrical instruments in rooms close to MRI (magnetic resonance imaging) equipment since their scanners produce very strong magnetic fields.

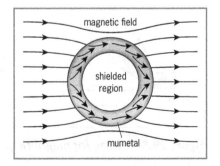

Figure 29.11 *Magnetic shielding*

Storing magnets

If magnets are bundled together, the **randomly directed** magnetic fields can **disarrange the alignment** of the atomic dipoles in neighbouring magnets. To avoid this, magnets should be stored by using soft-iron 'keepers' (see Figure 29.12). The magnetic field then forms a **closed loop** without disarranging the atomic dipole alignment within the permanent magnets; only the atomic dipoles within the soft-iron will change orientation.

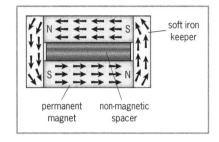

Figure 29.12 *Storing magnets*

Mapping magnetic fields

Using iron filings

- Attach a bar magnet to the work bench using tape and place a sheet of card over it.
- Gently tap the card as iron filings are lightly poured onto it.
- The filings will align with the magnetic field, revealing the pattern shown in Figure 29.13.

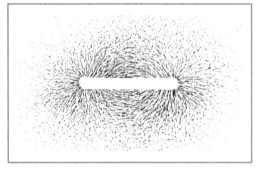

Figure 29.13 *magnetic field shown by iron filings lines*

Using a plotting compass

- Place the magnet at the centre of a large sheet of paper and outline its perimeter.
- Label the north and south poles of the magnet on the paper.
- Place a plotting compass near to the N pole and plot small dots to mark the position of the head and tail of its pointer.
- Advance the compass so that the tail of its pointer is now over the dot where the head was previously. Mark the new position of the head of the pointer.
- Continue the process until the dots reach the other end of the bar or extend off the paper.
- Join the dots by a smooth curve with an arrow to indicate the direction of the field as shown in Figure 29.14.
- Repeat the process to plot other field lines by placing the plotting compass in a slightly different starting position near the N pole of the magnet.

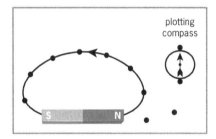

Figure 29.14 *Plotting the magnetic field of a bar magnet*

The Earth's magnetic field

The solid inner core and molten outer core of the Earth are composed mainly of the ferromagnetic elements iron and nickel. As heat passes from the inner core outward, convection currents consisting of moving electrical charges in the outer core generate the magnetic field of the Earth (see Chapter 30, Electromagnetism). The planet behaves as though it has a giant magnet at its centre with an axis slightly out of alignment with the *geographic axis* of rotation of the Earth (see Figure 29.15).

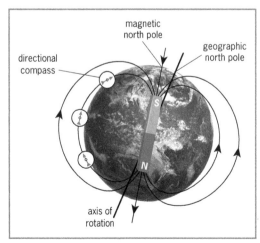

Figure 29.15 *The Earth's magnetic field*

The region in which this magnet field acts is known as the **magnetosphere**. Directional compasses placed in the magnetosphere will tend to align with their N pole pointing to the northern part of the planet. They will lie along a field line directed to the S pole **of the giant magnet** within the Earth.

Identifying the poles of a magnet using the Earth's magnetic field

A magnet hung from a string will align itself with the Earth's magnetic field (see Figure 29.16). It behaves as a directional compass and therefore has its N pole facing north.

In the northern hemisphere, the N pole of the suspended magnet declines (points downward) towards the Earth and in the southern hemisphere, it inclines (points upward) away from the Earth, as do the directional compasses shown in Figure 29.15.

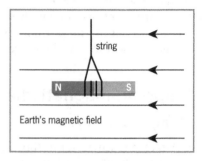

Figure 29.16 *Identifying the poles of a magnet using the Earth's magnetic field*

Recalling facts

 1 Draw a diagram showing the arrangement of atomic magnetic dipoles in a bar of:

   a unmagnetised iron

   b magnetised iron.

 2 **a** Define:

     i a magnetic field

     ii the direction of a magnetic field.

   b Sketch the magnetic field:

     i around a bar magnet

     ii around TWO bar magnets with ends of opposite polarity facing each other, and indicate the forces existing on the bars by means of arrows.

   c Sketch a magnetic field diagram to show how a uniform magnetic field can be produced. Shade the region in which the field is uniform.

 3 **a** Contrast the magnetic properties of iron and steel.

   b State which of iron and steel forms:

     i permanent magnets

     ii temporary magnets.

   c Name the strongest type of permanent magnet.

   d Other than iron, steel or neodymium, name ONE other material that forms:

     i permanent magnets

     ii temporary magnets.

   e State which of the following devices utilises permanent magnets, which utilises temporary magnets, and which utilises both permanent and temporary magnets:

     i electric bells

     ii directional compasses

    iii circuit breakers

    iv ammeters

    v moving coil loudspeakers

    vi moving coil microphones.

4 Draw magnetic field diagrams to show how:

  a magnets may be stored so that their magnetic poles are not affected by stray magnetic fields

  b devices can be protected from strong magnetic fields existing in the area.

Applying facts

5 **a** What is meant by the term *magnetic induction*?

   b Describe and explain, in terms of atomic magnetic dipoles, why a magnet placed on the door of a refrigerator remains attached to it.

6 **a** Why does the Earth behave as though it were a magnet?

   b Sunita is lost on her uncle's cattle ranch. She had walked north, away from the bank of the river, but now has lost all sense of direction. Her aunt had given her a large magnetised needle that morning with its N pole painted red. How can she use her shoelace, together with the needle, to determine the way back to the river?

7 Anita brings the N pole of a bar magnet close to one end of another bar, P, and notices that the bars attract each other. She then brings the S pole of the bar magnet near to the same end of P and notices that the bars repel.

The process is repeated by replacing P with another bar, Q, and she observes that the bars attract when either pole of the magnet approaches either end of Q.

State and explain what can be deduced of the nature:

  a of P

  b of Q.

8 Draw magnetic field lines to illustrate the magnetic field around the arrangement of bar magnets shown in Figure 29.17.

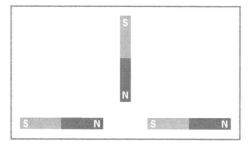

Figure 29.17 *Question 8*

30 Electromagnetism

Electromagnetism is the study of the forces produced by the interaction of electric and magnetic fields. As an electric charge moves relative to a magnetic field, an electromagnetic force is produced. Electromagnetic relays, loudspeakers, electric bells, ammeters and voltmeters, electric motors and generators, circuit breakers and security locks, are only some of the many devices which utilise electromagnetic forces in today's modern technology.

Learning objectives

- Investigate **magnetic field patterns** around **current-carrying conductors**.
- Apply the **right-hand grip** rule.
- Investigate the forces existing between wires carrying currents.
- Describe the structure of an **electromagnet** and explain the function of some of its applications.
- Investigate the **force on a current-carrying conductor** immersed in a magnetic field and describe the factors affecting the size of the force produced.
- Apply **Fleming's left-hand rule** to currents in wires and to charged particles moving in magnetic fields.
- Explain the function of the **moving coil loudspeaker**, the simple **DC motor** and the **moving coil ammeter**.

Magnetic fields around current-carrying conductors

Figure 30.1 (page 317) shows how we may observe the magnetic fields associated with electric currents in straight wires, short coils and long coils (solenoids). When **iron filings** are lightly sprinkled on the card, they align with the magnetic field, showing a unique field pattern in each case.

Notes:

- The magnetic field associated with the current in a solenoid is **identical** to that of a **bar magnet**.
- Magnetic field lines are directed **out of** the **north pole** and **into** the **south pole** of a current-carrying coil (*within the coil, the lines are directed from the south pole to the north pole*).
- The direction of the field can be obtained by application of the **right-hand grip rule** described next.

Relation between the direction of a current and its associated magnetic field

Right-hand grip rule

1. **Applied to a straight conductor** – Imagine gripping the wire with the right hand such that the thumb is in the direction of the current; the fingers will then be in the direction of the magnetic field.

2. **Applied to a coil** – Imagine gripping the coil with the fingers of the right hand in the direction of the current; the thumb will then indicate the direction of the magnetic field (the end of the coil that acts as a north pole).

Applying the right-hand grip rule (Figure 30.2) to each of the experiments shown in Figure 30.1 results in the field directions shown in the plan views.

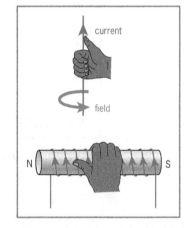

Figure 30.2 *The right-hand grip rule*

Right-hand screw rule (corkscrew rule)

If a right-hand drive screw (one that advances when turned clockwise) is rotated so that it advances in the direction of the current, then the associated magnetic field will be in the direction of rotation of the screw.

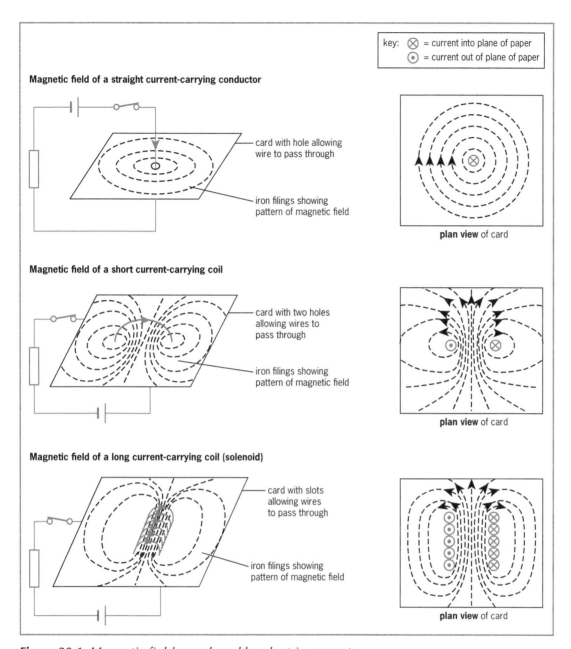

Figure 30.1 *Magnetic fields produced by electric currents*

Electromagnetic forces due to parallel current-carrying conductors

The magnetic fields due to currents flowing in two parallel wires can either cause **attraction** or **repulsion** of those wires. Figure 30.3 shows circuits in which these forces may be investigated.

In accordance with **Newton's 3rd law**, the forces, *F*, caused by each wire on the next should be equal in magnitude but opposite in direction (see Chapter 4).

By analysis of the field patterns shown in Figure 30.3, we see that:

* for Figure 30.3a, the wires attract each other due to the **longitudinal tension** within the field lines
* for Figure 30.3b, the wires repel each other due to the **lateral push** between the field lines.

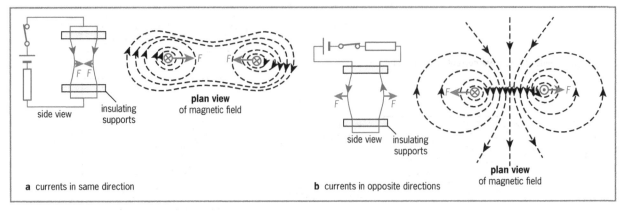

Figure 30.3 *Electromagnetic forces due to parallel current-carrying conductors*

Making a magnet using a solenoid

An unmagnetised steel bar can be magnetised by placing it along the axis of a solenoid (long coil) and passing a high **direct current** (DC) through it for a short period of about **10 seconds**. The high current in the coil produces a strong magnetic field along its axis, resulting in a N pole at one end and a S pole at the opposite end of the bar (see Figure 30.4).

Leaving the current flowing for a longer period may damage the coil. It will not increase the magnetic strength of the bar since the high current quickly **aligns** all the **atomic dipoles** to produce **magnetic saturation**. Use of the right-hand grip rule indicates that in this case, the N pole is as shown in Figure 30.4.

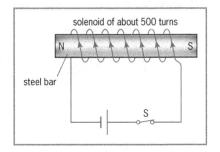

Figure 30.4 *Making a magnet using electricity*

Electromagnets

An electromagnet is a **temporary magnet** consisting of a conducting coil wrapped around a core of **'soft' magnetic material** (such as *soft-iron*), as shown in Figure 30.5. By switching the current on and off, the core can instantly be magnetised and demagnetised, respectively.

Electromagnets can be used to **lift heavy iron and steel objects** (such as damaged vehicles) with the help of a crane and then release them in another location. The **'soft' magnetic core increases the strength** of the magnetic field within the coil.

Electromagnets are also used in electric bells, electromagnetic relays, circuit breakers and door locks.

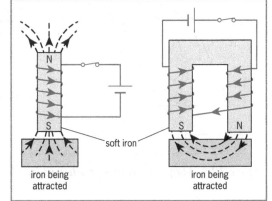

Figure 30.5 *Electromagnets*

The electric bell

Figure 30.6 shows a simple electric bell. The contacts are initially touching when the switch is open. On closing the switch, current flows through the coil and magnetises the electromagnet. The magnet then attracts the soft-iron armature, separates the contacts and causes the hammer to strike the gong. With the contacts open, the electromagnet is disabled, the hammer springs back, closes the contacts and repeats the process.

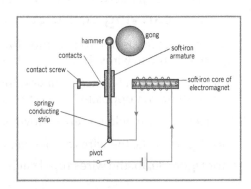

Figure 30.6 *The electric bell*

The electromagnetic relay

*An **electromagnetic relay** is an electrical switch which can enable or disable a second circuit.*

Figure 30.7 shows one type of electromagnetic relay. When the switch S is closed, the current in the coil creates a magnetic field that magnetises the two **soft-iron reeds** in accordance with the right-hand grip rule. The **adjacent ends** of the reeds become **opposite magnetic poles** and attract each other, completing the circuit connected to terminals T_1 and T_2. The reeds are protected from oxidation by surrounding them with an inert gas trapped in the glass enclosure. This type of relay is used in **electronic circuits** which open and close many times per second.

Figure 30.8 shows another type of electromagnetic relay. When the switch S is closed, the current in the coil magnetises the electromagnet, causing it to attract the soft-iron armature. The armature then rocks on its pivot, closes the graphite contacts and completes the circuit connected to terminals T_1 and T_2.

These terminals may be connected to the starter motor of a vehicle via **thick** wires that can carry **high** currents. When the **ignition key S** is turned on, the **low-current** circuit of the solenoid switches on the **high-current** circuit to the starter motor.

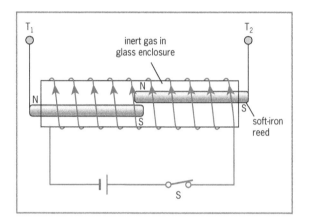

Figure 30.7 An electromagnetic relay

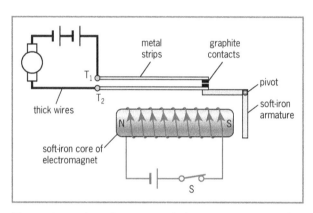

Figure 30.8 Another type of electromagnetic relay

Electromagnetic circuit breaker (trip switch)

Figure 30.9 shows a simple electromagnetic circuit breaker. The graphite contacts are normally closed, allowing current to flow to the device through the coil. When the current rises above the maximum permitted value, the magnetic field it creates in the coil is then strong enough to attract the soft-iron lever, causing it to rock on the pivot P. The lever forces past the catch on the flexible strip, causing the contacts to separate and the circuit to break. The high current is therefore prevented from damaging the device. After the problem is corrected, the trip switch is reset by manually repositioning the lever.

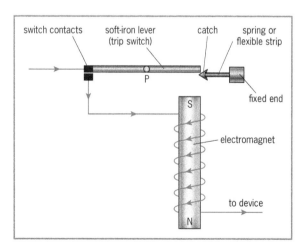

Figure 30.9 Electromagnetic circuit breaker

Electromagnetic door lock

Figure 30.10 shows one type of a simple electromagnetic door lock. Pressing the push-button switch allows current to flow in the coil and magnetises the electromagnet. The iron bolt is then pulled out of the slot in the door by the electromagnet and compresses the spring.

On releasing the push-button switch, the circuit breaks, the electromagnet demagnetises and the force in the compressed spring returns the bolt to the slot in the door.

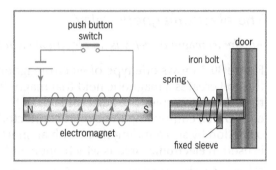

Figure 30.10 *Electromagnetic door lock*

Force on a current-carrying conductor in a magnetic field

Figure 30.11 shows a **stiff wire** suspended from a metal loop and immersed in a magnetic field. The lower end of the wire just touches the surface of **mercury** contained in a dish below. When the switch is closed a force acts on the wire which pushes it out of the mercury and breaks the circuit. The current then rapidly diminishes to zero, the wire falls back into the mercury and the process repeats.

The force, *F*, on the stiff wire shown in the field diagram of Figure 30.11 occurs due to the **longitudinal tension** within the magnetic field lines in which the wire is immersed.

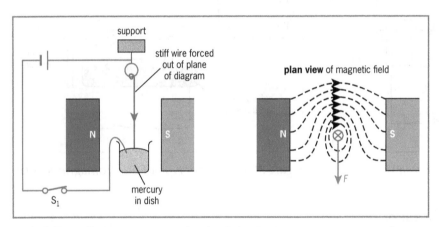

Figure 30.11 *Demonstrating the force on a current-carrying conductor placed in a magnetic field*

Fleming's left-hand rule

The direction of the electromagnetic force producing the motion of the wire in Figure 30.11 can be obtained by using Fleming's left-hand rule.

If the first finger, the second finger and the thumb of the **left** hand are placed mutually at right angles to each other (Figure 30.12), with the **F**irst finger in the direction of the magnetic **F**ield in which the current is immersed and the se**C**ond finger in the direction of the **C**urrent, then the Thu**M**b will be in the direction of the **T**hrust or **M**otion.

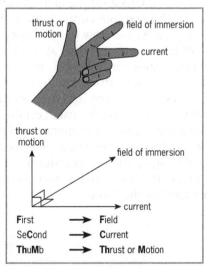

Figure 30.12 *Fleming's left-hand rule*

The strength of the thrust (force)

Increasing any of the following results in an increase in the magnitude of the electromagnetic force:

- the strength of the magnetic field in which the current is immersed
- the magnitude of the current
- the length of the conductor.

Notes:

- In cases where the current is not at right angles to the field, the **component of the current** which is at **right angles** to it is used with Fleming's left-hand rule to determine the direction of the thrust. This is illustrated in the final situation shown below in Example 1.
- The **maximum thrust (force)** for a given current and field strength is obtained when the direction of the **current** is at **right angles** to the direction of the **field**.
- **No force** is obtained when the **current and field** are in the **same direction** or in **opposite directions**.
- It is advisable to apply Fleming's left-hand rule to a 2D diagram rather than a 3D diagram.

Example 1

Figure 30.13 shows several situations of currents flowing in magnetic fields. The current in each case is indicated in red. Apply Fleming's left-hand rule to verify that the thrust on each of the current-carrying conductors is as stated. (Note: direction of field of immersion is from N to S.)

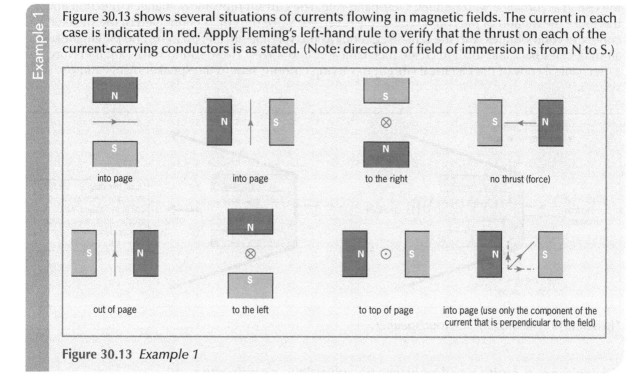

Figure 30.13 *Example 1*

Deflection of charged particles in magnetic fields

Since an electric current is a flow of charge, protons, electrons and other charged particles moving through magnetic fields also experience forces in accordance with Fleming's left-hand rule. If these particles enter a magnetic field which is at right angles to their motion, they will take a **circular path**.

Figure 30.14 shows the deflection of positive and negative charges as they enter magnetic fields.

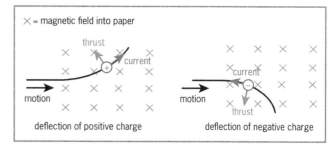

Figure 30.14 *Deflection of charged particles by magnetic fields*

Note the following when applying Fleming's left-hand rule to charged particles moving within such fields.

- Recall that the direction of the current is the direction of the flow of positive charge; for negative charges (for example, electrons) the **direction of current** is **opposite** to the **direction of motion**. See Chapter 24, *Electron flow and conventional current*.

- As the direction of the particle changes, the direction of the force on it also changes, so that it is always **perpendicular to its path** and produces **circular motion**.

- Recall that the work done by a force is the product of the force and the component of the displacement **in the direction of the force**. Since the force on the charged particle is always **perpendicular** to its motion, it does no work.

The moving coil loudspeaker

Figure 30.15 shows a moving coil loudspeaker. **An alternating current** from an amplifier flows in the **voice coil** which is wound on a hollow cylinder **attached** to the **speaker cone**. Since the current in the coil is immersed in the radial magnetic field of the permanent magnet, a force is created on the coil in accordance with Fleming's left-hand rule. The current repeatedly changes direction and therefore the motion of the coil is repeatedly reversed. Since the coil is connected to the speaker cone, the cone vibrates with the same frequency as the varying current, pushing and pulling nearby air molecules back and forth to create mechanical sound waves. These sound waves are therefore a mechanical copy of the electrical vibration (varying current) sent to the speaker by the audio system.

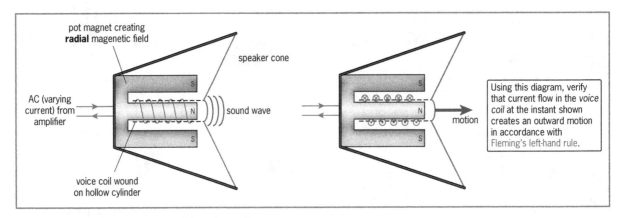

Figure 30.15 *The moving coil loudspeaker*

Forces on a rectangular current-carrying conductor immersed in a magnetic field

The current flowing in a rectangular current-carrying coil placed in a magnetic field produces forces in accordance with Fleming's left-hand rule (see Figure 30.16). Since the **current flows** in **opposite directions** on either side of the coil, the **forces** are also in **opposite directions**.

A pair of forces, equal in magnitude but opposite in direction and having different lines of action, constitute a **'couple'**, and will have a **turning effect** (moment) about the **axle** P of the coil. As the coil turns through **one half of a revolution** however, the current adjacent to any pole of the magnet changes direction, and the couple so formed causes the direction of rotation to **reverse**. The coil therefore will never undergo a complete rotation, but instead will **oscillate back and forth**, eventually coming to rest when its plane is perpendicular to the magnetic field (see Figure 30.16b).

For a complete rotation to occur, the connection between the coil and battery needs to be reversed every half revolution. This is achieved by adding a **split ring** (commutator) in the construction of the DC motor discussed next.

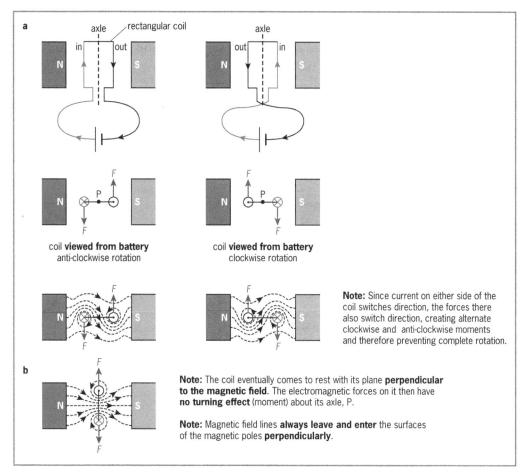

Note: Since current on either side of the coil switches direction, the forces there also switch direction, creating alternate clockwise and anti-clockwise moments and therefore preventing complete rotation.

Note: The coil eventually comes to rest with its plane **perpendicular to the magnetic field**. The electromagnetic forces on it then have **no turning effect** (moment) about its axle, P.

Note: Magnetic field lines **always leave and enter** the surfaces of the magnetic poles **perpendicularly**.

Figure 30.16 *Forces on rectangular coil in magnetic field*

Simple DC motor

Figure 30.18 shows a simple DC motor. One side of the coil has been coloured to aid in the explanation of the function of the device.

The **coil and commutator** are **fixed to the axle** and **rotate with the axle**. The current-carrying coil is immersed in a magnetic field and therefore forces are created on it in accordance with Fleming's left-hand rule. Since the current flows in opposite directions on either side of the coil, the electromagnetic forces, *F*, produced are in opposite directions and create a couple, causing rotation.

- The **concave faces** of the magnetic poles cause the forces producing rotation to be **perpendicular** to the **coil plane**, resulting in **maximum turning effect** about the axle.

- A **practical** motor has its coil wound on a **cylindrical, soft-iron core** (not shown in Figure 30.18) so that the **magnetic field** is more **concentrated** around its windings and therefore the **turning forces** created will be **greater**. Figure 30.17 shows that the core must be designed as a series of discs **insulated** from each other to prevent unwanted **eddy currents** (discussed in Chapter 31) developing within it as it moves through the magnetic field.

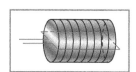

Figure 30.17
Laminated and insulated soft iron core with coil

The coloured side (*left-hand* side) of the coil in Figure 30.18a has a downward force on it, causing anti-clockwise rotation. Half a revolution later (Figure 30.18b) it is on the right-hand side and requires an upward force to maintain rotation. Since the split ring turns with the coil and axle, it switches connection between the coil and battery every half revolution as it rubs against the fixed graphite brushes. With the current switched, the force on the right-hand side will be upward, resulting in continuous rotation.

In Figure 30.18, current always enters the left side of the coil from the positive terminal of the battery and leaves the right side of the coil, returning to the negative terminal of the battery. The current, and therefore the forces, **on any given side of the coil** are always in the same direction within the coil and so it rotates continuously.

The strength of the thrust is proportional to:

- the magnitude of the **current**
- the strength of the **magnetic field**
- the **number of turns** of the coil (each turn of coil produces its own force).

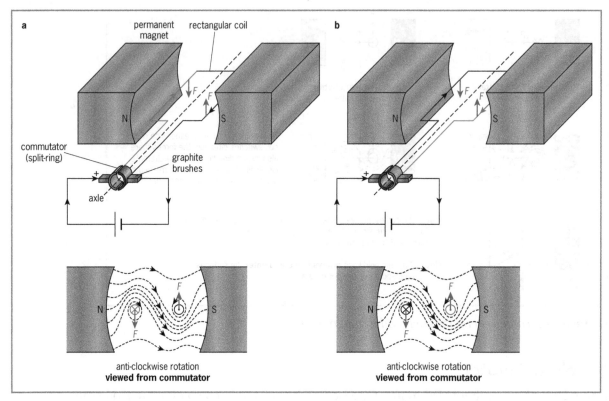

Figure 30.18 *Simple DC motor*

The moving coil ammeter

Figure 30.19 shows a plan view of a moving coil ammeter. The rectangular coil is wrapped on a light aluminium frame which is fitted over the fixed soft-iron cylinder such that it can rotate about the axis of the cylinder. Current from the circuit flows through the rectangular coil in the radial magnetic field produced by a pair of curved magnetic poles of a permanent magnet. The **electromagnetic forces** produced in accordance with Fleming's left-hand rule cause the coil to turn against the **torsional forces**

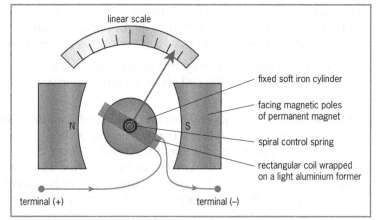

Figure 30.19 *Moving coil ammeter*

of the **spiral spring**. When the electromagnetic force creates a **moment**, **equal in magnitude** to the opposing moment created within the spring, the coil will remain stationary with its pointer indicating the strength of the current.

Recalling facts

1   **a**   With the aid of a diagram, explain the function of an electromagnetic relay. Your diagram should include current directions and polarities of magnetic dipoles.

      b   List TWO other applications which utilise electromagnetic forces.

2   Describe, with the aid of a diagram, an experiment that investigates the forces acting on parallel conducting wires carrying currents in the same direction. Indicate the direction of the forces acting on the wires in the diagram by using arrows labelled *F*.

3   Describe how a bar of steel can be magnetised using an electrical method.

Applying facts

4   Figure 30.20 shows wires carrying currents (indicated in red) in magnetic fields. The directions of the currents are as follows:

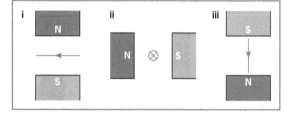

i   to the left

ii   perpendicularly into the page

iii   to the bottom of the page.

Figure 30.20   *Question 4*

      a   State the direction of the electromagnetic force that acts on each wire.

      b   Draw the magnetic field diagram for the situation shown in **ii**.

5   Figure 30.21 shows a plan view of TWO aluminium rails, R_1 and R_2, connected to the terminals of a battery. A cylindrical aluminium rod, A, rests across the rails and a magnetic field acts perpendicularly onto the system. Describe what will occur when:

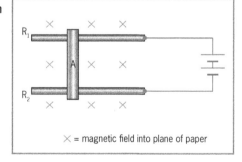

      a   the circuit is switched on

      b   a stronger battery is used

      c   the battery is replaced by a high frequency AC source.

Figure 30.21   *Question 5*

6   Figure 30.22 shows a horizontal electron beam about to enter a magnetic field, which acts vertically downward, perpendicular to its path.

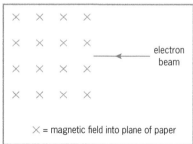

      a   Redraw the diagram to show the path of the beam when it enters the magnetic field.

          Show the direction of the electromagnetic force, *F*, on an electron in the beam.

      b   Name the rule used to determine the direction of the force.

Figure 30.22   *Question 6*

7 Figure 30.23 shows a simple DC motor.

a Identify the labels A, B, C and D.

b What is the polarity of D?

c Does the wire PQ move upward or downward at the instant shown?

d State how you determined your answer to **c**.

e What is the function of the part C?

f List THREE factors (other than those affecting efficiency) which can increase the rate of rotation.

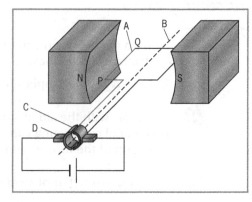

Figure 30.23 *Question 7*

8 **a** Redraw Figure 30.24 and sketch magnetic field lines between the magnetic poles.

b Indicate the forces acting on the current-carrying conductors, using arrows labelled *F*.

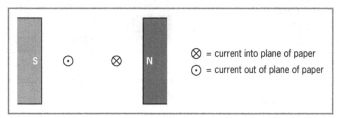

Figure 30.24 *Question 8*

9 Figure 30.25 shows a section through a rectangular coil carrying a constant current and pivoted on an axle P in a magnetic field.

a Show, by means of arrows labelled *F*, the direction of the forces acting on each side of the coil, when in each of the three positions.

b How do the *magnitudes* of the *forces* acting in position **i** compare with those in position **iii**?

c Give an expression for the *total moment* about the axle acting on the coil when in each of the three positions.

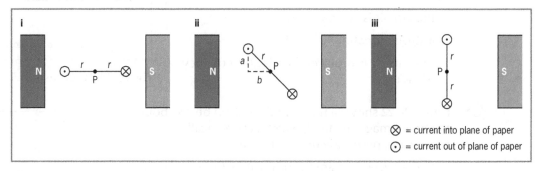

Figure 30.25 *Question 9*

31 Electromagnetic induction

Just as currents through conductors in magnetic fields can produce forces resulting in motion, the motion of conductors in magnetic fields can produce emfs resulting in currents. The generation of emfs in this way is known as **electromagnetic induction**. Two important scientists in this field are **Michael Faraday** and **Emil Lenz**. Electrical generators, transformers and microphones are a few examples of today's technology which utilise the science of electromagnetic induction.

Learning objectives

- Apply **Faraday's law** and **Lenz's law** of electromagnetic induction.
- Apply **Fleming's right-hand rule** and describe the factors affecting the **size of an induced current**.
- Describe and explain the function of the simple **AC generator**.
- Describe and explain the function of the **transformer**.
- Perform **calculations** involving the **transformer**.
- Discuss the reasons for long distance **AC distribution**.
- Describe and explain the function of the **moving coil microphone**.

Faraday's law and Lenz's law of electromagnetic induction

*Faraday's law – Whenever there is **a relative motion** of magnetic flux (field lines) linked with a conductor, an **emf** is induced, which is proportional to the **rate of cutting of flux**.*

If the circuit is complete, this emf will produce a current.

*Lenz's law – An **induced current** is always in a direction to **oppose the motion creating it**.*

Conducting rod moving through a magnetic field

Figure 31.1 shows a metal rod being pushed downward through a magnetic field.

- As the conductor **cuts** through the **magnetic field lines**, an **emf is induced** in it, which drives a current through the centre-zero galvanometer.
- If the rod is pulled upward, the induced current is in the **opposite direction**.
- **Increasing** the **speed** of the rod, **increases** the magnitude of the **current**.
- Using a **stronger magnet**, **increases** the magnitude of the **current**.
- If the rod is held **stationary** in the field, there is **no induced current**.
- The **direction of the current** can be obtained by application of **Fleming's right-hand rule** (see Figure 31.5).
- Like a cell, the **rod** is acting as a **source of electrical power**, and so its **positive end** is the end from which it **delivers current**.

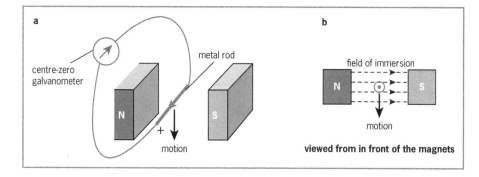

Figure 31.1
Demonstrating electromagnetic induction in a metal rod

Bar magnet moving into and out of a conducting coil

Figure 31.2 shows a pole of a magnet being pushed into and pulled out of a coil.

- An induced current is detected by the centre-zero galvanometer.
- The direction of the induced current changes each time the direction of the magnet changes.
- **Increasing** the **speed** of the magnet **increases** the magnitude of the **current**.
- Using a **stronger magnet increases** the magnitude of the **current**.
- If the magnet is held **stationary** in the coil, there is **no induced current**.
- The direction of the current in the coil of Figure 31.2 can be obtained by applying **Lenz's law**.
  - First, **determine the induced polarity** at the end of the coil. If the N pole of the magnet is pushed into the coil, it induces a N pole at that end of the coil, which opposes its motion. If the N pole of the magnet is pulled out of the coil, it induces a S pole at that end of the coil, which opposes its motion.
  - Then, using the **right-hand grip** rule on the coil, with the **thumb** in the direction of the induced **north pole of the coil**, the **fingers** give the direction of the **induced current**.

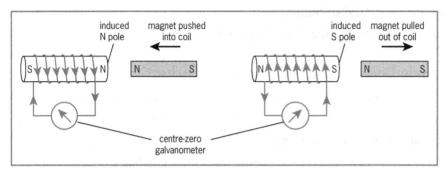

Figure 31.2 *Demonstrating electromagnetic induction in a coil*

Eddy currents and Lenz's law

As the copper plate of the pendulum of Figure 31.3 oscillates in the magnetic field, it cuts the field lines at differing angles and speeds, inducing changing emfs in accordance with Faraday's law. These emfs create induced 'eddy' currents forming closed loops within the plate. The motion of the pendulum is quickly **dampened** by the **opposing electromagnetic force** created in accordance with **Lenz's law**.

If the plate is **slotted** the oscillations occur for a much **longer period**. The damping effect is then greatly reduced since the slots break the circuits of the eddy currents that cause the opposing force.

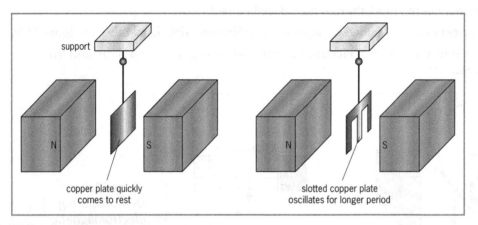

Figure 31.3 *Damping effect of eddy currents*

A very simple AC generator

One pole of a magnet is made to bob up and down inside the hollow coil of the circuit shown in Figure 31.4. Since the magnetic flux of the magnet repeatedly reverses its direction of motion as the magnet bobs into and out of the conducting coil, an alternating current is induced and the lamp glows. The oscillations will quickly **diminish** in accordance with **Lenz's law** due to the **opposing force** on the moving magnet. If the switch is opened, the circuit breaks, and so there is no longer an induced current and its opposing force, and the oscillations will continue for a much longer period.

Electromagnetic induction and the principle of conservation of energy

As the rod in Figure 31.1 is moved through the magnetic field and as the magnet of Figure 31.2 is moved into and out of the coil, the experimenter will experience a force opposing the motion in **accordance with Lenz's law**.

This is also in **accordance with the principle of conservation of energy**. The mechanical energy required by the experimenter to overcome the opposing force converts to electrical energy in the conductor.

mechanical energy → electrical energy

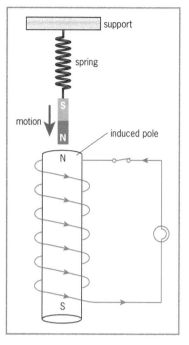

Figure 31.4 *A very simple AC generator*

Fleming's right-hand rule

If the first finger, the second finger and the thumb of the right hand are placed mutually at right angles (Figure 31.5), with the **F**irst finger in the direction of the magnetic **F**ield and the Thu**M**b in the direction of the **M**otion, then the se**C**ond finger will be in the direction of the induced **C**urrent.

Applying Fleming's right-hand rule to the rod of Figure 31.1b gives the direction of the induced current. It is advisable to apply this rule to a 2D diagram rather than a 3D diagram.

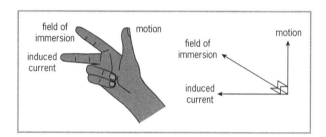

Figure 31.5 *Fleming's right-hand rule*

The AC generator

A simple AC generator is shown in Figure 31.6. The coil is rotated with its sides P and Q cutting through the magnetic field. This induces an emf that drives current around the circuit in accordance with **Fleming's right-hand rule**. After every **half revolution**, the side of the coil which was moving downward then moves upward through the field, **reversing** the **direction of the emf** and producing an alternating voltage.

The alternating voltage is transferred to the external circuit by a **pair of slip rings** which rub on a pair of **fixed graphite brushes** as the coil rotates. The slip rings ensure that each side of the coil is always connected to the same output terminal for any coil position, allowing the AC generated in the coil to be transferred to the external circuit. Figure 31.6 shows that the current through the lamp reverses direction every half revolution.

The slip rings are **rigidly connected** to the **shaft and coil** and therefore **rotate** together with it. They are necessary to **prevent tangling** of the wires connected to the external circuit.

- The **concave faces** of the magnetic poles cause the magnetic field to be **perpendicular** to the **coil plane**, resulting in **maximum current** being induced in the wires P and Q cutting through it.
- A **practical** generator has its coil wound on a cylindrical soft-iron core so that the **magnetic field** will be more **concentrated** around its windings and therefore the **induced current** will be **greater**. The core must be designed as a series of **insulated discs** (see Chapter 30 Figure 30.17) to prevent unwanted **eddy currents** developing within it as it moves through the field.

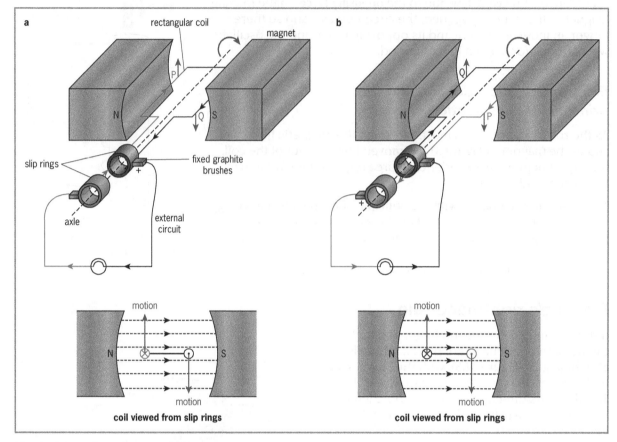

Figure 31.6 *The AC generator*

Given that the coil rotates in a clockwise direction, use Fleming's right-hand rule to verify that the currents in P and Q are as shown in Figure 31.6.

A graph of emf against time for one revolution of the coil is shown in Figure 31.7. For the positions shown, use Fleming's right-hand rule with respect to the 'coloured' side P of the coil (motion of P indicated by coloured arrow) to verify that the current alternates as illustrated by the graph.

Notes:

- In positions A, C and E, motions of the sides P and Q of the coil are **parallel** to the magnetic field and therefore there is **no cutting of field lines** and **no induced current**.
- In positions B and D, motions of the sides P and Q of the coil are **perpendicular** to the magnetic field and therefore there is **maximum cutting of field lines** and **maximum current**.
- Since each graphite brush is always in contact with the same ring, it can be on either side of the ring (see Figure 31.19).

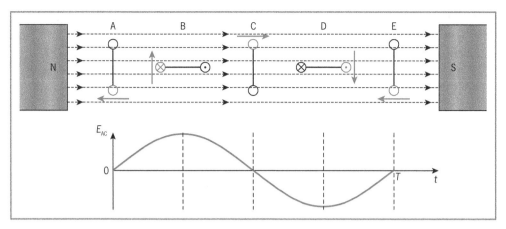

Figure 31.7 *Output of AC*

Factors which increase the emf induced in the coil

- The strength of the **magnetic field**.
- The **rate of rotation** of the coil.
- The **number of turns** of the coil (each turn of the coil produces its own emf and the total emf is the sum of these emfs since they are all in series).

Moving coil microphone

Figure 31.8 shows a moving coil microphone. Sound waves incident on the **flexible diaphragm** cause it to **vibrate**, and the attached coil therefore moves back and forth within the magnetic field of the permanent magnet. An **emf** that varies with the **same frequency** as the sound wave causing it is induced between the ends of the coil. The **stronger** the vibration of the incident **sound wave**, **the greater** is the **amplitude** of the **electrical signal**. An electrical copy of the incident sound wave is therefore created.

Compare this microphone with the loudspeaker in Figure 30.15.

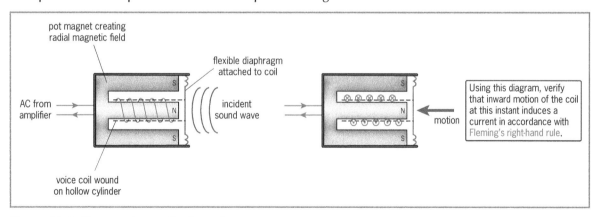

Figure 31.8 *The moving coil microphone*

Transferring electrical energy between coils

Coils A and B are connected in circuits arranged as shown in Figure 31.9. The following occurs when the switch connected in **A's** circuit is closed, remains closed, and is then opened:

Switch closed – On closing the switch, the current in coil **A** rapidly increases to some maximum value and then remains constant. As the current increases, the magnetic field associated with this current also grows to a maximum value. As it grows, the flux (field lines) **enters coil B** and **induces a current**

in it. The magnetic poles induced in **B** are in a direction such as to **oppose the entry** of the magnetic field of **A** and so a N-pole is induced at its right end.

Switch remains closed – When the current in **A** reaches its maximum value and remains constant, the magnetic field associated with it no longer grows, but **remains stationary** in **B**. **No current** is therefore induced in **B** at this time.

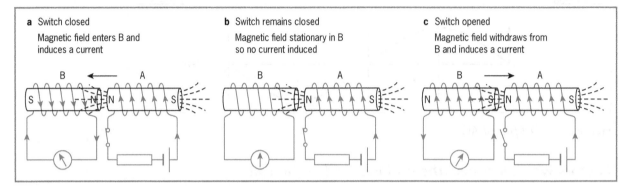

a Switch closed
Magnetic field enters B and induces a current

b Switch remains closed
Magnetic field stationary in B so no current induced

c Switch opened
Magnetic field withdraws from B and induces a current

Figure 31.9 *Transferring electrical energy between coils*

Switch opened – On opening the switch, the current in coil **A** rapidly diminishes to zero. The magnetic field associated with the current therefore also diminishes to zero. As it does so, the flux withdraws from **B** and once more induces a current there. The magnetic poles induced in **B** are now in such a direction as to oppose the withdrawal of the magnetic field of **A** and so a S-pole is induced at its right end.

Use the right-hand grip rule to confirm that the directions of current in coil **B** of Figure 31.9 are correct.

Figure 31.10 shows the variation in current within the coils during the process. The induced currents on opening and closing the switch are **oppositely directed** since the magnetic flux **moves into B** in the first case and **moves out of B** in the next.

Note: If the circuit of **coil A is powered by an AC supply** instead of the DC supply from the battery, the current and associated magnetic flux would be growing into B and then withdrawing from B repeatedly and therefore the **AC in A will induce an AC in B**. This is the basis of the transformer, described next.

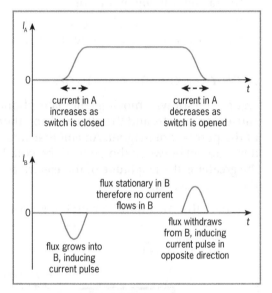

Figure 31.10 *Graphs showing variation of currents in coils A and B with respect to time*

The transformer

*A **transformer** is a device that increases or decreases an alternating voltage.*

Figure 31.11 shows a simple transformer. The primary coil is connected to an AC supply of voltage V_p (generally, the AC mains supply) and the secondary coil delivers a voltage V_s to the device to be operated. The changing current in the primary coil produces a changing magnetic

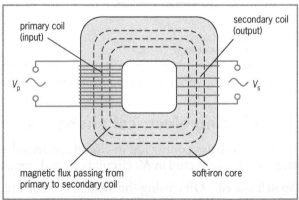

Figure 31.11 *The transformer*

field which repeatedly grows into, and diminishes from, the secondary coil, thereby inducing an alternating voltage within it.

Notes:

- The **soft-iron core** allows the **magnetic flux to pass easily** from the primary to the secondary coil.
- The current input to the primary coil of a transformer is never a constant DC. A varying current is needed to produce the varying magnetic field necessary to enter and withdraw from the secondary coil to induce a current there. The input to a transformer is generally the AC mains supply.
- The frequency of the input voltage is equal to that of the output voltage, since each oscillation of the input current induces an oscillation of the output.
- Placing a **fuse** in the **primary coil** of a transformer protects both coils from high currents, because current can only flow in the secondary coil if there is current in the primary.
- Step-up transformers increase the voltage and step-down transformers decrease it.
- A transformer that **increases voltage**, **decreases current**, and one that decreases voltage, increases current (**see calculations which follow**).
- If a transformer increases voltage, this does not mean that it increases power. $P = VI$ and any increase in V will have a corresponding decrease in I.
- The coil with the **least number of turns** carries the **lower voltage** and therefore **higher current** and **thicker** wires.

Factors affecting the efficiency of a transformer

Figure 31.12 *Laminated and insulated soft-iron core*

1. **Eddy currents** – The changing magnetic flux creates induced eddy currents in the core. This results in **energy lost as heat in the core** and therefore less energy is available from the secondary coil. To reduce these eddy currents, the core is **laminated** into thin sheets of soft-iron separated from each other by a film of **insulating varnish** (see Figure 31.12). This may be compared with the insulated discs of the soft-iron cylinders on which the coils of practical motors and generators are wound (see Figure 30.17).

2. **Hysteresis loss** – As the atomic magnetic dipoles within the soft-iron core constantly rearrange their direction with the changing magnetic field, there is internal friction that results in energy lost as heat. **Specially developed alloys** reduce this hysteresis loss to a minimum.

3. **Coil resistance** – Current through the coil's resistance results in energy being wasted as heat. By increasing the diameter of the wire, the resistance can be made smaller. Note that the coil which carries the larger current (smaller voltage or smaller number of turns) should be thicker.

4. **Incomplete flux linkage** – Some of the flux from the primary coil may not thread through the secondary coil due to either inadequate materials used for the core or inadequate shape of the core design and coil placement.

Reasons for power stations transmitting electricity as AC

1. **AC can be converted with minimum energy loss to different voltages by transformers.** Consumer appliances operate on several voltages, and transformers can be used to acquire these voltages when AC is applied to their primary coils. The efficiency of a transformer is typically above 95%!

2. **AC can be transmitted** from the station **at small currents** resulting in **minimum energy wastage** as heat due to the resistance of the long transmitting cables. Recall that the **power loss in the cables** is given by $P = I^2R$, so making the current smaller significantly reduces the power lost.

3. **Thinner cables can be used** to transmit the electricity at small currents and therefore the **cost of the wire** needed is less.

Transmission and distribution of electrical energy over long distances

Power stations generate electrical energy at alternating voltages and generally transmit it over long distances to the city where it is consumed. The power lost as heat in the transmission cables is given by $P = I^2R$.

To prevent large amounts of energy wastage, the transmission current I and the resistance of the cables R should be kept to a minimum.

- I is kept small by using a **step-up transformer at the power station** to increase the voltage and hence decrease the current.

- R is kept small by using **good conducting aluminium cables.**

  Copper is a better conductor than aluminium, but aluminium cables are preferred since aluminium is less dense and cheaper.

When the energy reaches the city or region of distribution, other transformers step the voltage down to an appropriate level. Figure 31.13 shows a network used to transmit and distribute electricity over long distances.

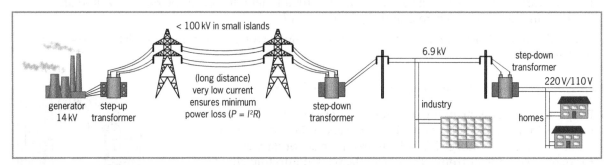

Figure 31.13 *Transmitting and distributing electricity*

Calculations involving transformers

The symbols below are used in the equations which follow.

V_p and V_s = voltage across the primary and secondary coils

N_p and N_s = number of turns in the primary and secondary coils

I_p and I_s = current in the primary and secondary coils

P_p and P_s = power in the primary and secondary coils

For all transformers:

- $$\frac{V_p}{V_s} = \frac{N_p}{N_s} \quad \text{... (i)}$$

- $$\text{efficiency} = \frac{\text{useful power output}}{\text{power input}} \times 100\% = \frac{V_s I_s}{V_p I_p} \times 100\% \quad \text{... (ii)}$$

For IDEAL transformers, efficiency = 100%, so power output is equal to the power input:

- $V_p I_p = V_s I_s$

  $$\therefore \frac{V_p}{V_s} = \frac{I_s}{I_p}$$

  $$\therefore \frac{V_p}{V_s} = \frac{N_p}{N_s} = \frac{I_s}{I_p} \quad \text{... (iii)}$$

Note: The **current ratio** is **inverted** relative to the voltage ratio and number of turns ratio.

Example 1

An ideal transformer having 2000 turns on its primary coil and 500 turns on its secondary coil is used to operate a 30 V, 20 Ω device. Determine:

a the current taken by the device

b the current in its primary coil

c the voltage across the primary coil.

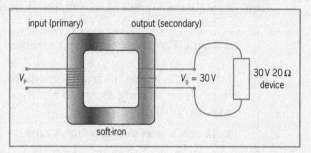

input (primary)   output (secondary)

V_P $V_s = 30$ V   30 V 20 Ω device

soft-iron

Figure 31.14 *Example 1*

a The current taken by the device is the current in the secondary coil:

$$V_s = I_s R \qquad 30 = I_s \times 20 \qquad I_s = \frac{30}{20} \qquad I_s = 1.5 \text{ A}$$

b Since the transformer is ideal:

$$\frac{I_p}{I_s} = \frac{N_s}{N_p} \qquad \frac{I_p}{1.5} = \frac{500}{2000} \qquad I_p = \frac{500 \times 1.5}{2000} \qquad I_p = 0.375 \text{ A} \quad (0.38 \text{ A to 2 sig.fig.})$$

c $\dfrac{V_p}{V_s} = \dfrac{N_p}{N_s} \qquad \dfrac{V_p}{30} = \dfrac{2000}{500} \qquad V_p = \dfrac{2000 \times 30}{500} \qquad V_p = 120 \text{ V}$

Example 2

A transformer of efficiency 96% is used to operate a 6.0 V, 30 W device from a 120 V_{AC} mains supply. Determine:

a the current in the secondary coil (windings)

b the resistance of the device

c the power input from the mains

d the current in the primary coil

e the turns ratio $\left(\dfrac{N_p}{N_s}\right)$ of the coils.

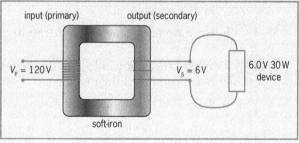

input (primary)   output (secondary)

$V_P = 120$ V $V_s = 6$ V   6.0 V 30 W device

soft-iron

Figure 31.15 *Example 2*

a $\qquad P_s = V_s I_s \qquad I_s = \dfrac{P_s}{V_s} \qquad I_s = \dfrac{30}{6} \qquad I_s = 5.0 \text{ A}$

b $\qquad V_s = I_s R \qquad \dfrac{V_s}{I_s} = R \qquad R = \dfrac{6.0}{5.0} \qquad R = 1.2 \text{ Ω}$

c efficiency $= \dfrac{P_s}{P_p} \times 100\% \qquad 96 = \dfrac{30}{P_p} \times 100 \qquad P_p = \dfrac{30 \times 100}{96} = 31.25 \dots \; (31 \text{ W to 2 sig. fig.})$

d $\qquad P_p = V_p I_p \qquad 31.25 = 120 \times I_p \qquad I_p = \dfrac{31.25}{120} \qquad I_p = 0.26 \text{ A}$

e $\qquad \dfrac{N_p}{N_s} = \dfrac{V_p}{V_s} \qquad \dfrac{N_p}{N_s} = \dfrac{120}{6.0} \qquad \dfrac{N_p}{N_s} = \dfrac{20}{1}$

Recalling facts

1   **a** State the laws of electromagnetic induction established by Michael Faraday and Emil Lenz.

     b Name TWO electrical devices which utilise electromagnetic induction.

2   Figure 31.16 shows an arrangement in which a magnet hangs from a spring above an electrical coil connected to a centre-zero galvanometer.

     a *Describe* and *explain* what occurs on the galvanometer if:

       i the lower magnetic pole is stationary *within* the coil as shown

       ii the magnet is pulled down and released so that its lower end oscillates into and out of the coil.

     b Redraw the coil as it is in the diagram and use arrows to indicate the direction of the current when the north pole *enters* it.

     c Explain why oscillation of the magnet quickly diminishes.

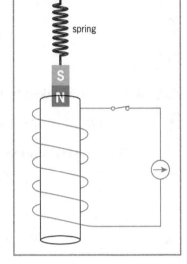

3   **a** Explain how each of the following affects the efficiency of a transformer:

       i eddy currents

       ii hysteresis loss

       iii coil resistance.

Figure 31.16 *Question 2*

     b How are eddy currents reduced in the core of a transformer?

4   **a** What causes power loss in the electrical cables through which electricity is delivered over long distances?

     b State TWO ways by which this loss can be minimised.

     c Why is AC preferred to DC as the form of electricity required for home and industry?

5   List THREE factors (other than those affecting efficiency) that can increase the output voltage of a generator.

Applying facts

6   Figure 31.17 shows a moving coil microphone.

     a Sketch the diagram and add arrows to it to indicate the direction of the current when a **rarefaction** reaches the diaphragm. *Explain* the reason for the chosen direction.

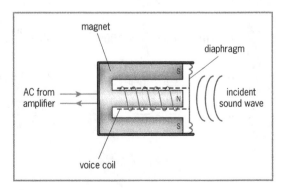

Figure 31.17 *Question 6*

b Describe the current induced in the coil, in terms of amplitude and frequency, when the microphone receives a sustained note of:

   i low volume and high pitch

   ii high volume and low pitch.

7 Figure 31.18 shows a plan view of a cylindrical metal rod P rolling to the right across metal rails R_1 and R_2. The rails are connected to an ammeter forming a circuit with the rod, and a magnetic field acts vertically downward on the system.

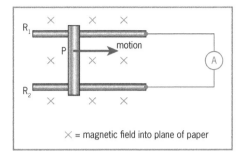

a Redraw the rod as shown in the diagram and show by means of an arrow the direction of the current within it.

b Mark the end of the rod which has positive polarity.

Figure 31.18 *Question 7*

8 A strong cylindrical magnet is allowed to fall vertically through a cylindrical PVC tube and then through a copper tube of the same internal diameter. Explain why the magnet falls much slower through the copper tube.

9 Figure 31.19a shows a simplified diagram of an AC generator. The crank is turned so that the axle, coil and slip rings rotate. Arrows on the coil indicate its **motion** at the instant shown in **a**.

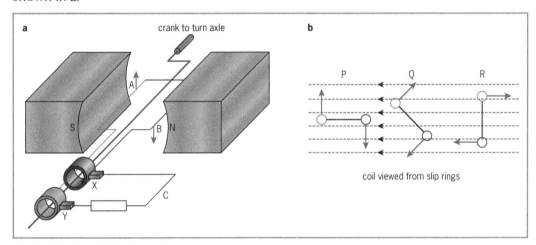

Figure 31.19 *Question 9*

a Redraw the circuit as shown in Figure 31.19a and indicate the direction of the current along sections A, B and C by means of arrows.

b Which graphite brush, x or y, is of positive polarity at the instant shown?

c Why are slip rings used?

d The coil should be wound on a soft-iron cylinder made of a series of insulated discs.

   i Why should the cylinder be made of a soft magnetic material?

   ii Why are the discs insulated from each other?

e Figure 31.19b shows THREE positions of the coil as it cuts through the magnetic field.

   State and explain the different emfs obtained within the coil in positions P, Q and R.

10   **a** An AC generator has a peak voltage of 6 V and a period of 0.02 s. Sketch a graph of voltage against time for ONE revolution, indicating the peak voltage and period.

     b Calculate the frequency of the AC produced.

     c The number of revolutions per second of the coil is now doubled. Determine the new:

       i frequency

       ii period

       iii peak voltage.

11   An ideal transformer has 4000 turns on its primary coil and 400 turns on its secondary coil. It is used to operate a 12 V device that draws a current of 5.0 A.

     a Calculate:

       i the power of the device connected to the secondary coil

       ii the current in the primary coil

       iii the voltage input to the transformer.

     b State, with a reason

       i if it is a step-up or step-down transformer

       ii which coil should have the thicker wires

       iii which coil which should be fitted with a fuse.

12   A poorly designed transformer of efficiency 80% has its primary coil of 5000 turns connected to a 120 V AC mains supply and its secondary coil attached to two 6.0 V, 15 W lamps in parallel. Draw a diagram of the arrangement and calculate:

     a the number of turns in the secondary windings

     b the current through each lamp

     c the current in the secondary windings

     d the current in the primary windings.

13   Figure 31.20 shows a transformer that has two secondary coils, A and B. The primary windings have 4000 turns and the secondary windings, A and B, have 400 turns and 100 turns, respectively. If an AC supply of 120 V is applied to the primary coil, calculate the voltage available at each of the secondary coils.

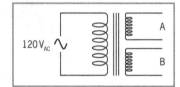

Figure 31.20 *Question 13*

14   Power is sent over cables of resistance 4.00 Ω from a supply of 15.0 kW. Figure 31.21 shows that some of the voltage of the supply provides a pd across the cables and the remainder is available to the consumer. Calculate the power lost in the cables if the electricity is sent from the supply at:

     a 250 V

     b 100 000 V.

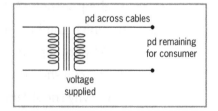

Figure 31.21 *Question 14*

15 Figure 31.22 shows an ideal transformer with its primary coil connected to a 120 V AC supply and its secondary coil to a variable resistor. Table 31.1 shows readings of the currents flowing in the primary and secondary coils as the resistance is varied.

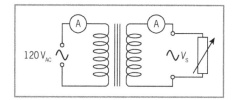

Figure 31.22 *Question 15*

Table 31.1 *Question 15*

Current in primary coil, I_p/A	0.10	0.30	0.50	0.70	0.90	1.10
Current in secondary coil, I_s/A	0.7	2.5	4.0	5.8	7.4	8.5

a Plot a graph of I_s against I_p using the readings of Table 31.1.

b Calculate the slope of the graph.

c Use the slope of the graph to calculate the voltage obtained from the secondary coil.

d Determine the current in the secondary coil when the current in the primary coil is 1.0 A.

e How many turns are in the primary windings if the secondary windings are made of 500 turns?

f What is the value of the load resistance when the current in the primary coil is 1.0 A?

Figure 31.22 shows an ideal transformer with its primary coil connected to a 230 V AC supply and its secondary coil to a variable resistor. Table 31.1 shows readings of the currents flowing in the primary and secondary coils as the resistance is varied.

Figure 31.22 Question 17

Table 31.1 Question 17

0.10	0.20	0.30	0.40	0.50	1.10
1.0	2.0	3.0	4.0	5.0	7.0	8

a. Plot a graph of ... against ...

b. Calculate the slope of the graph.

c. Use the slope of the graph to calculate the voltage obtained from the secondary coil.

d. Determine the current in the secondary coil when the current in the primary coil is 0.6 A.

e. ...

f. What is the value of the load resistance when the current in the primary coil is 0.4 A?

Section E
The physics of the atom

In this section

The atom

- Here you will learn of the evolution of the concept of the structure of the atom.

Radioactivity

- This chapter deals with natural radioactivity: the study of the spontaneous disintegration of unstable atomic nuclei.

Nuclear fission and nuclear fusion

- These are extremely energetic nuclear reactions: The splitting of large atomic nuclei and the joining of small atomic nuclei.

32 The atom

Humans have been interested in the study of matter from the beginning of modern civilisation, but up until near the end of the 19th century, little progress was made in understanding it. Since then, however, there has been a series of significant scientific developments in the field, and to this day, our concept of matter continues to evolve as new theories emerge to explain various phenomena.

Learning objectives

- Describe the work of **scientists** involved in the **evolution** of the concept of the atom.
- Describe and explain the **Geiger–Marsden** (or Rutherford scattering) experiment.
- Sketch a simple model of the **structure of the atom**.
- Compare the **relative mass** and **relative charge** of the electron and the neutron to the proton.
- Apply the relation $A = Z + N$ and use the nuclide notation $^A_Z X$.
- Explain the term **isotope**.
- Relate the **shell model** of the atom to the **Periodic Table**.

Evolution of the concept of the atom

In around 450 BC – The Greek philosopher, Democritus, postulated that all matter consisted of small indivisible particles. He called these particles **'atomos'**, from the Greek word for **indivisible**.

In 1897 – Joseph John Thomson discovered the **electron**. He observed rays of negatively charged particles emitted from cathodes at high negative potential. Since the atom is neutral, he assumed that these 'corpuscles' (later called electrons) existed as smaller particles **interspersed within a positive material**; the total charge of the electrons being equal in magnitude to the charge of the positive material. This became known as the **plum pudding** concept of the atom, where the electrons are the 'plums' scattered within the positively charged 'pudding'.

In 1911 – Ernest Rutherford proposed that **most of the atom** is **empty space** and that it contains a **concentrated positively charged nucleus**. He suggested that small, negatively charged particles (electrons) existed in a surrounding 'electron cloud', making the total charge on the atom zero. His proposition was made with the help of his assistants, Hans Geiger and Ernest Marsden, who carried out an experiment where they investigated the behaviour of alpha particles as they were shot at a thin sheet of gold foil (see below). Rutherford proposed his **planetary model** of the atom, which suggested that electrons revolve around the nucleus in orbits of arbitrary radii, just as planets orbit the Sun.

In 1913 – Niels Bohr proposed the Bohr model of the atom, which better defined the electron orbits of the Rutherford atom and became known as the Rutherford–Bohr atom, as follows.

- Electrons revolve around the nucleus in **discrete** orbits of a **particular radius**. These orbits exist in 'shells' and the electrons in a particular shell possess a **particular quantity of energy**.

- Electrons possess **more energy** if they exist in **shells of larger radii**.

- Electrons **emit photons (energy packets) of electromagnetic radiation** when they **jump inward** to a shell of a lower energy level, and can **absorb photons of electromagnetic radiation** when they **jump outward** to a shell of a higher energy level. The energy of the emitted or absorbed radiation is equal to the difference between energy levels of the shells. Figure 32.1 shows that an electron in the third shell and of energy E_3, jumping to the first shell where it would have energy E_1, would emit a photon of energy value $E_3 - E_1$.

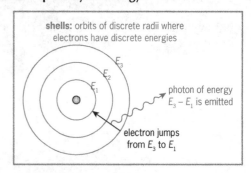

Figure 32.1 *Electron shells of the Bohr model of the atom*

In 1932 – James Chadwick discovered **neutrons**, uncharged particles that exist together with protons within the nucleus of an atom. Neutrons were difficult to detect since uncharged particles are **unaffected by electric and magnetic fields.**

Today's scientists are aware of other atomic models with many more subatomic particles, but an understanding of the atom in terms of protons, neutrons and electrons is sufficient to explain many of the phenomena we encounter. Figure 32.2 shows a summary of the different models mentioned above.

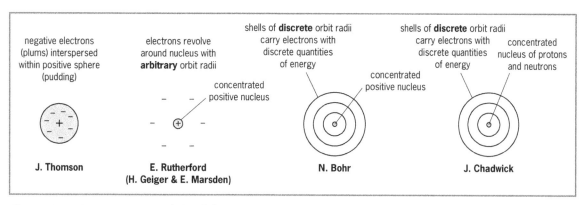

Figure 32.2 *Successive models of the atom*

The Geiger–Marsden experiment (or Rutherford scattering experiment)

Ernest Rutherford believed that most of the atom was empty space and performed a series of experiments to test his prediction. He worked with Hans Geiger, investigating the behaviour of alpha particles as they were shot through metal foils. Alpha particles consist of two protons and two neutrons, and are emitted from an atomic nucleus at high speeds (see Chapter 33).

In 1911, Geiger, together with a student, Ernest Marsden, carried out a research project in Rutherford's laboratory. They shot **alpha particles** at a thin sheet of **gold foil** (see Figure 32.3). The chamber was **evacuated** since alpha particles are stopped by just a few centimetres of air.

The **path** of the particles was detected by observing the **scintillations** they produced on crashing onto a **zinc sulfide screen** fixed to the front of a **rotatable microscope**. The microscope was rotated from the straight-through position of **0°** up to an angle of **150°** to receive scintillations at various angles of deflection.

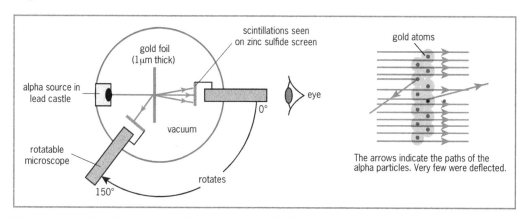

Figure 32.3 *The Geiger–Marsden (or Rutherford scattering) experiment*

Observations:

1. The scintillations indicated that most of the particles passed through the gold foil **without deflection.**

2. As the microscope was rotated, the number of scintillations observed **rapidly declined.**

3. As the microscope was rotated to angles greater than 90°, a small number (less than 1 in 8000 emissions) of **strong** scintillations were still detected.

Conclusions:

1. The first two observations indicate that **most** of the atom is **empty space.**

2. The third observation indicates that the **nucleus** of the atom must be an extremely **small, dense, positively** charged mass, capable of **repelling** the approaching **positively** charged alpha particles at **very high speeds.**

Rutherford found the strong deflections of the alpha particles to be very surprising and so commented:

"*It is almost as incredible as if you fired a fifteen-inch shell at a piece of tissue paper and it came back and hit you.*"

Structure of the atom

The **proton number** or **atomic number**, **Z**, of an element is the number of protons contained in the nucleus of an atom of the element.

The **neutron number**, **N**, of an element is the number of neutrons contained in the nucleus of an atom of the element.

The **mass number** or **nucleon number**, **A**, of an element is the SUM of the numbers of protons and neutrons contained in the nucleus of an atom of the element.

$$A = Z + N$$

A neutral atom contains the **same number** of **protons** as it does **electrons.** In a later section of this chapter, Figure 32.5 shows the arrangement of protons, neutrons and electrons in three neutral atoms.

Mass and charge of proton, neutron and electron

The Periodic Table (see later, Figure 32.4) is an ordered list of the number of protons in an element. The proton is therefore the reference particle to which the masses and charges of other subatomic particles are compared (see Table 32.1).

Table 32.1 *Masses and charges of subatomic particles*

	Proton	Neutron	Electron
Relative mass	1	1	$\dfrac{1}{1840}$
Relative charge	+1	0	−1
Absolute mass/kg	1.66×10^{-27}	1.66×10^{-27}	9.11×10^{-31}
Absolute charge/C	1.6×10^{-19}	0	-1.6×10^{-19}

Representation of an atomic nuclide

A **nuclide** is an atomic nucleus with a particular number of protons and neutrons. It is represented by:

$$^A_Z X \qquad \text{i.e.} \quad ^{Z+N}_{Z}X \quad \text{or} \quad ^{\text{protons + neutrons}}_{\text{protons}}X$$

Where X is the symbol of the element. For example:

$^{24}_{11}$Na contains 11 protons and 13 neutrons

$^{14}_{6}$C contains 6 protons and 8 neutrons.

Representation of a subatomic particle

$$^{Z+N}_{\text{relative charge}}Y \qquad \text{i.e.} \quad ^{\text{protons + neutrons}}_{\text{relative charge}}Y$$

where Y is the symbol for the particle. For the proton, neutron and electron:

$$^1_1\text{p: proton} \qquad ^1_0\text{n: neutron} \qquad ^0_{-1}\text{e: electron}$$

The Periodic Table

*The **Periodic Table** is a table of elements of increasing proton number, arranged to categorise their electronic configurations and chemical properties.*

Figure 32.4 shows the first 20 elements of the Periodic Table listed with their proton numbers.

		Groups (number of electrons in outer shells)								
		I	II	III	IV	V	VI	VII	0	
Periods	1	H 1								He 2
(number of shells)	2	Li 3	Be 4	B 5	C 6	N 7	O 8	F 9	Ne 10	
	3	Na 11	Mg 12	Al 13	Si 14	P 15	S 16	Cl 17	Ar 18	
	4	K 19	Ca 20							

Figure 32.4 *The first 20 elements of the Periodic Table: their symbols and their proton numbers*

Electronic configuration

*The **electronic configuration** of an atom shows the number of electrons in each of its shells.*

Periods

The period number indicates how many electron shells typically contain electrons in an atom of the element. Elements in the same row have the **same number** of **electron shells** in their atoms. Atoms of hydrogen (H) and helium (He) are in period 1 and therefore each have only one shell, whereas atoms of sodium (Na) and chlorine (Cl) are in period 3 and each have three shells.

Groups

With the exception of group 0, the group number indicates how many electrons are in the outermost shell of an atom of the element. Elements in the same column have the **same number** of **electrons** in the **outer shells** of their atoms. Atoms of sodium (Na) and potassium (K) therefore each have one electron in their outer shells, whereas atoms of fluorine (F) and chlorine (Cl) each have seven electrons in their outer shells.

- For the first 20 elements of the Periodic Table, the shells hold these maximum numbers of electrons:

    1st shell: 2 electrons   **2nd shell:** 8 electrons   **3rd shell:** 8 electrons

- The **innermost** shells are **filled first**.
- Elements in the **rightmost** column (group 0) have **full outer orbits**. So helium (He) has two electrons in its outer orbit, and neon (Ne) and argon (Ar) each have eight electrons in their outer orbits.

*Isotopes are variants of an element having the **same atomic** number but **different mass numbers**.*

or

*Isotopes are variants of an element having the **same number of protons**, but **different numbers of neutrons** contained in their atoms.*

Note: Isotopes of the same element have the **same chemical reactions**.

Examples: Isotopes of sodium are $^{24}_{11}$Na and $^{23}_{11}$Na   Isotopes of carbon are $^{12}_{6}$C and $^{14}_{6}$C

Figure 32.5 shows the structure of different neutral atoms. Their electronic configurations can be obtained by examining their electron shells.

The configurations can be represented by writing the number of electrons in each shell starting from the innermost shell. Figure 32.5 shows that the electronic configuration of $^{24}_{11}$Na is 2, 8, 1.

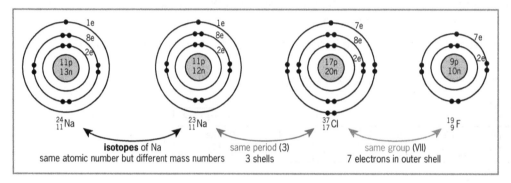

Figure 32.5 *Electronic configurations – isotopes, periods and groups*

Recalling facts

1 Name the scientist in each case.

  a Who first proposed that electrons orbit the nucleus of an atom in shells and must have a particular energy to exist in a given shell?

  b Who discovered the neutron?

  c Who first postulated that matter must be composed of fundamental particles which were indivisible?

2 Describe the atom proposed by each of the following:

  a Ernest Rutherford

  b Joseph Thomson.

3 The following questions refer to the Geiger–Marsden experiment.

  a What type of radiation was used in the experiment?

  b Describe the material that the radiation was directed towards.

c Why was the experiment carried out in a vacuum?

d Describe the detector used to determine the path of the radiation.

e Explain why it was concluded that:

   i   most of the atom is empty space

   ii the nucleus of the atom consists of a very concentrated positive charge.

f Name the scientist who played an important role in analysing the results of the Geiger–Marsden experiment and who proposed a model of the atom based on the experiment.

4 a Differentiate between atomic number and mass number.

b Define the term *isotopes*.

5 a A student says that the charge on an electron is –1. Explain why this is incorrect.

b By what *factor* is the mass of a proton greater than that of an electron?

Applying facts

6 a Sketch the structures of the neutral atoms having the following nuclide representations, showing for each, the numbers and positions of their protons, neutrons and electrons.

   i $^{24}_{12}Mg$ ii $^{34}_{16}S$ iii $^{18}_{8}O$

b By analysing their electronic configurations, determine which of the atoms mentioned in part **a** are in the same:

   i   group

   ii period.

c Using the same notation as in part **a**, give one possible example of an isotope of $^{210}_{82}Pb$.

7 Five nuclides are given below.

$^{24}_{12}Mg$ $^{33}_{16}S$ $^{34}_{16}S$ $^{32}_{15}P$ $^{16}_{8}O$

a What is the mass number of $^{24}_{12}Mg$?

b Which of the nuclides has/have 16 protons?

c Which of the nuclides is positioned immediately before $^{34}_{16}S$ in the Periodic Table?

d How many electrons are in a neutral atom of $^{32}_{15}P$?

e How many electron shells are in a neutral atom of $^{16}_{8}O$?

f How many neutrons are in $^{33}_{16}S$?

g Give the electronic configuration of a neutral atom of $^{24}_{12}Mg$.

33 Radioactivity

In 1896, **Henri Becquerel** accidentally discovered that an ore of uranium emits a form of radiation. This phenomenon was further investigated by **Marie Curie** and today plays a significant role in modern science, having several applications in the medical and industrial fields.

Learning objectives

- Describe **Marie Curie's work** in the field of radioactivity.
- Define **radioactivity** and be familiar with the various **properties of α-, β- and γ-emissions**.
- Understand the meaning of **background radiation** and be aware of its **sources**.
- Understand how to detect and measure radiation using a **GM tube and accessories**.
- Identify alpha, beta and gamma radiations using the **absorption** test, the **electric field** deflection test, the **magnetic field** deflection test and the **cloud chamber**.
- Write and interpret **equations of radioactive decay**.
- Describe the **random** and **exponential decay** of the radioactive process.
- Recall that the radioactive process is **independent** of conditions **external** to the nucleus.
- Describe an experiment which can **simulate** the determination of the **half-life** of a radioisotope.
- Perform **calculations** involving **half-life** including those involving **radio-carbon dating**.
- Discuss the **applications** of radioisotopes and the precautions to be taken when using them.

Marie Curie's work in the field of radioactivity

Marie Curie was born in Poland in 1867. Following up on Henri Becquerel's work on the radiation emitted from **uranium**, she realised that the **intensity** of the rays was **dependent** only on the **mass of the emitting sample**. Since neither environmental physical factors nor chemical factors could affect the rate of emission of the radiation, she concluded that it was dependent only on the properties of the atom itself and was therefore an **atomic phenomenon**. She called the phenomenon **radioactivity**.

In 1903 Marie and her husband, Pierre, shared the Nobel Prize in **Physics** with Henri Becquerel for their work on **radioactivity**. In 1911 she was awarded a second Nobel Prize, this time in **Chemistry**, for her **isolation of** the elements **polonium** and **radium**.

Marie Curie was the first woman to be awarded a Nobel prize, the first person to be awarded two Nobel Prizes, and remains the only person to be awarded Nobel Prizes in two different scientific fields!

During World War 1 she, together with her daughter, Irene, worked in the development of X-radiography. As a member of the Academy of Medicine in France, Marie Curie's work opened the field of **nuclear medicine** with its **applications** in **diagnosis** and **therapy**.

She died in 1934 from the cumulative effects of the radiation she encountered during her investigations.

The nature of radioactivity

Within an atomic nucleus there are strong **electrical forces of repulsion** existing between the positively charged **protons**. The system is still tightly bound, however, due to the strong **nuclear forces of attraction** existing between the **nucleons**.

Neutrons situated between the protons **reduce** the electrical forces of repulsion because this increases the distance between the positive charges. However, too many neutrons can also cause instability. The neutron-to-proton ratio in small nuclei is approximately 1:1, but increases to 1.6:1 in larger nuclei. If the nuclide of an isotope becomes unstable it can emit α-particles, β-particles and/ or γ-rays (see below) to become more stable.

Radioactivity is the spontaneous disintegration (decay) of unstable atomic nuclei.

Some elements of small atomic number have radioactive isotopes, but all isotopes of atomic number greater than 83 are radioactive.

Some properties of radioactive emissions are outlined in Table 33.1.

Table 33.1 *Some properties of radioactive emissions*

	Alpha particle (α)	Beta particle (β)	Gamma ray (γ)
Nature	2 protons and 2 neutrons tightly bound	1 electron	high frequency electromagnetic wave
Symbol	$_2^4\alpha$ or $_2^4$He	$_{-1}^{0}\beta$ or $_{-1}^{0}$e	γ
Relative mass	4	1/1840	Zero
Relative charge	+2	−1	Zero
Velocity in air or vacuum/m s⁻¹	2×10^7	up to almost the speed of light	3×10^8
Type of energy	kinetic energy	kinetic energy	electromagnetic energy (photon energy)
Ionisation of the air	strongly ionising on collision with neutral air molecules	weakly ionising on collision with neutral air molecules	very weakly ionising when the wave energy is absorbed by neutral air molecules
Absorbed by	a thin sheet of paper or about 5 cm of air	a few m of air or about 3–5 mm of aluminium	several m of concrete or 4 cm of lead absorbs most of it
Detection of presence by	GM (Geiger–Müller) tube and accessories	GM tube and accessories	GM tube and accessories
Effect of electric field	deflects less than β	deflects more than α	no deflection
Effect of magnetic field x = magnetic field perpendicularly into plane of paper	deflected in accordance with Fleming's left-hand rule (see Chapter 30) α-flow has the direction of conventional current deflection should be observed in a vacuum since α-particles are easily absorbed by a few cm of air	deflected in accordance with Fleming's left-hand rule β-flow has the direction opposite to conventional current deflected more than α-particles since the mass of an α-particle is more than 7000 times greater than that of a β-particle	no deflection

Background radiation is the ionising radiation within our environment arising from radioactive elements in the Earth and its surrounding atmosphere, together with X-rays from medical equipment and high speed charged particles from the cosmos.

Figure 33.1 shows that most of the background radiation we receive is from **radon gas** in the atmosphere.

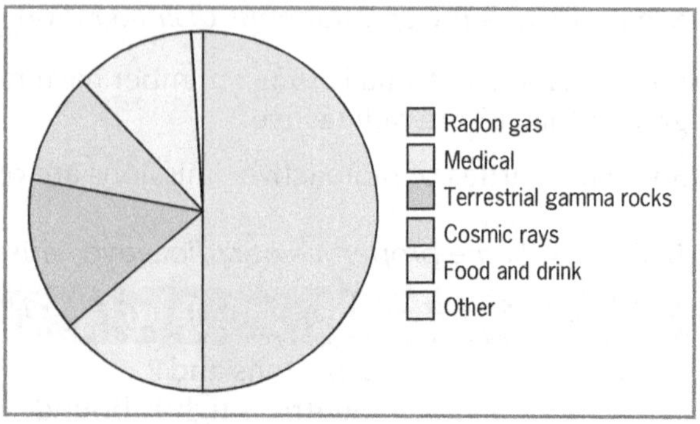

- Radon gas
- Medical
- Terrestrial gamma rocks
- Cosmic rays
- Food and drink
- Other

Figure 33.1 Background radiation

Detection using the Geiger–Müller detector

Radioactive emissions may be detected by a **Geiger–Müller tube** (GM tube) and accessories, which together are called a GM detector and is commonly referred to as a **Geiger counter**. Radiation entering the tube through a thin **mica** window **ionises** the gas within it, and produces an **electrical pulse** between a pair of electrodes which is passed to one or more of the following:

- a **ratemeter**, which gives the rate at which emissions occur

- a **loudspeaker**, which produces a 'click' on detecting an emission

- a **scalar**, which counts the emissions.

Figure 33.2 shows a GM detector receiving radiation from a radioactive source and producing an output via a speaker, ratemeter and scalar counter.

*The **activity** of a sample of radioactive material is the rate at which its nuclei decay.*

*The **becquerel**, Bq, is a rate of one nuclear disintegration per second.*

*The **count rate** of a sample of radioactive material is the rate of emissions detected from it.*

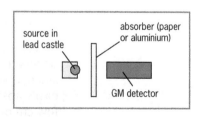

Figure 33.2 Detecting radiation from a radioactive source using a GM detector

Received count-rate

The **count-rate (C)** detected by a GM detector is inversely proportional to the square of the **distance (r)** from the source to the detector $\left(C \propto \dfrac{1}{r^2}\right)$; the greater the distance, the smaller is the amount of radiation detected. Experiments investigating the activity of a source should therefore maintain **a constant distance between it and the detector.**

The **observed count-rate** will include **background radiation**. This **must be subtracted from the received count rate** to obtain the correct portion detected from the source being investigated (see Example 6).

Testing for the type of radiation

1 – Absorption test

- The background count rate is measured by a GM detector.

- The radioactive source is then placed in front of the mica window of the detector and the count rate is again measured.

- A thin sheet of paper is placed between the source and detector (see Figure 33.3) and the count rate is measured.

source in lead castle

absorber (paper or aluminium)

GM detector

Figure 33.3 Absorption test

i   If the count rate is reduced to the background count rate, then the source is an α-emitter.

ii If the count rate is unaffected, then the source is either a β-emitter or a γ-emitter and the paper is replaced by an aluminium sheet of thickness 5 mm.

iii If the count rate now returns to the background count rate, then the source is a β-emitter; otherwise, it is a γ-emitter.

Note: Since α-particles are readily stopped by air, for experiments involving α-emission, the source must be placed **very close** to the detector or the apparatus should be set up in a **vacuum**.

2 – Electric field deflection test

- The apparatus is set up as shown in Figure 33.4 with the **switch in the off position** and the GM detector in position P, with its mica window directly facing the emissions from a radioactive source.

- The count rate is recorded.

- The **electric field** is then **switched on**.

    i   If the count rate is unaffected, the source is a γ-emitter.

    ii If the count rate falls and only returns to its initial value when the detector is shifted towards the positive plate, then the source is a β-emitter. The positive plate would have attracted the negative β-particles.

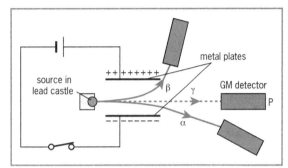

Figure 33.4 *Electric field deflection test*

iii If the count rate falls and only returns to its initial value when the detector is shifted towards the negative plate, then the source is an α-emitter. The negative plate would have attracted the positive α-particles.

3 – Magnetic field detection test

- A GM detector is positioned at P so that its mica window is directly facing the emissions from a radioactive source.

- The count rate is recorded.

- A magnetic field is then directed perpendicular to the path of the rays (see Figure 33.5).

    a   If the count rate is unaffected, the source is a γ-emitter.

    b   If the count-rate falls, the detector should be shifted until it returns. Current is a flow of charge and therefore the α- and β-particles will experience forces in accordance with **Fleming's left-hand** rule (see Chapter 30).

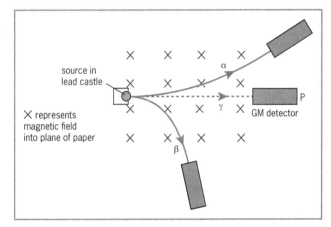

Figure 33.5 *Magnetic field deflection test*

When applying the rule, recall that the direction of the current is the direction of positive flow (α-flow), and is opposite to the direction of electrons (β-flow). The forces on the particles are perpendicular to their paths and therefore cause them to have **circular motion** (see Chapter 30, Figure 30.14). Note that the beta particles are deflected much more than the alpha particles since they are of much smaller mass.

4 – The diffusion cloud chamber test

Figure 33.6 shows a diffusion cloud chamber and the tracks which may be produced in it. The air in the chamber is cold due to the **black metal** surface, and is **supersaturated with alcohol vapour** from the soaked felt strip. As the radioactive emissions ionise the air, **condensation of the alcohol** occurs on the **cold ions**, producing cloud tracks characteristic of the type of radiation emitted by the source.

Note: Water may be used instead of alcohol to saturate the air, but alcohol is preferred since it condenses more readily.

- **α-tracks** – An α-particle has a mass of more than **7000 times** that of a β-particle and so is **strongly ionising** on collision with other particles, producing **straight, thick** tracks.

- **β-tracks** – A β-particle has a relatively small mass and so is **weakly ionising** on collision with other particles, producing **weak** tracks. The tracks are **randomly directed** since the particles are **easily deflected** on collision.

- **γ-tracks** – A γ-ray has no mass. Unlike an α-particle or β-particle, which have kinetic energies, a γ-photon has electromagnetic energy. A γ-photon, incident on an atom, may be absorbed by it. In so doing, it transfers its **electromagnetic energy** to an **orbital electron,** providing energy for it to **jump out of the atom** and so produce a **pair of ions**. Since this occurrence is relatively rare, γ-tracks are **extremely weak** and **dispersed**.

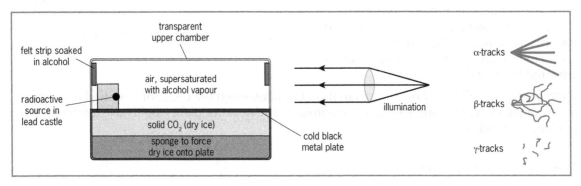

Figure 33.6 *Cloud chamber and cloud tracks of radioactive emissions*

Beta emission

Beta particles are electrons emitted from the nucleus of an atom. During beta emission, a neutron decays into a proton and an electron. The electron shoots out of the atom as a high-velocity β-particle, leaving the newly formed proton within the nucleus (see Figure 33.7).

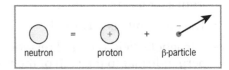

Figure 33.7 *Neutron decay and beta emission*

Equations of radioactive decay

Table 33.2 *Common symbols used in nuclear equations*

Alpha particle	Beta particle	Gamma ray	Proton	Neutron
$_2^4\alpha$ or $_2^4\mathrm{He}$	$_{-1}^{0}\beta$ or $_{-1}^{0}\mathrm{e}$	γ	$_1^1\mathrm{p}$	$_0^1\mathrm{n}$

When writing nuclear equations, always check the following:

- the sum of mass numbers on the left-hand side = the sum of mass numbers on the right-hand side

- the sum of atomic numbers (or relative charges) on the left-hand side = the sum of atomic numbers (or relative charges) on the right-hand side.

Section E: The physics of the atom

α-emission causes an isotope to become an atom of an element **two places back** in the Periodic Table.

Example: $^{226}_{88}Ra \rightarrow \, ^{4}_{2}He + \, ^{222}_{86}Rn$

β-emission causes an isotope to become an atom of an element **one place forward** in the Periodic Table.

Example: $^{14}_{6}C \rightarrow \, ^{0}_{-1}e + \, ^{14}_{7}N$

γ-emission does not alter the **mass number** or **atomic number** of the emitting isotope.

Example: $^{234}_{91}Pa \rightarrow \, ^{234}_{91}Pa + \gamma$

Some nuclei may remain in an **excited state** (more energetic state) after emitting an α- or β-particle. A γ-photon is then emitted, returning the nucleus to its more stable, **ground state**.

Examples: $^{238}_{92}U \rightarrow \, ^{234}_{90}Th + \, ^{4}_{2}\alpha + \gamma$
$^{90}_{38}Sr \rightarrow \, ^{90}_{39}Y + \, ^{0}_{-1}\beta + \gamma$

During radioactive decay, the decaying element is called the **parent** and the nuclide produced is the **daughter**.

Random decay of the radioactive process

Radioactive decay is **random** and can be compared to tossing a coin, throwing dice or shooting at a target. If a radioactive source is placed in front of a GM detector the following will occur, depending on the instrument connected to the GM tube:

- the **ratemeter** needle will **flicker**
- the **loudspeaker** will produce 'clicks' at a **random** rate
- the **scalar** will indicate a **random** increment of the detected emissions.

Half-life and the exponential decay of the radioactive process

Although the activity of a radionuclide is random, there is a **definite period** for which half of a given sample of it will decay.

*The **half-life** of a radioisotope is the time taken for the mass (or activity) of a given sample of it to decay to half of its value.*

The half-lives of some radioisotopes are very short, and some others are very long. For example:

- The half-life of hydrogen-7 ($^{7}_{1}H$) is only 2.3×10^{-23} s !
- The half-life of bismuth-209 ($^{209}_{83}Bi$) is 1.9×10^{19} years; over 1 billion times longer than the estimated age of the universe!

The half-life of a radioisotope is **not affected by conditions external to the nucleus**. This includes:

- **physical conditions**, such as temperature and pressure
- **chemical conditions**, such as its existence in the free state or chemically combined with some other element in a compound.

Figure 33.8 shows that radioactive decay is **exponential** with time. Note that the curve never meets the time axis but gets closer and closer to it. The activity (rate of decay) at any given time is proportional to the mass, the count rate and the number of nuclei at that time. The **y-axis** of the graph can therefore be activity, **mass**, **count rate** or **number of nuclei remaining**.

The half-life is found by choosing any point on the graph and finding the time for its activity, mass, count rate or number of nuclei remaining to fall to half of its value at that point.

An **average** from **three decay periods** (see Figure 33.8) gives a more accurate result. See also question 22 at the end of this chapter.

Experiment simulating radioactive decay

- About 300 dice are placed in a bucket. The number of dice, D, is recorded.

- The bucket is emptied (the dice 'thrown') and the dice with the value 6 facing upward are removed. The new value of D is recorded.

- The process is repeated using the remaining dice, recording the number of throws, N, and the new value of D, until fewer than 25 dice remain.

- A graph is plotted of D against N as shown in Figure 33.9.

Notes:

- The points on the graph are not exactly on the smooth curve drawn through them because the generation of the value 6 occurs at **random**.

- The **exponential** nature is indicated by the curve.

- As D decreases, the outcome of each throw becomes statistically less accurate.

- To obtain the **'half-life'** of this decay:

    i read from the graph THREE values of D and determine the corresponding period taken for each to fall to half of its value (see Figures 33.8 and 33.9)

    ii calculate the mean value of the three periods.

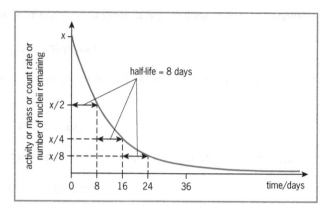

Figure 33.8 *Graph showing exponential decay and half-life*

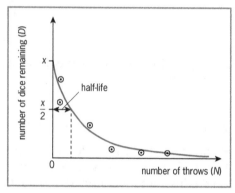

Figure 33.9 *Experiment simulating radioactive decay*

Calculations involving half-life

Two methods to calculate the half-life of a radioisotope are outlined below. The examples which follow demonstrate the use of the methods.

Arrow method

- Each arrow represents a half-life transition and points to half the value it originates at.
- The total time of decay is the sum of the half-lives.

Table method

- Particularly useful if a graph is to be plotted.
- The *Time* column always **starts at 0** and increases by one half-life for each new row.
- The *Activity, Count rate, Mass remaining* or *Nuclei remaining* column decreases to half its value for every new row.

Example 1

A sample of the radioisotope $^{137}_{55}$Cs gives a count rate of 1000 Bq. Determine the count rate after 90 years if its half-life is 30 years.

Arrow method:

1000 Bq → 500 Bq → 250 Bq → 125 Bq

    30 years   30 years   30 years

After 90 years, the count-rate is 125 Bq

Table method:

Count rate /Bq	Time/years
1000	0
500	30
250	60
125	90

Example 2

$^{60}_{27}$Co is a radioisotope with a half-life of 5.3 years. How long will it take for a mass of the isotope to decay from 400 g to 50 g?

Arrow method:

400 g → 200 g → 100 g → 50 g

    5.3 y   5.3 y   5.3 y

Time taken to decay to 50 g = 3 × 5.3 = 15.9 years

Table method:

Mass remaining/g	Time/years
400	0
200	5.3
100	10.6
50	15.9

Example 3

A radioisotope of iodine, $^{125}_{53}$I, decays to 1/16 of its mass in 240 days. Determine its half-life.

Arrow method:

With fractions, start with 1.

1 → ½ → ¼ → ⅛ → ¹⁄₁₆

————————————————→

    240 days

The 4 arrows indicate that there are 4 half-lives in the 240 days.

Therefore, one half-life = 240/4 days = 60 days.

Table method:

Fraction remaining	Time/days
1	0
1/2	60
1/4	120
1/8	180
1/16	240

Example 4

A scientist has 0.8 g of the radioisotope C-14. Given that the half-life of $^{14}_{6}$C is 5700 years, what would have been the mass of the C-14 sample 17 100 years ago?

Arrow method:

$$\frac{17\ 100}{5700} = 3$$

So it has decayed for 3 half-lives.

Work backwards from 0.8 g for 3 half-lives.

6.4 g → 3.2 g → 1.6 g → 0.8 g

————————————————————→

    17 100 years

17 100 years ago, the mass was 6.4 g.

Table method:

Mass remaining/g	Time/years
6.4	0
3.2	5 700
1.6	11 400
0.8	17 100

Example 5

The radioisotope $^{24}_{11}$Na has a half-life of 15 hours. How long will it take for its mass to fall by 87.5%?

Arrow method:

With percentages, start with 100%

It falls **BY** 87.5% and therefore falls **TO** 12.5%

$100\% \rightarrow 50\% \rightarrow 25\% \rightarrow 12.5\%$

$\quad$ 15 hours $\quad$ 15 hours $\quad$ 15 hours

It takes 3 half-lives, therefore 3×15 hours = 45 hours, to fall by 87.5%.

Table method:

% remaining	Time/days
100	0
50	15
25	30
12.5	45

Example 6

A radioisotope of half-life 10 minutes is placed in front of a GM detector. The received count rate is 165 counts per second when the background count rate is 5 counts per second. Determine:

a the received count rate after 20 minutes

b the time taken for the received count rate to be 25 counts per second.

a The initial count rate from the radioisotope is $(165 - 5)$ s^{-1} = 160 s^{-1}

Background count rate does not decrease with time, so only the count rate from the source in front of the detector will diminish.

Arrow method:

160 s$^{-1} \rightarrow 80$ s$^{-1} \rightarrow 40$ s^{-1}

$\quad$ 10 min $\quad$ 10 min

The count rate after 20 minutes from the radioisotope placed in front of the tube is 40 s^{-1}

The received count rate includes the background and is therefore $(40 + 5)$ = 45 s^{-1}

b When the received count rate is 25 s^{-1}, the count rate from the radioisotope would be $(25 - 5)$ = 20 s^{-1}

Arrow method:

The arrow chain should therefore finish at 20 s^{-1}

160 s$^{-1} \rightarrow 80$ s$^{-1} \rightarrow 40$ s$^{-1} \rightarrow 20$ s^{-1}

$\quad$ 10 min $\quad$ 10 min $\quad$ 10 min

It would take a total time of 30 minutes for the received count rate to fall to 25 s^{-1}.

Uses of radioisotopes

Medical uses – cancer therapy

1 – External beam radiotherapy

Gamma radiation from an **external source** of cobalt-60 (**Co-60**) can be used to destroy malignant growths (cancerous tumours). The treatment is only used if the cancer is **localised** in a small region, because the radiation also destroys good cells.

This type of therapy has some drawbacks.

- Normal tissue is irradiated between the surface and the tumour.
- The beam cannot be sharply focused, so good tissue surrounding the target becomes damaged.
- Bone may shield the tumour from radiation.

External beam therapy using linear accelerators is replacing the use of Co-60. The high-energy X-ray beams from these accelerators are easier to control and can be more sharply focused.

2 – Brachytherapy (sealed therapy)

Short-range radiation sources (α, β) and low energy γ are placed directly at the cancerous site where their emissions are absorbed. The sources may be:

- stuck to the patient's skin to treat **skin** cancers
- placed into the tumour using **needles** or a **catheter**, as in the treatment of **breast** cancer
- placed into **sealed capsules** and inserted into body cavities, as in the treatment of **cervical** cancer
- placed into **titanium seeds** and inserted at the site of the tumour, as in the treatment of **prostate** cancer.

Iridium-192 (**Ir–192**) and caesium-137 (**Cs-137**) are used for treating **cervical cancer**.

Iodine-125 (**I-125**), palladium-103 (**Pd-103**), caesium-137 (**Cs-137**) or iridium-192 (**Ir-192**) can be used to treat **prostate cancer**.

- Unlike external beam therapy where the beam may damage tissue in front of the tumour, this method mainly reaches the target since the sealed radioisotopes are placed there.
- Unlike radiopharmaceutical therapy (see below), the source is **not dissolved in body fluids** and so does not move around inside the body.
- The capsules, needles or seeds move with the patient or tumour so the irradiation is **always on target**.
- Unlike other therapies, a **high dose** of radiation can be given for a **short period**, so the cancer has less time to develop. The source can then be **removed**.
- The radioactive implants are usually only **temporary** and so the sources can have long half-lives, since the needles, capsules, seeds or wires **may be removed** from the patient at any time. Some implants using seeds may be **permanent** and therefore these must be of low-emitting γ-sources.

3 – Radiopharmaceutical therapy (drug therapy)

This form of treatment is used when a pharmaceutical can carry the radioactive source to the target site. The radioactive source is **injected** or given **orally** to the patient. An example is in the treatment of thyroid cancer. If a patient is given a dose of iodine it is targeted to the thyroid where it is needed. I-131 emits β- as well as γ-radiation; its β-radiation is particularly good at destroying cancer cells.

- For this form of therapy, the sources should have **short half-lives** so as not to expose the patients to excessive doses of radiation for long periods.
- To protect others, patients should **remain in the hospital** until a safety level of the radiation they emit is reached.

Medical uses – diagnostic

As in radiopharmaceutical therapy, radioisotopes used for diagnosis are given to patients orally or by injection and so should have the following properties.

- **Weakly ionising** – preferably **γ-emitters** (but may also be β-emitters) because **γ-radiation** is **least absorbed** by body tissue and is therefore **least damaging** to it.
- **Non-toxic.**
- **Short half-life** – so that the patient will not be exposed to the radiation for long periods.

1 – Radioactive tracing

*A **radioactive tracer** (radioactive label) is a chemical compound in which one or more elements have been replaced by radioisotopes so, due to its radioactive decay, its path can be followed to investigate chemical reactions.*

A suitable tracer is given to the patient and a detector is used to follow its path through the body. Iodine-123 (**I-123**) is used as a tracer to investigate the functioning of the thyroid gland. Unlike I-131, which emits β- and γ-radiation, I-123 only emits γ-radiation and is almost **harmless to thyroid cells**. If **I-131** is used it must be administered in low doses, since its β-radiation can be **damaging** to body tissue.

Solutions of salts containing sodium-24 (**Na-24**), a β-emitter, are injected into the bloodstream to detect blockages.

Technetium-99m (**Tc-99m**), a γ-emitter, tends to concentrate in **damaged tissue** of the **lungs, heart and liver** and so is used to investigate such regions. It does not remain in the body for long and has a half-life of just 6 hours. The 'm' in its symbol indicates that it is a metastable isotope (its nucleus is in an excited state for a longer period than is typical).

Thallium-201 (**Tl-201**) concentrates in **good tissue** of the **heart** and so is used to identify its healthy parts.

2 – Photography (scanning)

The patient is given a **γ-emitting** radioisotope which can be targeted to a particular region within the body. A **scanner** or **gamma camera** is positioned outside the patient to detect the emitted radiation. Tumours present will appear as a shadow in the image since they absorb the radiation. **Tc-99m** is used in approximately **85%** of all radioactive scanning.

- **Thyroid scans** – I-123 is suitable for the same reason it is used as a thyroid tracer.
- **Bone scans** – A **phosphate** compound containing **Tc-99m** will collect in the bone where it will produce increased phosphate metabolism in cancerous regions. The radiation emitted from these regions then casts the image which shows up in the scan.

Aging specimens using radioisotopes

Living plants absorb carbon dioxide from the atmosphere by the process of **photosynthesis**. A very small amount of the carbon absorbed is of the radioisotope C-14 (carbon-14 or ^{14}C). Animals eating the plants and other animals eating those animals will then also absorb the C-14 atoms. The **ratio** of non-radioactive **C-12 to radioactive C-14** in the atmosphere is therefore the **same** for all living plants and animals.

When a plant or animal **dies**, decaying **C-14 is no longer replaced**, so the **ratio of C-14 to C-12 decreases** with time. The half-life of C-14 is 5700 years, therefore every 5700 years the mass of C-14 in a dead specimen falls to half the value it had at the beginning of that period.

If the amount of C-14 per gram of carbon in a living organism is known, together with the amount per gram of carbon in an old specimen, the age of the specimen can be calculated.

In natural carbon there is only **ONE** atom of **C-14** for every 8×10^{11} atoms of **carbon**! Due to this very small proportion of C-14 in a sample of carbon, even from a living organism, the count rate obtained from very old specimens is very low. This type of dating is therefore not useful for determining the age of samples which have been dead for over 60 000 years (about 10 half-lives).

Table 33.3 shows radioisotopes of half-lives much greater than C-14 which may be used to date ancient materials.

Table 33.3 *Other isotopes used for radioactive dating*

Isotope	Half-life
Uranium-238	4500 million years
Potassium-40	1300 million years
Uranium-235	700 million years

Example 7

An explorer discovers an old axe with a wooden handle. 5.0 g of carbon extracted from the wood gives a count rate of 7.5 min⁻¹. Given that the half-life of C-14 is 5700 years, calculate the age of the axe if 5.0 g of carbon extracted from a living plant gives a count rate of 30.0 min⁻¹ when the source is placed the same distance from the detector.

With questions on carbon dating always **start with the living** specimen and **end with the dead** specimen. The **masses used** in the calculation for both specimens **should always be the same**.

Comparing activities of 5.0 g of the living and of 5.0 g of the dead specimens.

Arrow method:

(5.0 g: living)   (5.0 g: dead)

$30 \text{ min}^{-1} \rightarrow 15 \text{ min}^{-1} \rightarrow 7.5 \text{ min}^{-1}$

         5700 y   5700 y

2 arrows implies 2 half-lives

Therefore age of the specimen is 2×5700 y = 11 400 years.

Table method:

Count rate/min⁻¹	Time/years
30	0
15	5700
7.5	11 400

Example 8

An archaeologist discovers an old wooden stool and decides to investigate its age. 12 g of carbon extracted from it has an *activity* of 45 per minute. Given that the activity of 1 g of carbon from a living plant is 15 per minute and that the half-life of C–14 is 5700 years, determine the age of the specimen.

Note: Unlike example 7, the masses of the living and dead samples are different. We need to compare the activities of **similar masses**; either determine the activities of 1 g or of 12 g of the living and dead specimens. It is advisable to compare the larger masses to reduce the possibility of fractional values.

Say we compare activities of 12 g of the living with 12 g of the dead specimens.

Living: 12 g of the living specimen has an activity of $12 \times 15 \text{ min}^{-1} = 180 \text{ min}^{-1}$

Dead: 12 g of the dead specimen has an activity of 45 min⁻¹

Arrow method:

(12 g: living)   (12 g: dead)

$180 \text{ min}^{-1} \rightarrow 90 \text{ min}^{-1} \rightarrow 45 \text{ min}^{-1}$

          5700 y   5700 y

2 arrows implies 2 half-lives

Therefore the age of the stool = 2×5700 years = 11 400 years.

Table method:

Count rate	Time/years
180	0
90	5700
45	11 400

Sterilisation

The objects to be sterilised are packed and **sealed** in plastic containers which are then irradiated by γ-radiation from a cobalt-60 (**Co-60**) source to kill germs within them.

Diagnostic photography used in industry

Cobalt-60 is a **powerful γ-source** and is used in industry to produce images that detect flaws in metal castings and weak spots in welded joints in a similar way to which X-rays are used to obtain images of our bones.

Tracers used in industry

Natural gas, trapped below layers of rock, can be released by a process of hydraulic fracturing known as **fracking**. Pressurised liquid containing dissolved radioactive tracers are forced into the rock structure, causing it to break. Detectors are used to measure the emissions at various points. The positions of the fractures are indicated at locations from which there is a high concentration of radiation emitted from the tracers that have settled there.

Leaks in underground water pipes can be found in a similar way by using dissolved salts of **sodium-24 (Na-24)**.

Tracers should generally be **gamma** sources so that their radiations are **strongly penetrating**, and they should have **short half-lives** so that they will emit at **safe levels** after a **short period**.

Phosphorus-32 (**P-32**) is a β-emitter used to trace fertiliser uptake in plants.

Thickness measurement and control

The thickness of various materials can be determined by examining the amount of **β- or γ-radiation** that can penetrate them.

Figure 33.10 shows how paper thickness can be measured and controlled during manufacture. Measurements from the detector are fed back to a computer, which then instructs the roller controls on how to alter the pressure on the paper and thereby adjust the thickness to the desired value.

For this application, a β-source is used. Radiation from an α-source would be absorbed before reaching the detector and radiation from a γ-source would pass through the paper almost unaffected.

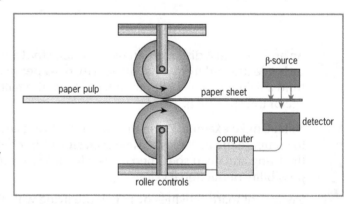

Figure 33.10 *Measuring and controlling the thickness of paper during manufacture*

These sources should have **long half-lives** so that changes in detected radiation are only due to absorption by the materials through which the radiation is passed. The emitted radiation should also be of **moderate energy** so that the risk of exposure to it is reduced.

Smoke alarm

Alpha radiation from americium-241 (Am-241) **ionises** the air between two oppositely charged metal plates, causing a small current in an electric circuit. If smoke enters the region it absorbs some of the ions and therefore reduces the current. The circuit detects the fall in current and this triggers the alarm.

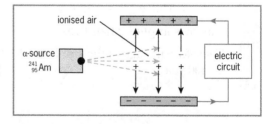

Figure 33.11 *Smoke alarm*

Experiment to measure the thickness of metals using β-radiation

- Record the count rate received from a β-source placed directly in front of a GM detector.
- Place a sheet of the metal, 0.5 mm thick, between the source and detector (see Figure 33.12) and record its **thickness**, together with the **new count rate**.
- Repeat the procedure several times, **adding** sheets of the metal of the same thickness (0.5 mm) between the source and detector each time.
- Plot a graph of count rate against thickness to act as a **calibration curve**.
- The sheets are then replaced by a single sheet of the same metal of unknown thickness and the new count rate (y_1) is measured.
- The unknown thickness of the sheet is determined by reading the x-coordinate (x_1) of the calibration curve corresponding to y_1.

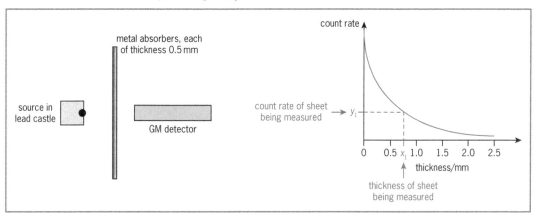

Figure 33.12 *Experiment to measure the thickness of metals using β-radiation*

Choice of half-life in applications

Long half-life

Radioisotopes used for **thickness measurements** have long half-lives so that their activity remains steady during the measurements they carry out. Those used in radioactive **dating** have long half-lives so that they still emit, even after thousands of years and so can provide useful readings.

Short half-life

Radioisotopes given to patients **orally or by injection** have short **radioactive half-lives** so that they are not in the patient at dangerous levels for long periods. However, the half-life **cannot be too short** because time is required for the radioisotope to be transported by the blood to the target site.

A short **biological half-life** (a measure of the time for the source to be excreted from the body) is also required to prevent long-term contamination.

Note: Radioisotopes used for **external beam therapy** or **brachytherapy** (sealed therapy) can have long or short half-lives since they may be removed from the patient's environment at any time.

Table 33.4 shows the emissions and half-lives of some radioisotopes used in the fields of medicine and industry. You need not memorise the table but you should relate the type of emissions and the half-lives to the uses of the isotopes.

Table 33.4 *Half-lives and emissions of some radioisotopes used in the fields of medicine and industry*

Radioisotope	Symbol	Half-life	Emission/s	Uses
Cobalt-60	Co-60	5.3 years	γ	external beam therapy (radiotherapy)
Technetium-99m	Tc-99m	6 hours	γ	tracer, scans
Iodine-123	I-123	13.2 hours	γ	tracer, scans
Iodine-131	I-131	8 days	β, γ	tracer, radiopharmaceutical therapy
Sodium-24	Na-24	15 hours	β, γ	tracer, radiopharmaceutical therapy
Iridium-192	Ir-192	74 days	β, γ	tracer, sealed therapy
Caesium-137	Cs-137	30 years	β, γ	sealed therapy
Strontium-90	Sr-90	29 years	β	sealed therapy, thickness measurement
Phosphorus-32	P-32	14 days	β	tracer, radiopharmaceutical therapy
Americium-241	Am-241	432 years	α, γ	smoke alarms

Hazards of ionising radiations

Ionising radiations are high energy particles or electromagnetic waves which can dislodge electrons from atoms to produce ion pairs.

α-particles, β-particles, γ-rays and X-rays are examples of ionising radiations.

A body exposed to radiation is said to be **irradiated**. The cells of a living organism that has been irradiated by alpha, beta, gamma or X radiations can be severely damaged due to ionisation on absorbing the emissions.

The damage can result in the following outcomes.

- **Radiation burns** – Very concentrated doses of radiation can burn the body tissue, killing the cells of a particular region.
- **Cancer** – Radiation affecting the DNA in the nucleus of a cell can cause the cell to divide uncontrollably to form a tumour.
- **Mutations** – Radiation affecting a sperm or egg cell can damage the DNA of its genes, causing future generations to be burdened with genetic problems.

Sources outside the body

α-sources are not dangerous once they are a few centimetres from the body. Their emissions are readily absorbed by air and cannot penetrate dead skin cells.

β-sources, several metres away, are also safe since their emissions will be absorbed by that distance of air.

γ-sources produce the least ionising emissions, but these can penetrate the body; if the emissions are of high energy, the damage can be substantial as they are absorbed by the body cells.

Sources inside the body

Sources can enter the body by **inhalation** of radioactive gases (such as radon and thoron) or radioactive dust, or by **ingestion** of foods contaminated with radioactive materials.

α-sources are extremely dangerous, since their emissions are strongly ionising and can cause much cellular damage.

β-sources are also dangerous, since much of the energy of their emissions will ionise body cells in their paths.

γ-sources are the least damaging since they are the least ionising. Most of their energy passes straight through the body and into the air. However, some γ-sources, especially in high concentrations, can be dangerous.

Safety precautions against the hazards of ionising radiations

* The **tri-foil** hazard sign (Figure 33.13) must be clearly visible wherever ionising radiation is hazardous.
* Radioactive material should always be **labelled** by name and as being dangerous.
* Persons working with radioisotopes should **wear protective**, **lead-lined clothing** or be **shielded by lead-lined walls.**
* The use of **robotic arms, tongs** and **gloves** should be employed when handling radioisotopes.
* Radioactive samples should be **stored** in **lead** containers.
* **Food and/or drink** should **never** be **consumed** in locations where radioactive material is present.
* Hands should be **washed** or **baths** should be taken after working with radioactive material.
* One should comply with **current regulations** on the **disposal** of radioactive waste.
* Nuclear weapon testing should be done **underground.**
* High-altitude airline staff must work in shifts to reduce the amount of cosmic radiation they receive.
* Staff constantly working with radioactive material need to wear **badges** that **monitor** the radiation they receive.

Figure 33.13
Tri-foil

Recalling facts

1. **a** Name the female scientist who shared a Nobel prize for her work on radioactivity.
   b What led her to conclude that radioactivity from an atom was dependent only on the atom itself?
   c What area of medical science has benefited from her work?
   d What is believed to have led to her death?

2. **a** What is meant by the term *radioactivity*?
   b How does the atomic number of a radioisotope change on emission of:
      i a beta particle
      ii an alpha particle
      iii a gamma wave?

3. **a** Name the instrument that is commonly used to count and measure the rate of radioactive emissions received at a location.
   b Name and give the symbol of the unit used for the rate of radioactive emissions per second.
   c What is meant by the term *activity of a sample of a radioisotope*?

4 For each of the following, state the type/s of nuclear radiation pertaining to the particular characteristic.

   a Can penetrate a sheet of paper but is stopped by 5 mm of aluminium.

   b Is least ionising.

   c Has the weakest penetrating capability.

   d Is not deflected by electric or magnetic fields.

   e Is stopped by a sheet of paper.

   f Consists of two protons and two neutrons.

   g Is an electron.

   h Is emitted at the greatest speed.

   i Is a particle of relative charge -1.

   j Is a particle of charge -1.6×10^{-19} C.

   k Is comprised of particles of total charge $+3.2 \times 10^{-19}$ C.

   l Is comprised of particles of total mass more than 7000 times that of an electron.

5 **a** What source provides approximately half of all background radiation received on Earth?

   b List THREE other sources of background radiation.

6 $\times$ represents the direction of a magnetic field acting perpendicularly into the plane of Figure 33.14.

What type of radiation is emitted from the source if the magnetic field causes it to take the circular path shown?

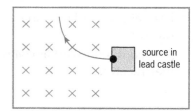

Figure 33.14 *Question 6*

7 **a** Sketch diagrams showing the different types of tracks produced in a cloud chamber by α- and β-radiations.

   b Explain the difference in shape and strength of the tracks mentioned in **a**.

8 Explain how it is possible for electrons to shoot out of an atomic nucleus as beta particles.

9 **a** What property of radiation causes it to be classified as *ionising radiation*?

   b Name FOUR ionising radiations.

   c Describe THREE ways in which ionising radiations may be harmful.

   d List THREE precautions which should be taken when working with ionising radiations.

10 **a** Explain why an α-source is not dangerous to a person when it is more than a few centimetres outside of his or her body, but can be extremely dangerous if inhaled or ingested.

   b Why is it less dangerous to consume a β-source than it is to consume an α-source?

11 **a** Explain why a half-life of 15 hours is suitable for an isotope used as a medical tracer, but half-lives of 15 seconds or 15 years are not.

   b Why is a half-life of 6 hours unsuitable for an isotope used to measure the thickness of metal sheets in a factory?

12 Explain why the needle of a ratemeter connected to a GM detector flickers when the window of the tube is placed to face a radioactive source.

13 a What is meant by the half-life of a radioisotope?

b Sketch a graph of mass against time showing the decay of a radioisotope of initial mass M. Label the x-axis t_1, t_2 and t_3 to indicate the times taken to complete the first, second and third half-lives, and label the y-axis to show the masses at those times.

Applying facts

14 Use Table 33.4 to explain each of the following:

a Tc-99m is used as a tracer

b Sr-90 is used to measure the thickness of paper

c I-131 is used in treating cancer of the thyroid

d Co-60 is used in external beam therapy

e Cs-137 can be used on patients receiving brachytherapy (sealed therapy)

f I-123 is not as damaging to body tissue as is I-131

g Am-241 can produce a consistently high level of ionisation in a smoke detector.

15 Radon gas is responsible for much of the background radiation received on Earth. During decay, its isotopes and daughter nuclides emit alpha particles. Explain how and why this can be a danger to people.

16 Write an equation for each of the following radioactive emissions:

a $^{222}_{86}Rn$ emitting an alpha particle to become an isotope of Po

b $^{32}_{15}P$ emitting a beta particle to become an isotope of S

c $^{99m}_{43}Tc$ (Tc in an excited state) emitting a gamma photon to become Tc in its ground state.

17 Find the values of p, q, r, s, t and u in the following decay chain.

$$^{238}_{92}U \rightarrow {}^{4}_{2}\alpha + {}^{p}_{q}Th$$

$$^{p}_{q}Th \rightarrow {}^{0}_{-1}\beta + {}^{r}_{s}Pa$$

$$^{r}_{s}Pa \rightarrow {}^{0}_{-1}\beta + {}^{t}_{u}U$$

18 a The activity of a sample of pure iodine-131 falls by 93.75% in 32 days. Determine its half-life.

b What is the effect on the half-life if:

i the temperature is doubled?

ii the pressure is halved?

iii the same quantity of iodine-131, in the form of the compound sodium iodide, is examined?

iv the mass of the iodine is doubled?

19 A radioactive source of half-life 6 hours is placed in front of a GM detector and the ratemeter indicates a count rate of 68 Bq. If the background count rate is 4 Bq, determine the received count rate after 24 hours.

20 **a** 3 g of carbon extracted from a fossilised organism gives a count rate of 2.5 min^{-1} when placed in front of a GM tube. When 3 g of carbon extracted from a living organism is placed the same distance from the front of the tube, the count rate is 20 min^{-1}. Determine the age of the fossilised organism if the half-life of C-14 is 5700 years.

b Explain why C-14 is unsuitable to measure the age of an organism that is 70 000 years old.

21 Figure 33.15 shows how the masses of three radioisotopes A, B and C vary with time. State with reason:

a which of A and B has the shorter half-life

b what can be said of the half-life of C.

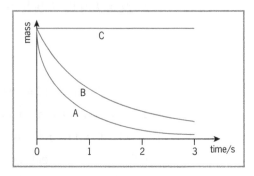

Figure 33.15 *Question 21*

Analysing data

22 Figure 33.16 shows the apparatus used to determine the half-life of radon-220 (Rn-220), a radioactive gas of short half-life which emits α-particles. A freshly prepared sample of the gas is forced into the vessel to take the place of air which is forced out, and the taps are immediately closed as the Rn-220 immediately begins to ionise the air. A pair of oppositely charged electrodes connected to a power source and a milliammeter register the ionisation current. As the Rn-220 decays, readings of the ionisation current, *I*, and the corresponding time, *t*, are taken at intervals of 20 s. Table 33.5 shows the readings taken.

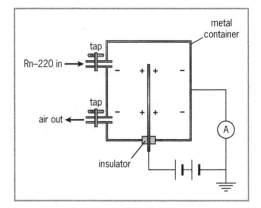

Figure 33.16 *Question 22*

Table 33.5 *Readings of ionisation current, I, and corresponding times, t*

Ionisation current I/mA	1.40	1.10	0.82	0.68	0.50	0.44	0.28	0.22	0.22	0.16	0.08
Time t/s	0	20	40	60	80	100	120	140	160	180	200

a Plot a graph of ionisation current versus time, using the graph paper in landscape format.

b Determine the time t_1 for the current to fall from 1.40 mA to half its value.

c Choose two other points on the graph and find the times t_2 and t_3 for their corresponding currents to fall to half their values.

d Determine the half-life of Rn-220 by calculating the mean value of t_1, t_2 and t_3.

e Why are the data points not falling exactly on the curve although the milliammeter and power source are not faulty?

f Read from the graph the current when $t = 50$ s.

g Why do the experimental points deviate more from the curve for larger values of *t*?

34 Nuclear fission and nuclear fusion

During the process of natural radioactivity, an atomic nucleus emits alpha, beta or gamma radiation to become more stable. Two other ways by which the nucleus can **increase its stability** are the processes of nuclear **fission** and nuclear **fusion**. During natural radioactivity, nuclear fusion and nuclear fission, nuclides **release energy** undergoing a transition to a **lower energy state**, resulting in **their nucleons** becoming **more tightly bound**.

Learning objectives

- Understand the processes of **nuclear fission** and **nuclear fusion** reactions.
- Write **equations** representing nuclear fission and nuclear fusion reactions.
- Define the **unified atomic mass unit** and use the relation 1 u = 1.66 × 10⁻²⁷ kg.
- Apply Einstein's equation $\Delta E = mc^2$ to solve problems on nuclear reactions.
- Cite **arguments for and against the utilisation of nuclear energy**.

Nuclear fission

Nuclear fission is the splitting of a heavy atomic nucleus into two or more atomic nuclei, resulting in a large output of energy and a decrease in mass.

Note: In natural radioactivity the atomic nucleus is also split on emission of alpha or beta particles. However, alpha and beta particles are much lighter than the daughter nuclides produced by nuclear fission.

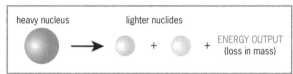

Figure 34.1 *Nuclear fission*

Nuclear power plants

Nuclear fission **chain reactions** (Figure 34.2) are utilised in **controlled nuclear reactors** (Figure 34.3) to provide electricity. Uranium and plutonium are the elements used as the fuel in such reactors.

Two possible fission reactions of uranium-235 (U-235) are shown below.

$$^{235}_{92}\text{U} + ^{1}_{0}\text{n} \rightarrow ^{141}_{56}\text{Ba} + ^{92}_{36}\text{Kr} + 3^{1}_{0}\text{n} + \text{energy}$$

$$^{235}_{92}\text{U} + ^{1}_{0}\text{n} \rightarrow ^{144}_{56}\text{Ba} + ^{90}_{36}\text{Kr} + 2^{1}_{0}\text{n} + \text{energy}$$

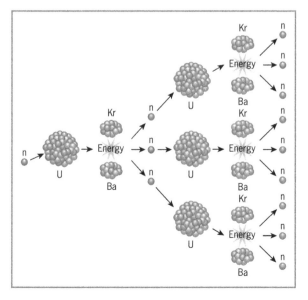

Figure 34.2 *Nuclear fission chain reaction*

- A **graphite moderator** is used to **reduce the speed** of stray neutrons, making it more likely that they can trigger nuclear fission reactions.
- A **chain reaction** is produced, since the fission of each uranium atom produces two or three neutrons which can trigger further fission reactions.
- **Boron control rods** are inserted between the **uranium fuel rods** to absorb some of the neutrons and so prevent a catastrophic explosion (**atomic bomb**) due to an exponential chain reaction.
- **Thermal energy** is produced in the graphite core and in the boron rods as they absorb **kinetic energy** from the neutrons.
- A **coolant** of **liquid or gas** is pumped through the system to absorb the heat and carry it to a **heat exchanger**, where it warms water to produce steam. The pressure of the steam is then used to turn the turbines of electrical generators.
- The **concrete shield** protects against ionising radiation produced from the daughter nuclides of the fission reaction and provides a strong casing for the high pressures within the reactor.

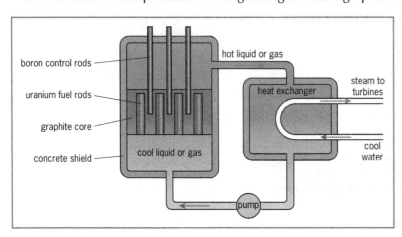

Figure 34.3 *Nuclear reactor*

Nuclear fusion

Nuclear fusion is the joining of two light atomic nuclei to produce a heavier atomic nucleus, resulting in a large output of energy and a decrease in mass.

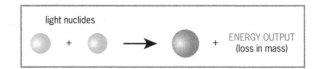

Figure 34.4 *Nuclear fusion*

Nuclear fusion is the process by which **radiant energy** is produced in the **Sun** as two hydrogen nuclei join to form a single helium nucleus. The **hydrogen bomb** also utilises nuclear fusion.

Two possible fusion reactions of hydrogen nuclei are shown below.

$$^2_1H + {}^2_1H \rightarrow {}^3_2He + {}^1_0n + energy$$

$$^2_1H + {}^3_1H \rightarrow {}^4_2He + {}^1_0n + energy$$

These reactions occur in the **core of the Sun** where the temperature is close to 15.7 million Kelvin and provides the positive hydrogen nuclei (protons) with very high velocities. The resulting high kinetic energy helps in the process of overcoming the extreme forces of repulsion between the positive nuclei and is able to produce a slow nuclear fusion reaction.

The difficulty of **sustaining** and **controlling** these **high temperatures** has been the main **obstacle** in the **design** of **fusion reactors** on Earth.

Einstein's equation of mass-energy equivalence

When nuclides undergo **fission, fusion** or **natural radioactivity**, the resulting nuclides become **more stable** as **matter** is converted **into energy**.

This **loss in mass**, Δm, is related to the energy output, ΔE, and to the speed of light in a vacuum, c, by the following equation proposed by **Albert Einstein**:

$$\Delta E = \Delta mc^2$$

The **speed of light in a vacuum is 3.0×10^8 m s^{-1}**.

Some students have problems inputting $(3.0 \times 10^8)^2$ to the calculator, especially when it is in the denominator of a fraction. It may be useful to remember that you can replace it with 9.0×10^{16} (see examples 1 and 2).

Example 1

Determine the energy output from a nuclear reaction in which the loss in mass is 2.0 g.

$\Delta E = \Delta mc^2$

$\Delta E = 2.0 \times 10^{-3} \times (3.0 \times 10^8)^2$

$\Delta E = 2.0 \times 10^{-3} \times 9.0 \times 10^{16}$

$\Delta E = 1.8 \times 10^{14}$ J

Example 2

Nuclear fusion reactions within a certain star convert energy at a rate of 4.5×10^{25} J s^{-1}. Determine the loss in mass of the star each second.

a $\qquad \Delta E = \Delta mc^2$

$\qquad \therefore \Delta P = \dfrac{\Delta mc^2}{t}$

$\qquad 4.5 \times 10^{25} = \dfrac{\Delta m(3.0 \times 10^8)^2}{1}$

$\qquad 4.5 \times 10^{25} = \dfrac{\Delta m \times 9.0 \times 10^{16}}{1}$

$\qquad \Delta m = \dfrac{4.5 \times 10^{25}}{9.0 \times 10^{16}}$

$\qquad \Delta m = 5.0 \times 10^8$ kg

Example 3

The nuclear fusion reaction involving nuclides of H-2 and H-3 has an energy output of 2.88×10^{-12} J. Use the information below to answer the question which follows.

$^2_1\text{H} = 3.345 \times 10^{-27}\text{kg} \qquad ^3_1\text{H} = 5.008 \times 10^{-27}\text{kg} \qquad ^1_0\text{n} = 1.675 \times 10^{-27}\text{kg}$

$^2_1\text{H} + ^3_1\text{H} \longrightarrow ^4_2\text{He} + ^1_0\text{n} + \text{energy}$

Note: If we are just balancing the mass numbers, atomic numbers and relative charges, it is acceptable to write nuclear equations without including the term *energy* (see question 5 at the end of this chapter, and Chapter 33). However, for questions that involve the energy output we need to include the energy term for its use in *mass-energy* conversions.

a Calculate the equivalent mass of the energy output.

b Calculate the mass, M, of the He-4 nucleus.

a $\quad \Delta E = \Delta mc^2$

$\quad \therefore \Delta m = \dfrac{\Delta E}{c^2} = \dfrac{2.88 \times 10^{-12}}{(3.0 \times 10^8)^2} = 3.2 \times 10^{-29}$ kg

---

b We must rewrite the equation using only masses.

$$3.345 \times 10^{-27} + 5.008 \times 10^{-27} = M + 1.675 \times 10^{-27} + 3.2 \times 10^{-29}$$

$$3.345 \times 10^{-27} + 5.008 \times 10^{-27} - 1.675 \times 10^{-27} - 3.2 \times 10^{-29} = M$$

$$6.646 \times 10^{-27} \text{ kg} = M$$

Unified atomic mass unit

Since the masses of nuclides and their atoms are very small it is typical to express them in a special unit known as the unified atomic mass unit, u.

1 u is equal to $\frac{1}{12}$ of the mass of the nuclide of carbon-12.

Since a nuclide of $^{12}_{6}C$ has 12 nucleons, **1 u** is equal to the mass of **1 nucleon**.

When using the equation $\Delta E = \Delta m c^2$ we must express Δm in kg. We can convert 'u' to 'kg' using the relation:

$$\mathbf{1 \text{ u} = 1.66 \times 10^{-27} \text{ kg}}$$

Note for the examples which follow: When working with equations involving energy output due to radioactivity, nuclear fission and nuclear fusion, if the masses are expressed in 'u' they can be either all expressed as **nuclear** masses or all as **atomic** masses. This is so since when using atomic masses, the masses of the included electrons on either side of the equations will effectively cancel.

Example 4

Determine the energy released during the nuclear fission reaction of U-235, given the following nuclear masses: $^{1}_{0}n = 1.0087$ u $^{235}_{92}U = 235.0439$ u $^{148}_{57}La = 147.9327$ u $^{85}_{35}Br = 84.9156$ u

$$^{235}_{92}U + ^{1}_{0}n \longrightarrow ^{148}_{57}La + ^{85}_{35}Br + 3^{1}_{0}n + \text{energy}$$

Interpret the equation in terms of only masses, assigning Δm as the equivalent mass of the energy output.

$$235.0439 \text{ u} + 1.0087 \text{ u} = 147.9327 \text{ u} + 84.9156 \text{ u} + 3(1.0087) + \Delta m$$

$$236.0526 \text{ u} = 235.8744 + \Delta m$$

$$0.1782 \text{ u} = \Delta m$$

We can now convert the mass to kg and determine the equivalent energy in J:

$$\Delta E = \Delta m c^2$$

$$\Delta E = (0.1782 \times 1.66 \times 10^{-27})(3.0 \times 10^8)^2$$

$$\Delta E = 2.7 \times 10^{-11} \text{ J}$$

Example 5

Thorium-228 (Th-228) is a radioisotope of atomic number 90 and emits an alpha particle to produce a daughter nuclide radium (Ra).

a Write the nuclear equation for the decay.

Given the following atomic masses: Th-228 = 228.0287 u   Ra-224 = 224.0202 u   He-4 = 4.0026 u

b Calculate the equivalent mass of the energy output in unified atomic mass units.

c Determine the energy released by a Th-228 nucleus as it decays.

---

a $^{228}_{90}\text{Th} \rightarrow {}^{224}_{88}\text{Ra} + {}^{4}_{2}\text{He} + \text{energy}$

b Interpret the equation in terms of only masses, assigning Δm as the equivalent mass of the energy output.

$$228.0287 \text{ u} = 224.0202 \text{ u} + 4.0026 \text{ u} + \Delta m$$

$$228.0287 \text{ u} - 224.0202 \text{ u} - 4.0026 \text{ u} = \Delta m$$

$$5.9 \times 10^{-3} \text{ u} = \Delta m$$

c $\Delta E = \Delta mc^2$

$\Delta E = (5.9 \times 10^{-3} \times 1.66 \times 10^{-27}) \times (3.0 \times 10^8)^2$

$\Delta E = 8.8 \times 10^{-13} \text{ J}$

Arguments for and against the utilisation of nuclear power plants

Arguments for

- **Availability** – A large supply of U-235 is available.
- **Efficiency** – A small amount of nuclear fuel produces an enormous amount of electricity. This efficiency is thousands of times greater than that in fossil fuel power plants.
- **Inexpensive** – Since small quantities of fuel are required, delivery and storage are relatively cheap when compared to conventional power plants.
- **Reliability** – Does not depend on weather or climatic conditions.
- **No greenhouse gases** – The nuclear reactions do not produce greenhouse emissions. However, if the energy used to mine and refine the ore uses fossil fuels, there will be a small amount of greenhouse emissions per unit of electricity produced.
- **Clean** – In the absence of natural disasters, they do not contaminate the environment if carefully managed.
- **Medical isotopes** – Many radioisotopes used in medicine are produced in nuclear power plants.
- **Suitable back-up** – Nuclear power can provide a suitable energy **back-up to green energy sources** when weather conditions are unsuitable. This reduces the reliance on fossil fuels and therefore reduces the damage that may be caused by global warming.

Arguments against

- **High construction costs.**
- **Danger of acquiring radioactive materials** – Mining and transportation can be dangerous.
- **Nuclear war** – Advancements in this technology can lead to rising dangers of nuclear warheads.
- **Danger to staff** – Staff may be irradiated by radioisotopes which contaminate their surroundings and may be contaminated themselves if they ingest or inhale the material. This can lead to radiation burns, cancer or genetic mutations as discussed in Chapter 33.
- **Danger of spent radioactive fuel** – Spent radioactive fuel contains radioactive material that is hazardous to the environment. It is usually **stored under water** at the reactor site to **remove the excess heat** it produces. The water also acts as a **radiation shield** in protecting the surroundings. After a decay period, some of the spent fuel can be transfered to dry concrete shielded casks at the reactor site or in a remote area.
- **Possibility of a catastrophic event** – A critical malfunction at a power station can lead to huge explosions that can cause **radioactive fallout** (the spreading of radioactive materials over large regions). This **large-scale contamination** can have a severe negative impact on the planet.

- **Costly plant closures** – Nuclear power stations must be shut permanently after several years because the plant and machinery become heavily contaminated.

Data which may be used in answering the following questions:

speed of light in a vacuum $= 3.0 \times 10^8$ m s^{-1} $\qquad\qquad$ 1 u $= 1.66 \times 10^{-27}$ kg

Recalling facts

1 Define the terms:

   a nuclear fission $\qquad\qquad$ **b** nuclear fusion.

2 Which of the two processes, nuclear fission and nuclear fusion, is associated with:

   a the hydrogen bomb?

   b the atomic bomb?

   c nuclear power plants?

   d energy from the Sun?

3 **a** Write Einstein's mass-energy equation and give the meaning of each of its symbols.

   b How does the unified atomic mass unit (u) relate to the mass of a C-12 nuclide?

4 List FOUR arguments for and FOUR arguments against the generation of electricity by nuclear power plants.

Applying facts

5 **a** Rewrite and complete the following nuclear equation:

$$^{235}_{92}U + {}^{1}_{0}n \rightarrow {}^{144}Xe + {}_{38}Sr + 2^{1}_{0}n$$

   b Determine the value of X and complete the nuclear equations in each reaction below:

    i $\quad ^{235}_{92}U + {}^{1}_{0}n \rightarrow {}^{140}_{56}Ba + {}^{93}_{36}Kr + X^{1}_{0}n$

    ii $\quad ^{239}_{94}Pu + {}^{1}_{0}n \rightarrow {}^{204}Au + {}^{31}_{15}P + X^{1}_{0}n$

6 Nuclear fusion reactions on a certain star release energy at a rate of 3.6×10^{27} J s^{-1}. Determine the loss in mass each second.

7 Calculate the energy released by a nuclear fission reaction that results in a loss in mass of 4.0 kg.

8 Use the following masses to determine the energy output during the nuclear fusion reaction below:

$$^{2}_{1}H = 3.345 \times 10^{-27} \text{ kg} \quad ^{3}_{2}He = 5.007 \times 10^{-27} \text{ kg} \quad ^{1}_{0}n = 1.675 \times 10^{-27} \text{ kg}$$

$$^{2}_{1}H + {}^{2}_{1}H \rightarrow {}^{3}_{2}He + {}^{1}_{0}n + \text{energy}$$

9 Use the following masses to determine the energy output during the nuclear fission reaction below:

$$^{235}_{92}U = 235.0439 \text{ u} \qquad ^{144}_{55}Cs = 143.9321 \text{ u} \qquad ^{90}_{37}Rb = 89.9148 \text{ u} \qquad ^{1}_{0}n = 1.0087 \text{ u}$$

$$^{235}_{92}U + {}^{1}_{0}n \rightarrow {}^{144}_{55}Cs + {}^{90}_{37}Rb + 2^{1}_{0}n + \text{energy}$$

Index